GEOMORPHOLOGY AND ENVIRONMENTAL SUSTAINABILITY

About the Editors

Prof. S.C. Kalwar, Professor and Ex-Head, Department of Geography, University of Rajasthan, Jaipur, is a specialist in Agricultural Geography, Rural Development, Land Use Planning Applied Economic Geography and Surveying. Professor Kalwar has written/edited 13 books related to resources, wasteland, ecology, environment, poverty and socio-economic development and has so far participated in about 25 conferences/seminars/symposia of national and international levels, and has contributed 50 research papers in various journals, supervised 32 Ph.D. scholars, more than 100 M.A./M.Sc./M.Phil. dissertations and conducted several research projects sponsored by U.G.C., I.C.S.S.R. and D.R.D.O., New Delhi. He is a member of Editorial Board of 'Studies in Geography', a research journal published by the Department of Geography, University of Rajasthan, and is the organiser of refresher courses and other academic conferences/seminars. He is also an external member of "Board of studies" and Research Degree Committee of various universities.

Prof. Kalwar is a life member of National Association of Geographers of India, Rajasthan Geographers Association, Asian Environmental Council and Indian Institute of Geomorphology. He was also a member of National Committee, Government of India for selecting Hindi books for National Award (2002-2003).

He was selected for a Research Award (Career Award) in 2002 by the University Grants Commission, New Delhi.

Dr. M.L. Sharma (born on 3rd September, 1942) is presently working as Principal Shri Bhawani Niketan College, Jaipur and had a long (35 years) teaching and administrative experience. Dr. Sharma graduated from University of Rajasthan, Jaipur, did his M.A. in Geography from University of Udaipur in 1966 with first division and Ph.D. from University of Rajasthan. Hailing from a small remote village of Phagi Tehsil, District, Jaipur. Dr. Sharma joined as a lecturer in Geography in a Government College in July, 1966. Dr. Sharma published six books and 70 research papers in the reputed national journals. Apart from the academic achievements, he served as N.C.C. officer (Major) and Leader Trainer (Rover scouting) and guided the youth in becoming good citizens. He is also a life member of N.A.G.I. R.G.A., I.I.G. and Rajasthan State Bharat Scout & Guides. Besides, he has organised a number of geographical conferences, seminars and published souvenirs.

Dr. R.D. Gurjar, Associate Professor in Geography is presently Convener, Board of Studies in Geography, member of Academic Council, Senate and Research Board as well as Chief Warden Campus hostels (boys) in the University of Rajasthan, Jaipur. Prior to it, he has held the administrative positions of Director, B.R. Ambedkar Administrative Service Examination Institute; Additional Chief Warden, University hostels; Vice-Principal and Principal of University Rajasthan College, Jaipur.

Dr. Gurjar has also been member, Telecom Advisory Committee Rajasthan Circle, Jaipur (1992-1994) under the Ministry of Communication; member, National Road Safety Council (1986-88, 1992-94) Ministry of Surface Transport; member, Expert Committee for Thermal Power Projects (1994-96) Ministry of Environment and Forest, Government of India.

Dr. Gurjar has several research papers as well as a book 'A Hand Book for Geography Teachers' at his credit.

Dr. M.K. Khandelwal, Head, Department of Geography, University of Rajasthan, Jaipur in a specialist in Livestock and Dairy Development, Integrated Rural Area Development, Urban Geography and Research Methodology. He has number of research papers and books to his credit. He has supervised many Ph.D. theses and edited Changing Contours of Arid Ecology, Rajasthan University Studies in Geography, Arid Ecology-Resources, Hazards and Rural Development Policies. He has attended and organised many academic conferences, and refresher courses. He is a life member of Asian Environmental Council, National Association of Geographers India, Indian Institute of Geomorphologists, Geographical Society of India, Institute of Indian Geographers and Rajasthan Geographical Association.

Dr. S.K. Wadhawan, a winner of Gold Medal of Delhi University, hold Ph.D. from University of Rajasthan. He has been a professional geoscientist in the Geological Survey of India for over 25 years. He has successfully completed several Quaternary geological mapping and applied geomorphological investigations, geoenvironmental and geotechnical appraisals in aridity affected dry lands of Thar Desert in Rajasthan and Gujarat. Dr. S.K. Wadhawan is presently posted as Director in GSI.

As Convener of Indian National Working Group of UNESCO-IGCP Projects, Dr. S.K. Wadhawan has conducted and coordinated several frontline research programmes on Late Quaternary Coastal Records of Rapid Changes; Desert Margins and Palaeomonsoons; Understanding Future Dryland Environmental Changes from Past Dynamics, etc. He has over 40 internationally acclaimed scientific publications to his credit.

GEOMORPHOLOGY AND ENVIRONMENTAL SUSTAINABILITY

(Felicitation Volume in Honour of Professor H.S. Sharma)

Edited by

Prof. S.C. Kalwar

Dr. R.D. Gurjar

Dr. M.L. Sharma

Dr. M.K. Khandelwal

Dr. S.K. Wadhawan

CONCEPT PUBLISHING COMPANY, NEW DELHI-110059

Cataloging in Publication Data—DK
Courtesy: D.K. Agencies (P) Ltd. <docinfo@dkagencies.com>

Geomorphology and environmental sustainability :
felicitation volume in honour of Professor H.S. Sharma / edited by S.C. Kalwar ... [et al.].
p. cm.
Hari Shanker Sharma, b. 1945, Indian geographer; contributed articles.
Includes bibliographical references.
Contents: section 1. Geomorphic process — section 2. Applied geomorphology — section 3. Environmental change — section 4. Environmental impact and sustainability — section 5. Bio-diversity.
ISBN 8180690288

1. Geomorphology—India. 2. Sustainable development—India. 3. Environmental management—India. 4. Biological diversity—India. I. Sharma, H. S. (Hari Shanker), 1945- II. Kalwar, S. C. (Sugan Chand), 1947-

DDC 551.41 21

ISBN : 81-8069-028-8

First Published 2005

Published and Printed by
Ashok Kumar Mittal
Concept Publishing Company
A/15-16, Commercial Block, Mohan Garden
NEW DELHI-110059 (India)

Phones : 25351460, 25351794
Fax : 091-11-25357103
Email : publishing@conceptpub.com

CONTENTS

SECTION-II
APPLIED GEOMORPHOLOGY

SECTION-III
ENVIRONMENTAL CHANGE

SECTION-IV
ENVIRONMENTAL IMPACT AND SUSTAINABILITY

SECTION-V
BIODIVERSITY

LIST OF FIGURES

LIST OF TABLES

Prof. H.S. Sharma

Prof. H.S. Sharma and Family

Profile of Professor H.S. Sharma

Dr. Hari Shankar Sharma was born on 5th January, 1945 in a small remote village Aliyabad in Newai tahsil of Tonk district, Rajasthan. In his childhood, he was a domicile, disciplined and dedicated boy with twin ability of compassionate understanding and reasoning. By virtue of these qualities, he rose up to the level of Professor in spite of his immemorable hardship, adversities and atrocities in background. He did his early schooling from small village near Aliyabad, High School from Newai town and embarked in higher education in 1961 in the prestigeous Maharaja's and Rajasthan- constituent colleges of University of Rajasthan, Jaipur. In the childhood, Nature was not benign to Dr. Sharma, as he lost his father at a growing age of 10 years. He was nurtured, and brought up by his elder brother, Shri Hari Narayan Sharma and mother Smt. Shanti Devi who shaped and sharpened the abilities of Dr. Sharma for about 20 years. Dr. Sharma got married in the early age and therefore, the devotion and sacrifice of Mrs. Maya Sharma cannot be overlooked in shaping his academic career.

After graduating from Rajasthan University in 1964, he joined the prestigeous Department of Geography, University of Sagar (M.P.) on the wise advice of his venerable teachers, Dr. Abul Hassan Sahir and Miss A.E. Dass of the Department of Geography, University of Rajasthan. Here, at Sagar he came into contact with Professor S.M. Ali, an intellectual giant heading the Department of Geography, who really sharpened his wisdom, kindness, the cupidity in his mind about study, learning and speech. He did post graduation in 1966 with I division from Sagar University, Sagar (M.P.). Prof. Ali's intellectual and emotional impress initiated his thoughts to pursue for the Doctoral degree in 1966 in geomorphology under his guidance. The untimely demise of Professor S.M Ali in December 1966 was a great shock to Sharma. However, he persued his research on the same topic and completed Ph.D. work under the able guidance of Professor Pramila Kumar, a reputed teacher and scholar of great eminence.

On the basis of his meritorius career, Dr. Sharma was selected as an Assistant Professor in the Department of Geography, University of Rajasthan, Jaipur in July 1969, and elevated as Associate Professor and Professor in 1985 and 1996 respectively.

Cutting a career across the varied academic cross roads, Dr. Sharma notched numbers of landmarks, which few geographers in India have been able to surpass. He is the first recipient of Career Award (U.G.C., Delhi, 1979) as a young scholar at the University of Rajasthan, Jaipur and also recipient of ISHEER Award (2001) of Indian Society of Health, Environment, Education and Research, Bikaner. He also acted as visiting lecturer at Jawaharlal Nehru University, New Delhi (1980), visiting fellow at Salford and East Anglia Universities of U.K. under the scheme of Alis link programme (1993), Panjab University, Chandigarh (2001) and visiting Professor at Mizoram University, Aizawl. He delivered extension lectures, attended seminars and conferences as well as delivered Key-note addresses in India and abroad like U.K. (1985, 1986, 1993), Germany (1989, 1995), Netherlands (1989), Canada (1993), Norway (1996), and U.S.A. (2001).

Dr. Sharma is a prolific writer. Although his interest is well known for geomorphology, yet he wrote immensely on varied aspects of geography with special reference to environmental issues of Rajasthan. He has special inclination about environmental geomorphology, desertification and human impact on environment, arid resources, sustainable development, soil erosion, human ecology, environmental hazards landuse changes etc. He has written and published 15 books on different themes of geography. His books Physiography of the Lower Chambal Valley and its Agricultural Development (1979), Tropical Geomorphology (1987), Ranthambhore Sanctuary: A Dilemma of Eco-development (2000), Geography of Rajasthan (2003), Ozone Depletion and Environmental Consequences (2004) are worth mentioning. Apart from this, he is the author of about 75 research papers published in journals of national and international repute including *Nature* (1973) and *Journal of Geomorphology* (1989) Germany. His books gave him a name and fame, and confirms garnered popularity. His monumental work *Perspectives in Geomorphology* in four volumes brought him wide acclaim. Amidst his multifaceted works, Professor Sharma did not abandon his duty to plough books, research articles and guiding the research scholars to nourish the discipline of geography.

Dr. Sharma is a charismatic geographer of Rajasthan State. His flair for language, captivating extempore speeches, power to generate brilliant ideas and enjoying the respect and confidence of a galaxy of royal students, researchers, friends, colleagues and superiors rose him to this status. He is the only geographer of the state, who laboured and worked in accordance with his ideas for the benefit, development and architecting the sphere of geography in the state and outside. He has guided as many 43 research scholars for the degree of Ph.D. in varied aspects of geography especially geomorphology and environment and thus he commands respect from a large number of student community. He is very popular teacher not only in the University of Rajasthan but also at any place of his working either as a visiting professor, guest of honour in conferences, seminars, work-shop etc. Due to this scholarly posture and academic apex, he has been accepted as a fellow of National Association of Geographers India (NAGI), Institute of Indian Geographers (IIG) Pune, Rajasthan Geographical Association (RGA), Geographical Society of India (GSI), Calcutta, Indian National Cartographic Association (INCA) and the like. He also enjoyed the prestigious position of RGA Presidentship; member, British Geomorphological Research Group, London (U.K.); Secretary, Institute of Indian Geographers; President, Indian Human Ecology Council; Vice-President of NAGI; member editorial board, District Gazetteers, Government of Rajasthan, Jaipur; President, Indian Institute of Geomorphologists etc. Hailing from a small village, Dr. Sharma established a corona of intellectuals, students of Geography, Geomorphology, Geology, Ecology etc. and disseminated the rays of scholarly light throughout his academic career and thus gave a distinctly dominated field of teaching, study, research, administration and organisation capacities and capabilities in Rajasthan, India and abroad. May it be in the capacity of a nominee for selections and evaluation or invitee to deliver lectures, Dr. Sharma's reach in the department of geography extends far and wide in India.

Professor Sharma was aware that to safeguard the study of discipline a smooth functioning of the university administration was a critical necessity. Hence by virtue of his popularity amongst teachers he was elected as President of Rajasthan University Teachers Association (RUTA) in 1992-93. Sharma's wise counsel and friendly cheer to dissolve vexing situation won him a place in numerous administrative post of the University. He worked with vigour not only in a number of committees of the University of Rajasthan but also took on various administrative posts in the University such as Principal, University Rajasthan College (1998-99), Director Pre-entry IAS Coaching Institute for S.C and S.T. 1995, Director Indira Gandhi Centre for Human Ecology, Environmental and Population Studies (1999-2003), Chairman of the Department of Geography (1989-92 and 1996-99). He served as Dean, Faculty of Science (2001-2004), member, Syndicate and also acted as a Convener, Dean's Committee looking after the office of the Vice-Chancellor.

Universities are kept alive merely on the shoulders of researchers and teachers, and it is only when the structure of the entire system is healthy can scholarship be upheld. Further it is not mere books and research, but the faceted experiences of life that channels wisdom, Sharma enjoyed the University in wide spectrum of ways. He epitomises the best in a university man.

The Department of Geography since its inception was lodged in University Rajasthan College Campus. Professor Sharma has been instrumental in establishing a separate building of the Department on the main Campus of the University. Research Assistance, Grants and Projects flowed to the department from diverse quarters like UGC, Ministry of Defence, Ministry of Environment and Forest and Space Application Centre, under the guidance of Professor Sharma.

Indeed, what is admirable is that despite a bed of internal uncertainties and strifes Professor Sharma carries on his professional duties with pleasant, easy and harmony to nature. He never turns a cynic and never cast a caustic comment. His flaouring as well as lucid voice and amicable temperament always disallow him to loose a friend. A mild heart-stroke in 2003 and open heart surgery clouds his health. Dr. Sharma maintains a compassionate composure despite dissatisfying. One fact is clear that from his village Aliyabad, Sharma has indeed come a long way.

Jaipur **Dr. M.L. Sharma**

Major Research Publications of Prof. H.S. Sharma

Books

1. Physiography of the Lower Chambal Valley and its Agricultural Development Concept 1979.
2. Ravine Erosion in India, Concept, 1980.
3. Tropical Geomorphology : A Morphogenetic Study of Rajasthan, 1987, Concept Publishing Company, New Delhi.
4. Perspectives in Geomorphology (Ed.) Concept International Series in Geography, No. 2, 1981. Volume I : Recent Trends, Volume II : Quantitative Fluvial Geomorphology, Volume III : Applied Geomorphology, Volume IV: Essays on Indian Geomorphology.
5. Environmental Design and Development, Conflict or Harmony (Ed.) Scientific Publisher, Jodhpur, 1987.
6. Ecology of Land and Water Management (Ed.) Kuldeep Prakashan, Ajmer, 1989.
7. Indian Geomorphology (Ed.), Concept Publishing Company, New Delhi, 1991.
8. Facets of Geography of Rajasthan (Ed.) 1992, Kuldeep Publications, Ajmer.
9. Hamara Paryavaran - Sampadakiya - Rajya Sarkar Paryavaran Vibhag Ke Sojanya Se 1995.
10. Bhuakriti Vigyan (in collaboration with Prof. Pramila Kumar), Madhya Pradesh Hindi Granth Academy, Bhopal.
11. Sustainable Development : Concepts and issues (Ed.) 1999, Concept, New Delhi.
12. Ranthambhore Sancturary - Dilemma of Eco-Development, Concept, New Delhi, 2000.
13. Environmental Conservation, Depliting Resources and Sustainable Development, Aviskar Publishers, Jaipur, 2003 (in collaboration with TI. Khan).
14. Physical Geography, Panchseel Prakashan, Jaipur-II Edition (Hindi), 2003.
15. Geography of Rajasthan, Panchseel Prakashan-III Edition (Hindi), 2003.
16. Ozone Depletion and Environmental Consequences, (in collaboration with T.I. Khan), Aviskar Publishers, Jaipur 2004.

Research Papers

1. Geomorphological Investigations of the Areas Around Jaipur. Journal of Saugar University, *Natural Science*, Vols. 15 & 16, Year 1967.
2. Morphometric Analysis of Jaipur Region, Uttar Bharat Bhoogol Patrika, Vol. VII, No. 2, Year 1967.
3. Land Capability Classification of the Lower Chambal Valley, India, Proceedings of the 21st International Geographical Symposium on Landuse in Developing Countries, 1968.
4. Evolution of the Chambal Scrap, *Uttar Bharat Bhoogol Patrika*, Vol. V, No. 2, 1969.

5. Genesis of Ravines of the Lower Chambal Valley, India, 21st International Geographical Congress, Selected Papers, Vol. I, pp. 114-118.
6. Erosion Surfaces in the Lower Chambal Valley, India. *International Geography* 1972, Vol. I, pp. 61-62, Canada.
7. Ramgarh Structure, India, *Nature, Mac Milian Journal Limited*, London, Vol.242, March 2, 1973, No.5392.
8. Some Applications of Geomorphology with Special Reference to India, University of Rajasthan, Studies in Geography, Vol. I, 1969-70.
9. Genesis and Pattern of Ravines of the Lower Chambal Valley Geographical' Review of India, Calcutta, Vol. 30, 1968. Special Issue, 21st International Geographical Union.
10. Geomorphology of the Ramgarh Dome, Rajasthan, Geographical Review of India, Calcutta, Vol. XXXII, 1970.
11. Micro-geomorphic Regions of the Lower Chambal Valley, *Indian Geographical Journal*, Madras, Vol. XIV 1970.
12. Problem of Ravine Erosion and its Effects on Landuse in the Lower Chambal, Valley, Essays in Applied Geography, Professor S.M. Ali Memorial Volume, 1975.
13. Morpho-agricultural Regions of the Lower Chambal Valley. *Geographical Review of India*, Calcutta, Vol. 34, No.2, pp. 109-122, 1972.
14. Some Preliminary Observations on the Weathering Process in the Lower Chambal Valley, University of Rajasthan, *Studies in Geography*, Vol. 2, pp. 69-79,1972.
15. Geomorphic Evidences of Rejuvenation in the Lower Chambal Valley, University of Rajasthan, *Studies in Geography*, Vol. IV, 1972, 1972-73, pp. 27-36.
16. Morphology of Ravines of the Morel Basin, *International Geography*, 1976, Moscow, Vol. I, pp. 219-223.
17. Quantitative Fluvial Characteristics of Streams of the Morel Basin [in collaboration]. *Geographical Review of India*, Calcutta, Vol. 39, No.4, 1977, pp. 365-366.
18. Quantitative Fluvial Characteristics of Streams of the Mej Basin [in collaboration]. Quantitative Fluvial Geomorphology. Edited Volume II, Concept Publishing Company, 1981.
19. The Study of Slopes in the Mej Basin [collaboration], *Geographer*, Aligarh, 1980.
20. Hill Side Slopes Evolution in the Morel Basin [Rajasthan], Proceedings of Symposium on Morphology and Evolution of Landforms, December 22-23, 1978, pp. 177-189.
21. Some Morphometric Techniques with relation to Discharge of Masi River Basin [AP], Proceedings of Symposium on Morphology and Evolution of Landforms, 22-23 December, 1978, pp. 168-179.
22. Some Aspects of Geomorphology of Menali Basin [Rajasthan] [collaboration], Professor R.P Felicitation Volume, Bhagalpur.
23. Trends of Geomorphic Research in India, *Annals of Association of Rajasthan Geographers*, Vol. 11, 1982.
24. Quartzitic Tors, A Case Study of Jawaharlal Nehru University, New Delhi, Perspective in Geomorphology, Vol. I, Concept Publishing Company, 1981.
25. Environmental Degradation in Jaipur Urban Complex. Proceedings of Symposium on Environmental Management, Allahabad, pp. 307-315, 1983.
26. Origin of Boulders and Gravels in Rajasthan Desert with Special Reference to Bap Boulders and Jayal Gravels. Paper presented before IV Nagi, Bombay, 1982. *Published in Annals* of RGA, 1983.

27. Lineaments and Drainage in Rajasthan, *Annals of National Association of Geographers*, India, Vol. IV, 1985.
28. Unprecedented Rainfall and its Aftermath at Jaipur [1981]. Presented before NAGI III, New Delhi, December 1981, *Annals of National Association of Geographers*, Vol. IV, 1984.
29. Inland Drainage of North Eastern Rajasthan, and Its Possible Link with GhaggarYamuna System. Transaction Institute, *Indian Geographers*, Pune, Vol. 5, November 1983.
30. Morphogenetic Regionalization of Rajasthan. *Annals of the Association of Rajasthan Geographers,* Vol. 3, 1984.
31. The Study of Palaeo-drainage of Rajasthan. *Annals of Association of Rajasthan Geographers,* Vol. IV, 1984.
32. Rajasthan: Identification of Eco-system, Environmental Problems and Recommendations for the Development of Areas under Desert Development Programme and Drought Prone Areas Programme, Occasional Paper, EMBS and TC, Jaipur, 1984 [collaboration].
33. Climate and Morphometry, A Case Study of Rajasthan. Paper presented at Ist International Conference of Geomorphology at Manchester [UK], 15-21 September 1985 [now published in *International Geomorphology,* Vol. 2, John Wiley & Sons, London. Ed. by V Gardiner.
34. The Problem of Desertification in Rajasthan, Paper presented at National Seminar of Resource Base and Pattern of Development, Hyderabad, March 1985 [published in proceedings].
35. Granitic Inselbergs: A Case Study of Rajasthan. Paper presented at VII Indian Geography Meet, Poona, February 1986, Explorations in Tropical Geomorphology, 1987.
36. Origins of Gravel and Boulders in Rajasthan in the Light of Palaeochannels. Paper presented at Desert Sediment Meeting held at London [UK] Published in proceedings.
37. The Origin of Interdunal Flats in the Rajasthian Desert, India. *Annals of Association of Rajasthan Geographers*, Vol. VI, 1987.
38. The Study of Planer Surfaces in Rajasthan. Special Volume of *National Geographers,* Vol. XXI, No. 1, 1986.
39. Aeolian Geomorphology of the Sand-dunes of Rajasthan. Paper presented at International Symposium on Geomorphology and Environmental Management held at Allahabad from 17-21 January; 1987 [Abstract].
40. The Problem of Soil Erosion in Rajasthan. Paper presented at the Annual Conference of INCA. held at Trivandrum [Kerala], 26-28 November 1986. [Abstract].
41. The Concept of Morphogenetic Regions and Land Systems: A Case Study of Rajasthan Annals of R.G.A., 1986.
42. Trends of Coastal Geomorphological Research in India. Key Note Address delivered at Mangalore, 21 February 1985 Published in proceedings.
43. Geography in Rajasthan, India Book Chronicle, March 1986.
44. Geomorphology and Environmental Assessment. Environmental Design and Development: Conflict or Harmony [Ed.] 1987.
45. Origin of Ravines, Lead Paper presented at National Symposium on Land and Water Management in Ravine Lands held at Agra, 20-22 March, Indian Association of Soil and Water Conservationists.
46. Climate, Drought and Desertification. *Presidential Address* delivered at the 19th

All Rajasthan Geographical Conference held at Sambhar Lake, 26 September, 1987.

47. The Lost Saraswati: An Application of Remote Sensing in Palaeo-Channel Study, Paper accepted for publication in Proceedings of the COSPAR [Committee on Space Research] held at Espoo, Finland [18-29 July 1988].
48. The Problem of Desertification and Strategy for Combating with Reference to Rajasthan Desert for Publication in the Proceeding of the Indo-Soviet Seminar held at New Delhi from 10-12 October, 1988.
49. Urban Geomorphology of Jaipur, Paper presented at Ist Conference of the Institute of Indian Geomorphologists held at Waltair from 28-30 October, 1988. Published in *India Geomorphological Diversity,* Edited, by K.R. Dikshit *at al.*, 1994.
50. Geornorphological Aspect of Process of Desertification in North Eastern Rajasthan, Paper presented at IInd International Conference held at Frankfurt, West Germany during 23-29 September, 1989. *Published in J. Geomorphology N.F Supplement. Bd. 87, Berlin* 1993, 61-69.
51. Waterlogging in Ganganagar-Hanumangarh Area, Paper presented at IInd Conference on Geomorphology held at Jaipur, 2-4 November, 1989, and published in Indian Geomorphology, 1991.
52. The Problem of Soil Erosion in Rajasthan—Survey and Monitoring by Remote Sensing Techniques. Paper presented at Asian Workshop organised by Indian Human Ecology Council, 2-12 April 1991, Published in proceedings.
53. Development of Desert with Reference to Rajasthan International Seminar on Regional Policies and Development in the Third World, 10-15 December 1991, BHU, Varanasi.
54. Landscape Changes in the Part of Indira Gandhi Canal Area, A Study based on Satellite Data presented at an International symposium on Evolution of Desert from 10-15 February 1992.
55. Land Transformation in Jaipur City, 5th Indo-British Seminar held at Mussouri, 6 12 April 1993. [published in a book on Environment and Development by Concept Publishing Company, New Delhi, 1993].
56. Environmental Hazard in a Semi-Arid Region, A Case Study of the Bandi River Basin, Paper presented at National Seminar held at Aligarh Muslim University, Aligarh, from 23-25 October, 1994.
57. Canal Irrigation and Environmental Degradation, A Case Study of Indira Gandhi Canal Area in Ganganagar District, Paper presented at 3rd International Seminar on Rajasthan at Jaipur from 14-18 December, 1994, Journal of Land and Water Management, Vol. 6, 1995.
58. Quaternary Stratigraphy and Morphology of Desert Ranns and Evaporite Pans in Central Rajasthan, India, *Man and Environment*, Vol. 22 (2), 1997, pp. 1-10.
59. Palaeo-climatic Changes in Rajasthan, A Review Paper presented at First International Science Conference under GAIM from 25-29 September 1995, Garmisch, Germany, *National Gographical Journal,* Varanasi, 2001.
60. Status of Water Resource and Management for sustainable development in Rajasthan, Invited Paper in Indian Science Congress held at Pusa, New Delhi, 4-7 January, 2001, Published in *Rajasthan Review*, 2002.
61. Geomorphology of The Ranthambhore National Park and its Environs, Published in *Recent Advances in Geomorphology* Edited by S.K. Tandon and B. Thakur, 2002, Manisa Publications, New Delhi.
62. Impact of Population Growth on Environment, in *Environmental Conservation* Edited by H.S. Sharma and TI. Khan, 2003, Aviskar Publishers, Jaipur.

63. Rajasthan Tribal sub-plan area—A Eco Development of Data Set (1 991) in *Environmental Conservation*, Edited by H.S. Sharma and TI. Khan 2003, Avishkar Publishers, Jaipur, 2003.
64. Geomorphology, Environmental Hazards and climatic change, *14th I.G.I. Presidential Address,* Chidambarm, Dec. 2001, Published in *Indian Journal of Geomorphology* Vol. 7, pp. 89-98, 2002.
65. Evaluation of Master Plan and changing character of Fringe Areas of Jaipur Urban Complex accepted for publication for proceeding of National Seminar on Managing The Urban Fringe of Indian Cities, University of Pune, 27-29 Nov., 2002.
66. Global Environmental Changes and its consequences: Three Prof. R.N. Dubey Memorial Lectures delivered on 15 and 16 Dec. 2002 at Allahabad University, Allahabad, Published by Prof. R.N. Dubey Memorial Foundation Edited by Prof. L.R. Singh, 2003 pp. 1-94.
67. Water Borne Disease—Flurosis-A silent killer—A Case study of Dausa District, Rajasthan, Transactions of I.I.G., Pune, Volume 25 (No. 1 & 2), 2003, pp. 108-1 17.
68. Drought and Depletion of Water Resource in Rajasthan : A Strategy of Sustainable Development in the book *Water Crisis and Sustainable Management* by D.N. Singh, J. Singh and K.N. Prudhu Raju, Tara Book Agency, Varanasi, 2003, pp. 122-139.
69. Ecological Perspective of Sustainable Development Keynote address in valedictory session- of 31st Annual conference of RGA, Dec. 2003, published in *Geographical Aspects,* Vol. 6, 2004.
70. The Indira Gandhi Canal Project : A Critical Review of the Institutional Arrangements for Resource Management, in *Prof. Majid Hussan Felicitation volume*, Jamia Millia University, New Delhi.
71. Heirarchy & Spatial Distribution Pattern of Urban Centers in Rajasthan in *Professor Ashok Datta (Akron) Felicitation Volume,* Delhi School of Economics, University of Delhi, Delhi (Editor Prof. Baleshwar Thakur) in press.

Research Projects

1. Morphogenetic Regionalisation of Rajasthan, UGC sponsored project (1980-83).
2. Land System, land capability and Landuse, Case Studies from Rajasthan, UGC sponsored Project (1986-89).
3. Coordinated an International Project on Sustainable development of Arid Lands, Indo- 04- Interinstitutional Project between University of Rajasthan and Noragric, AS Norway sponsored by NORAD.
4. EcoPerception—A Man Environment Relationship, A case study of Ranthambhore Sanctuary, Rajasthan, Project sponsored by Ministry of Environment and Forest, Government of India, New Delhi, 1998.
5. Morphometric Study of Sand dune Type Areas of Rajasthan Desert, Project sponsored by Defence Research and Development Organisation (DRDO) Delhi in May 1999 for one year completed in June 2000.
6. Status of Desertification in Dausa District, project sponsored by Space Application Centre, Ahmedabad, completing on 31th Dec., 2004.
7. An Integrated Approach for Watershed Management—A case Study of Mej Watershed of South Eastern Rajasthan, Project Sponsored by UGC, New Delhi from 1st July, 2003 to 30th June, 2005.

INTRODUCTION

The discipline of geomorphology has under gone sea change in content, methodology and approaches to the study of landforms and related process since 1945 when R.E. Horton introduced quantitative methods for the analysis of morphometric characteristics of fluvially eroded drainage basins. The post-1955 geomorphology witnessed a sharp divide between evolutionary approach evolving cyclic concept of landform development characterized by progressive changes through time and process response approach involving non-cyclic concept of equilibrium model. Thus the need of hour is to integrate the cyclic concept involving long term historical evolution of landforms and non-cyclic concept involving dynamic equilibrium, functional and process response models on the one hand and micro geomorphology involving smaller spatial and shorter temporal scale and mega geomorphology involving larger spatial and longer temporal scale on the other hand (Savindra Singh, 1998, 2000).

Research work in geomorphology has focused on landforms, erosional surfaces and morphometric analysis. There has been an overwhelmingly large number of case studies of landscape evolution which utilize the techniques of identification, relative dating and interpretation of erosional surfaces. In the last few years, new themes have attracted the attention of geomorphologists, including channel forms and process, sea-level changes, quaternary geomorphology and weathering process.

Geomorphic understanding is crucial for scientific landuse planning and environmental reconstruction. Multi-disciplinary studies in the earth sciences have, in the past decade, clearly demonstrated the importance of looking at the Earth-Ocean-Atmosphere system in an integrated manner. New technologies are available now for the study of the earth's surfacial processes. These include those related to satellite data, geodetic measurements and computational methods. Globally, there is a conscious efforts by geomorphologists to move towards quantification and determination of the rates of surface and near-surface-processes.

India has one of the most varied landform systems in the earth. They range from Thar desert to Sundarban Delta, and Himalayan heights to the monotonous plains of the Ganga basin. The Deccan tableland, the oldest mountain ranges denuded to form Aravali hills, thousands of kilometres of coastal land, apart from rivers, lakes, continental shelves, oldest to youngest mountains and great variety of climates and vegetation covers, offer a great opportunity for Indian researchers to make immense contribution to scientific study of evolution of surface features with various facets of development studies. In India there has been a considerable effort by quaternary geoscientists and geomorphologists to keep pace with the import of new methods and technology in this area of knowledge.

It was realized very early in the history of the development of geomorphology that the landscape of Peninsular India is of considerable antiquity and erosional surfaces form a prominent component of the landscape. Therefore studies on the geomorphic history and denudation chronology of the subcontinent became fashionable and have been by far the most popular themes of research among physical geographers. The very well know concepts developed by Davis. Penck and Budel provided the major stimulus to research in this field. Field observations and topographical maps were used for drawing superimposed profiles, hypsometric and altimetric curves for the area under consideration. However, the attention was mainly on spatial forms and patterns, with virtually no attention given to processes. For instance, the control of lithology, nature and intensity of geomorphic processes responsible for shaping the landforms and the chronology of development of landforms are still far from understood. In view of above statements, it is essential to encourage inter-disciplinary studies to have a holistic view of a region.

Prof. H.S. Sharma had striven during his academic career in the University of Rajasthan to establish physical geography and to encourage inter-disciplinary studies throughout his career he has been working and writing research papers on geomorphology and geoenvironmental issues of dry and subhumid regions of Rajasthan. Therefore, it would be a fitting tribute for his endeavours could be through the preparation of volume dealing with various aspects of geomorphology and environmental geo-sciences.

The present volume **Geomorphology and Environmental Sustainability** is dedicated to Prof. H.S. Sharma. It is a collection of 29 papers which have been organised in the following parts :

(1) Geomorphic Processes, (2) Applied Geomorphology, (3) Environmental Change, (4) Environmental Impact and Sustainability, and (5) Bio-diversity.

Amal Kar provides a cogent account of the relevance of processes, morphology and sand control measures in the Great Indian Desert. Amal Kar discusses the various types of sand movements, sand dunes and spatial pattern of aeolian activity with the help of wind erosion index. He has also classified the wind erodibility (classes) in western Rajasthan, stability of dunes, pattern of wind erosion and sand stabilizing programme. S.R. Basu, Anasuya Ghosh and Sunil Kumar Dey highlight the fluvial geomorphic processes in the light of meandering and cut-off of the holy river Bhagirathi in its lower reaches as river Ganga. Authors have studied the sinuosity, its relation with variance and predicted about the breaking point of a meandering river which leads to the formation of a cut-off. P.S. Dhinwa, Ajay and Bhavsar have discussed the process to combat the process of desertification in cold desert of Ladakh district (J&K) using RS and FIS. The author's have concluded that the anthropogenic factors and vegetal degradation are the important factors for desertification in this area of scanty winter rainfall. Neotectonic activity and seismicity have been discussed by J.K. Sharma and M.L. Sharma in their paper on Sambhar Lake and Environs. The problem of flood and water logging in Ghaggar river bed of Hanumangarh district have been analysed by S.C. Kalwar, H.V. Sharma and Avinash. Spatial pattern of land degradation in the Northern highland of Orissa has been discussed by *Punyatoya Patra* and Baleswar Thakur. Interpretation of ravine lands in the sub-watersheds of the Morel river using remote sensing technique has been done by R.D. Doi.

Surendra Singh has discussed the application of geomorphology in the appraisal and integration of natural resources for sustainable land use planning in an arid environment of Rajasthan. S.C. Mukhopadhyay has discussed applied geomorphology of eastern coastal zone with particular reference to the Sunderbans. S. Padmaja and G. Somiah have studied the Aestheticism and landscapes with a few issues concerning the scenic evaluation of landforms for recreation taking Khammam district of A.P. The applied aspect of geomorphology has been discussed by S.K. Sharma in the paper perspective of land resource development in Madhya Pradesh and Chhatisgarh. L.C. Agarwal has shown the relationship between distribution of settlements and correlation with morphometric attributes like slope and relief in the Parbati basin of south-east Rajasthan.

S.K. Wadhawan and A.K. Singhvi propose a text of material on environmental changes in drylands of India based on past dynamics and concluded that the timing of major millennium scale of dryland environmental changes was not synchronous from place to place owing to the roles of different climatic forcing factors that operate from decadal to multi-millennial time scales. Savindra Singh empasises the role of geomorphological indicators in climate change. He is of the opinion that on the basis of climochronology it can safely be inferred that climates have changed in the geological past, are changing now and they are bound to change in future. Greenhouse effect, global warming and climate has also been discussed by T.I. Khan. Azra Akhtar and Kalyan Hironi warns against the deadly consequences of acid rains on plants and ultimately to human race.

Santosh Shukla discusses the environmental impact assessment as a tool for environmental management. D.N. Sati also emphasises the impact of process of urbanisation on the environs of Ajmer city. R.K. Rai has evaluated the mining activities in Meghalaya and their impact on environment. The environment impact of Kota thermal power station on Kota city and environs has been assessed by B.L. Sharma, L.C. Agarwal and Pooja Pour. Jayanti Das and Lakshmi Sivaramakrishnan have thrown some light on the geo-environment and its influence on social factors of a rural and urban areas in West Bengal. Natural resource management for sustainable development—causes and concerns—have been discussed by S. Padmaja and Kausalya Ramachandran. Rajesh Bhakar has evaluated the sustainability of irrigated agriculture in arid zones taking a case study of Indira Gandhi Nahar Pariyojana. Meenakshi Vijay, Suresh C. Sharma, Ashu Rani and Kalyan M. Hironi have discussed the ground water quality of villages near fertilizer plant in Kota and Baran districts. R.B. Singh and S. Sen Roy have shown the influence of water availability on poverty in India. Utility of integrated approach of ground water management in depleted areas for sustainable development has been studied by D.D. Ojha and F.M. Golani

Ved Prakash Meena has discussed the salient features of desert national park and pointed out its uniqueness for conservation and development of bio-diversity in India. Prakash P. Bakre in his paper has highlighted that pesticides are killing out birds.

In the preparation of this felicitation volume, the editors have received cooperation and encouragement from colleagues, friends, students and admirers of Prof. H.S. Sharma to whom we are beholden and grateful.

The publisher, M/s. Concept Publishing Company, New Delhi not only undertaking the task of publication but doing fairly good job in a stipulated time. The editors would like to thank Shri Ashok Mittal of Concept Publishing Company for meticulous execution of this publication.

The editors also express sincere thanks to all those who have directly or indirectly helped in bringing this book in the present form.

Editors

SECTION-I

GEOMORPHIC PROCESS

1

Sand Dunes of Thar Desert

Relevance of Processes and Morphology in Sand Control Measures

AMAL KAR

Introduction

Sand dunes and other aeolian sand bodies in desert often create problems for regular human activities. Human settlements, roads and other communication lines, canal networks, agricultural and allied activities, are all affected when strong sand-shifting wind erodes sand from certain piece of land and deposits it elsewhere. Since many of the menacing sand bodies are in the form of sand dunes, the view of a sand dune usually provides an awe-inspiring vision of invading sand masses. All types of sand dunes in the Thar Desert of Rajasthan, India, and adjoining areas of Pakistan are, however, not so menacing unless their ecology is severely disturbed. Many dunes are well stabilised and have a significantly low rate of movement, while some are so stable that those may be termed as fossil dunes. There are some dunes, however, which move fast and pose the maximum threat to human activities. Therefore, a knowledge of different sand dune types, their morphology and genesis, as well as an understanding of the spatial pattern of wind erosion system here, can help in proper sand control measures. We provide here a summary view of the major findings on these aspects, as relevant to sand control measures.

Sand Moving Wind and Wind Erosion

Sand-moving winds are those that have sufficient velocity to move sand and create the dunes. Bagnold (1941) calculated the average impact threshold velocity of wind for dune sand as 16 km h^{-1} at 1 m height (or 4 m sec^{-1} at 1 cm height). The rate of sand movement varies approximately as the cube of drag velocity. Therefore, sand transportation capacity per hour increases sharply with per unit increase in wind velocity. For example, sand transportation by a 50 km h^{-1} wind for one hour is equivalent to that by a wind of 30 km h^{-1} for 14.3 hours and by a wind of 20 km h^{-1} for 611 hours. Since the sand moving winds in our desert blow from one or two directions and during some months of the year only, it is possible to find out the potential sand drift and drift directions from wind records.

Aeolian activity in the Thar Desert is usually restricted to the period of strong summer southwest monsoon wind. Sand and dust raising winds start blowing from March onwards when the surface is sufficiently dry after the previous year's monsoon rainfall (July-September). The mean maximum wind speed is reached during June. May and July are also very windy. Since this is also the period when much of ground flora is dry, conditions are appropriate for aeolian activities. The wind dissipates with the arrival of monsoon rains and sand blowing activity ceases. The north-eastern wind of winter months is weak and is a poor agent for such activity.

The spatial pattern of aeolian activity can be better appreciated from the wind erosion index map (Kar, 1993), which was constructed with the help of modified Chepil formula (Yallon and Ganor, 1966):

$$C' = (100*V^3)/[2.9\,(PE)^2]$$

where C' is climatic wind erosion index,
V is mean wind speed at 10 m height,
PE is the Thornthwaite's measure of precipitation effectiveness, and
2.9 is the annual average climatic index at Garden City, Kansas; values at other stations are expressed as a percentage of this figure.

The index values were classified into six groups, ranging from very low to extremely high (Table 1.1). The maximum potential wind erosion occurs roughly to the southwest of Jaisalmer in Rajasthan, and adjoining parts of Pakistan. It decreases gradually towards east and northeast, concurrent with the increasing rainfall and decreasing wind speed. This pattern is reflected in the spatial pattern of sand dune formation in the desert (Fig. 1.1).

Table 1.1 : Wind erosion index (March-July) for the meteorological stations in Thar Desert

Wind erosion index	*Category*	*Station*
480 and above	Extremely high	Jaisalmer, Chhor
120-479	Very high	Phalodi
60-119	High	-
30-59	Moderate	Bikaner, Jodhpur, Barmer, Pachpadra, Hyderabad, Sukkur, Khanpur
15-29	Low	Ganganagar, Churu, Nagaur, Jaipur, Badin
1-14	Very low	Hisar, Sikar, Sambhar, Ajmer, Erinpura, Bahawalpur

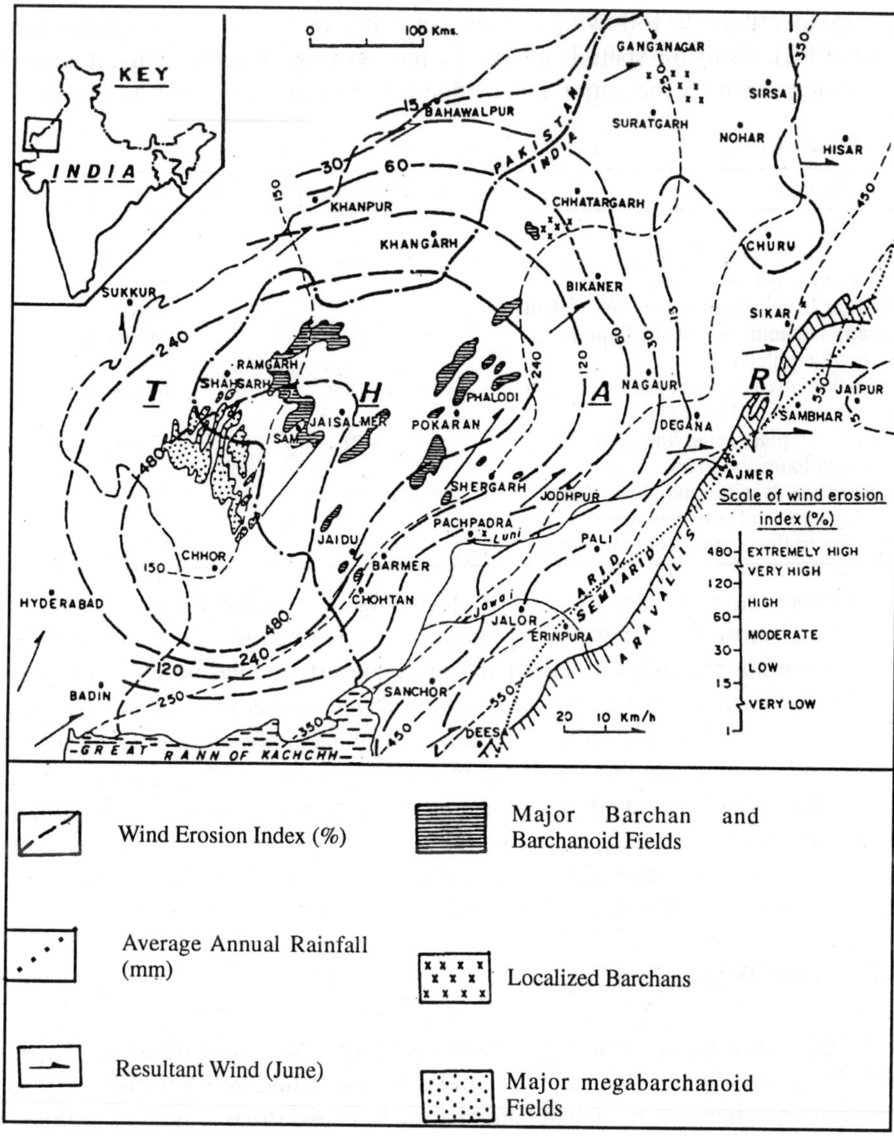

Fig. 1.1 : Thar Desert-Wind Erosion Index, Annual Rainfall & Mobile Crescentic Dunes

The most vulnerable landforms to wind erosion/deposition are the sand dunes and other sandy landforms. A closer look, however, indicates that the sandy landforms in the east are more stable than the sandy landforms in the west. While erosivity is a major factor for such spatial pattern, the erodibility of sediments is another major determinant. Using the per cent stable aggregates in a typical soil as the criteria for erodibility (Skidmore, 1988), Kar and Joshi

(1995) identified the following classes of wind erodibility in western Rajasthan (Table 1.2). Using the spatial pattern of wind erosivity and erodibility, it is now possible to identify the major areas vulnerable to wind erosion (Kar, 1996).

Table 1.2 : Wind erodibility classes in western Rajasthan, using soil aggregate method

Soil character	*Average stable aggregate (%)*	*Erodibility class*
Very deep, fine sand of dune complex	1-5	Very severe
Deep, loamy fine sand and fine sand of sandy plains and some dunes	5-10	Severe
Shallow loamy fine sand on sandy plains	5-8	Severe
Loamy medium sand of small dunes on flood plains	7-10	Moderate
Deep brown sandy loam and loamy fine sand on flat alluvial plains	10-15	Moderate
Deep flood plain soils, fine sand, sandy loam and loam	30-40	Slight
Seirozem soils with sandy loam, loam, clay loam and silty clay loam	20-50	Slight
Medium to fine textured gray brown soil	40-50	Negligible

Introduction of tractor for deep ploughing in recent decades, instead of traditional animal-driven wooden plough, has increased the sand load manifold for the aeolian processes in large parts of the desert, and has accelerated sand mobility. Increased destruction of the natural plant cover in common grazing lands for fuel and fodder, and cultivation in the marginally suitable sandy areas is also the responsible factors. Field measurements in western Thar during a sandstorm in 1987 revealed that the deep ploughed sandy plains lost more than 3000 tonnes of soil per hectare, while the sandy plains with 10-12 per cent plant cover, or with higher percentages of soil clods and some control measures, suffered almost negligible erosion (Dhir *et al.,* 1992).

Sand Dune Types and Systems

The major dune types in the Thar Desert are: parabolic, linear, transverse, star, as well as a variety of network dunes, major obstacle dunes along the hill slopes, barchans, barchanoids, megabarchanoids, and low sand streaks, including shrub-coppice dunes, zibars, source-bordering dunes, etc. (Singh, 1982; Kar, 1993, 1996; Fig. 1.2). The types reflect variations in shape and morphology of the dunes. Barchans, barchanoids and megabarchanoids have cresecentic shape and are devoid of vegetation. These are the fastest moving sand bodies in the desert.

Pandey *et al.* (1964) classified the sand dunes of the Thar on the basis of their age and stability. They recognised two 'systems' of dunes: the dunes of 'old system' and 'new system' (Singh, 1982). The new dunes are forming today and consist of the crescentic dunes, sand streaks and sandy hummocks. The old

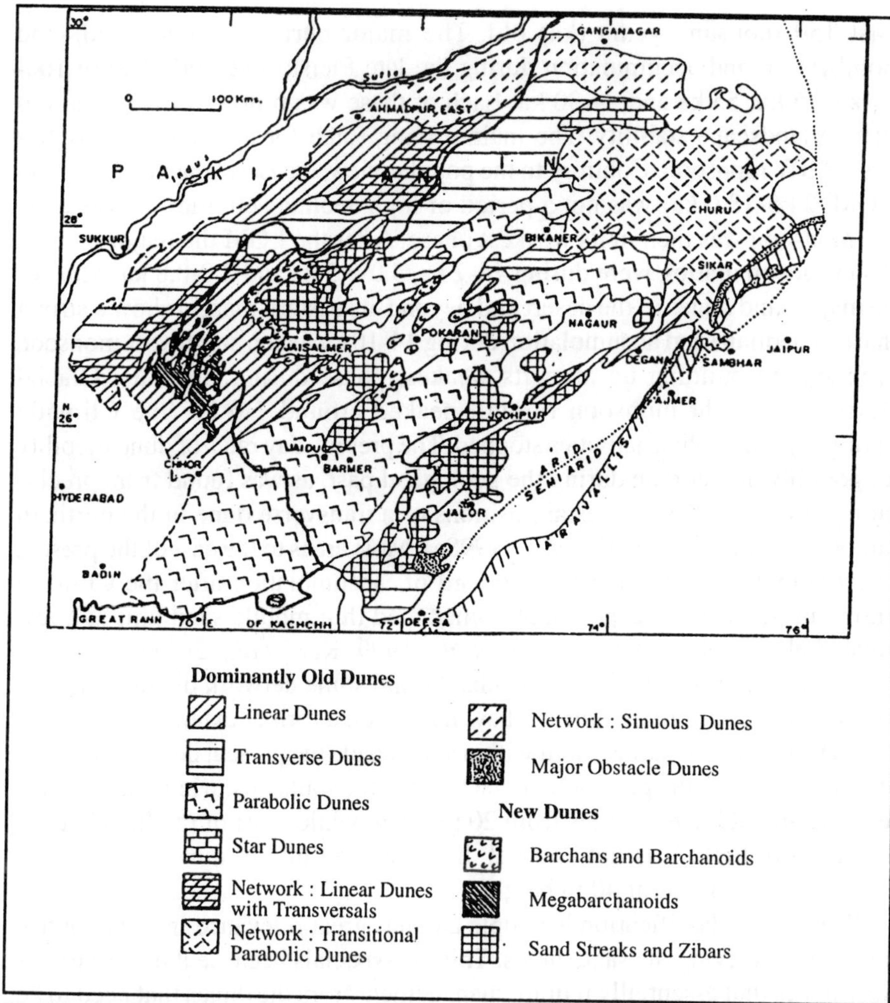

Fig. 1.2 : Thar Desert—Sand Dunes

dunes include the parabolic, linear, transverse, star, major obstacle and most network dunes.

Majority of the old dunes were last formed during 13 ka to 18ka BP, when the desert extended beyond the Aravalli hill ranges in the east (Goudie *et al.*, 1973; Wasson *et al.*, 1983 Singhvi and Kar, 1992). The sedimentation history of the old dunes *vis-a-vis* climate change is now better understood through a recently-concluded multi-institutional and multi-disciplinary research on a number of deep and profiles across the desert and its eastern fringe areas, which addressed the problems of process geomorphology, granulometry, sedimentology, geo-archaeology and absolute dating of sediments. It has been revealed that the aeolian history of the desert and its eastern fringe areas is at least 150 thousand years (ka) old. The major periods of enhanced sand mobilization and accumulation during the late Pleistocene period were 100-114ka, 75 ka, 55 ka and 24-30 ka. Each of these was followed by a period of relative landscape stability. The main driving force for the aeolian activities was the SW monsoon. Contrary to the popular view, the Last Glacial Maximum (LGM, 24.5-18.0 ka) was not a period of high aeolian activities, because the SW monsoon was then in its weakest phase. After the LGM the major periods of sand accumulation were during 14-7 ka, 5.0-3.5 ka and 2.0-0.8 ka. Most of the major sand dunes in the desert and beyond its eastern margin show a strong phase of aeolian sand accumulation during 14-10 ka. when southwest monsoon wind started building up after its weakest phase during the Last Glacial Miximum. As sht monsoon rainfall peaked around 7-6 ka (the Climatic Optimum), the aeolian activities stopped. The present-day rates of dune mobility are possibly higher than during the geological past, as was found from precise dating of the successive crestal positions of a transverse dune in the northern part of Jaisalmer district (Kar *et al.,* 1998a). Most areas to the east of the present desert boundary do not have any record of accumulation after the Climatic Optimum, save and except a human-mediated thin mobile sand unit at a few isolated places (Kar, 1999; Thomas *et al.,* 1999; Kar *et al.,* 2001)

Dissected obstacle dunes, whalebacks and some network dunes along the desert's wetter eastern margin and beyond mark the former extent of the desert. The trend of the 'old' dunes, both within and outside the present desert boundary, correlate well with the present regional trend of the southwest wind. The average height of the old dunes varies from 20 to 40 m, while most of the 'new' dunes are 2 to 8 m high. It is not unusual, however, to find old dunes of 30-60 m height and new dunes of 40 m height.

The above classification is simple and has great relevance in sand control measures, involving the sand dunes. This is esspecials because it differentiates those dunes that essentially require management from the dunes that need to be controlled.

Compound parabolic dunes cover the largest area in the Thar Desert. The dunes look like a coalescing U form, and their curved nosal part appears in the down wind direction. A three to four tier arrangement of dunes is noticed in almost every cluster (Singh, 1982; Wadhawan, 1988, 1998; Kar, 1993, 1996). Although the high dunes are mostly stabilised with copious vegetation and abundant carbonate root casts within them, increased human and livestock pressures on the them are reactivating many and are leading to the formation of secondary, low crescentic bedforms atop them. A gradual change in dune shape is noticed from the south-western part of the field to the north-eastern part. In the south-west the dune arms are usually 3 to 4 km long and join at sharp angles to form the hair-pin and chevron patterns, while towards the nort-east the arms become shorter and the sharp junction at the nose become more curved. The arm length is gradually reduced to about 500 m (Kar, 1987, 1999). Detailed measurements with Global Positioning System (GPS) revealed that in the central part of the desert the parabolic dunes contains 37 per cent of the sand volume in their nosal part (Kar *et al.,* 1998b).

Liner dunes are aligned in the direction of dominant and sand transport. These dunes are less numerous in the Thar than the parabolic dunes. The dunes occur mostly in the western part of the desert, although other smaller occurrences are noticed in the central and eastern parts. Unlike the linear dunes in the Sahara and the Middle East where these are called seif (sword edged or sharp crested) dunes, the linear dunes of the Thar are broadly convex. Natural vegetation is an integral part of many of these dunes. In their main field in the sest the recently forming and the reactivated linear dunes are barchan-crested. The dune form changes from 10 to 12 km long, unjoined and short-feathered ones in the extreme west to inverted Y-junctured ones (open upwind), and then to clusters of 5 to 8 linears joining at acute angle, but without much donwwind prolongation from the zone of convergence. This represents transition to the field of hair-pin parabolic dunes further east. A gradual decrease in wind energy eastward and a change from the narrow unimodal to a wider unimodal wind pattern in that direction should create such changes in form and explains the coexistence of the linear and the parabolic dunefields (Kar, 1987).

Transverse dunes occur mostly along the north-western part of the Thar and also in the east-central part. Unlike the linear dunes, which are aligned in the direction of the sand-moving south-west wind, the transverse dunes are oriented across the path of the sand-moving wind. The crest line is everywhere characterised by broad crescent. While the 20 to 40 m high old dunes have copious vegetation and sub-parallel linear ridges along their windward slopes, the new ones of lower height do not possess such linear ribs.

Star dunes occur mainly in the Suratgarh-Rawatsar-Nohar tract in the north, and mostly in chains (Kar, 1993; Wadhawan, 1996). Peaks of 10 to 25 m height are noticed along the central ribs where arms from two or more directions join. These dunes have formed at the convergence of southwesterly, westerly and northwesterly wind currents. Network dunes of different kinds are found especially along the transitional zones between the fields of major dune types. The most dominant are the compound parabolic dunes with transverse forms at right and the large parabolic forms with network patterns within. Both the patterns are noticed in the nort-eastern part of the Thar and are transitional to the compound parabolic dunes. Their average height varies from 8 to 15m. A less energetic wind regime, a greater directional variability of the wind and an increased vegetation cover might have contributed to their formation. In parts of the Saraswati-Krishadvati palaeo-flood plain in the north, many of these dunes were formed along streamlets, and are now a good indicator of palaeochannels.

Among the new dunes, barchans and barchanoids occur mostly to the west of the 120 contour of the wind erosion index (Kar, 1998, 1999). This runs approximately along the 200 mm isohyet in the sourthern part of the desert, but in the northern part it does not follow that isohyet because of a lower wind regime. The dunes are ideally formed along the margin of the rocky desert of Jaisalmer-Pokaran where 2 to 3 m high dunes move at an average speed of 27 to 32 m per annum. Barchanoids, or the coalesced barchans, form under many situations. In the low wind erosion index zones the dunes form in the interdune plains, mainly due to high human-induced reactivation of the old dunes and the sandy plains. Between 120 and 240 contours of the index there are more chances of such formation due to a drier envitronment and easy availability of movable material. Compound megabarchanoids occur in the relatively inaccessible westermost part of the Thar where the average annual rainfall is less than 100 mm and the wind erosion index is more than 480. A number of closely spaced elliptical fields of these 20-40m high dunes could be noticed amidst fields of vegetated linear dunes. Although not as fast moving as the low barchans, the barchanoids and the megabarchanoids are also very mobile and dangerous to human activities.

Among the other dunes the shrub coppice dunes (or the nebkhas) are the most dominant non-crescentic recent aeolian bedforms. Bushes of *Capparis decidua* and *Ziziphus nummularia* provide the best environments for their growth and dune-plant symbiotic relationship. *Calligonum polygonoides*, *Leptadenia pyrotechnica*, *Haloxylon salicornicum*, *Lasiurus sindicus* grass clumps, *Acacia jacquemontii* and degraded bushes of *Prosopis cineraria* also encourage their formation in different environments. Low linear and transverse sand streaks are common in the areas of intensive cultivation in the deep sandy terrain, especially to the east of 250 mm isohyet. The sand streaks usually form along the peripheries of agricultural fields when the farmers fence their fields to avoid sandy soil being blown out during the sandstorms (*Aandhi*). The fences, however, become the loci of sand accumulation.

The Thar sand dunes have mean sand grain sizes of 0.15 mm (2.74 phi) to 0.20 mm (2.30 phi). Like in other deserts, quartz is the most common constituent, but feldspar and some heavy minerals are also found (Singh, 1977). The dominance of quartz is due to the resistance it provides to mechanical abrasion and chemical weathering. Nearer the rock outcrops many small and unaltered rock fragments are also noticed within the sand population. These are related to the local geological set-up. The other major source of sand is the alluvium in the pre-existing landforms. The earlier view that the Great Rann of Kachchh provided the sand has been proved wrong (Pandey *et al.*, 1964). The dunes with

Fig. 1.3 : Western Rajasthan—Wind Erosion

stable crest (old dunes) have generally a finer grain size and better sorting than the dunes with mobile crest (new dunes and old dunes with reactivated crest) (Kar, 1993, 2002). Much of the high level suspended load, moving out of the desert and settling along the desert margin and beyond as dust fall, is composed of particles smaller than 0.06 mm size.

Stability of Old Dunes

The greater stability of old dunes is attested by the calcified nodules and root casts of plants within them. These dunes also support copious vegetation under natural conditions, with a basal cover of 4-8 per cent or more. Such a cover can reduce sand transport on the dunes by 30-50 per cent of that on bare sand, assuming a wind speed of 36 km h^{-1} (Buckley, 1987). The broadly convex vegetated linear dunes commonly possess a lesser basal cover of vegetation than the transverse and the compound parabolic dunes, which are essentially the sand accumulating forms. Apart from providing stability to the dune surface, vegetation also influences dune morphology, like anchorage and elongation of the parabolic dune arms and restricting sand movement along the linear dune crests to the strongest ridge-parallel winds. However, fewer plant species survive in the crestal part of the dunes in the very windy western part of the desert. The best colonisers of the dune crests include *Leptadenia pyrotechnica, Calligonum polygonoides, Panicum turgidum, Acacia jacquemontii* and *Crotalaria burhia* (Prakash and Gupta, 1976; Saxena, 1977).

Present-day Pattern of Wind Erosion

Based on field data and remote sensing signatures, the spatial pattern of wind erosion in western Rajasthan has been delineated (Singh *et al.,* 1994; Narain *et al.*, 2000). Broadly, the extreme western part of Jaisalmer district, bordering Pakistan, is under very severe erosion, where the wind crosivity index is greater than 480. Human activities are negligible in the area and, therefore, the severity is related to high wind and extreme aridity. The moderate to severe erosion in the central part are related mostly to the high human and livestock pressures on a vulnerable sandy landscape, although the pattern also partly follows the erosivity contours, especially in the southern part. The north-eastern part and the eastern fringe areas are affected by very less erosion, except in some pockets where, despite a low erosivity, the high erodibility and high human and livestock pressures cause moderate to severe wind erosion problems (e.g., Ratangarh-Churu-Sikar area; Pushkar area). Overall, 15 per cent area is affected by severe to very sever erosion, 35 per cent by moderate erosion and 25 per cent by slight erosion (Table 1.3; Fig. 1.3).

Table 1.3 : Wind erosion/deposition classes in western Rajasthan

Erosion/deposition class	*Area (km^2)*	*Percentage of total area*
Very severe	5800	2.78
Severe	25540	12.23
Moderate	73740	35.32
Slight	52690	25.24
Negligible	50981	24.43

Relevance in Sand Stabilization Programme

A knowledge of the air and sand flow patterns in the dune-covered areas and across other kinds of barrier help in finding out the possible vulnerable sites around those structures. For example, the old and vegetated transverse dune chains are generally associated with narrow deflation plains slightly downwind of their leeward slope, because of the recirculating captive airflow there. Such is also the case with the chains of vegetated compound parabolic dunes and the complex dunes involving the linear and the transverse dune ridges. Therefore, such sites in the interdune corridors are less prone to sand accumulation. However, once the cultivation and grazing pressures on the dunes increase, the dune slopes become more reactivated. The net sand movement then takes place from the windward slope and dune arms towards the crest, and from there into the erstwhile zone of deflation through sand avalanche and new bedform migration. Such information may be vital for sand control measures.

While the western part of the desert, especially the areas to the west of 120 contours of wind erosion index has high mobility of sand, the sand mobility decreases sharply in the eastern part, especially because of higher moisture status in the sand and a lower wind regime. There is, however, remote sensing evidence of very high reactivation of old dunes in some areas in the east (e.g. Nagaur-Churu tract) due to more pressure of tractor ploughing and cultivation. Many reactivated old dunes of the east have a stable surface at a depth of 1 to 3 m below the loose sand apron. Such dunes could be provided their earlier stable and vegetated status through better management practices everywhere in the desert. Not all the sand dunes are, however, mobile and, hence, not all the different dune types require the same kind of stabilisation treatment. Some dunes like barchans may even turn more dangerous if their regular pattern of movement is interfered with. The sand-passing dune forms may react differently to a specific kind of treatment than a sand-accumulating form.

Sand stabilisation programme on the new crescentic dunes, which are devoid of vegetation, need to be taken up very carefully. Studies in many deserts have indicated that it is much wiser and economical in the long run to avoid the paths of barchans, rather than to interfere with their natural movement. While biological control of such dune has not met with much success, the chemical stabilisation methods like oil spray have some success. However, once the surface is breached by wind a huge amount of sand becomes suddenly available for reworking and the problem multiplies (Watson, 1990). Moreover, the oil spray is costly and may spoil the land permanently in countries like India where even the barchan slopes are put to mono-cropping of pearl millet whenever there is monsoon rainfall. Therefore, it may be worthwhile to attempt a reshaping of the dune, or to stabilise the dune horns by mechanical means, so that its optimum aerodynamic form is disrupted and the barchan is forced to degenerate into other less dangerous forms. Trenching may also be tried. However, these techniques do not have any permanent success. As soon as the sand passes the zone of induced airflow it comes again under the grip of the prevailing aerodynamic system and the problem restarts. Moreover, the barchans are not some isolated dunes in the desert plains. They usually occur in a long series. So, the immobilization of one or two barchans, or disrupting their aerodynamic shape may be of little consequence when new barchans start arriving at the site and increase the sand load there.

Construction activities within the desert often lead to some sand reactivation, which, in due course of time, threatens the structures constructed. In order to minimise such hazard it is desirable to follow some precautions during the construction activities. For example, at the request of the Gas Authority of India Limited (GAIL) during construction of a pipeline for natural gas from Gamnowala Tar and Ramgarh to the west of Jaisalmer, Kar *et al.* (1994) suggested that while back-filling of the dug out material the heavier particles, including soil aggregates and gravels, should be spread sparingly over the loose sandy surface to protect them from faster movement. Care should be taken to protect the natural vegetation cover along the route. As far as practicable, the ground flora including the small shrubs, should not be uprooted, but may be cut or scrapped. This would help in regeneration of the vegetation. In the deflated interdune plains between the linear dunes additional protection measures were suggested, like compaction of the back-filled material in the trench, first with a mat of *Panicum turgidum* grass and then with a layer of *kankar*. This would provide greater resistance to the high wind in the narrow interdune corridors and along the flanks of the dunes.

Most sand dune stabilization programmes are directed towards the old dunes, so that the production potentials of these lands can be restored or improved upon. The activities include (a) protection of the area from human and livestock encroachment; (b) creation of micro-wind breaks on dune slopes, using locally available shrubs either in a checker board pattern or in parallel strips; (c) direct seeding or transplantation of indigenous and exotic species; (d) plantation of grass slips or direct sowing of grass seeds on leeward side of micro-wind breaks; and (e) management of revegetated sites (Muthana, 1982; Harsh *et al.*, 1995).

Plantation of trees and shrubs at every location may not be beneficial. In some areas plantation may even become counter-productive. For example, the highways in the drier western part of the Thar get more problems of sand drift in the areas where roadside plantation was taken up along both sides of the roads in the sandy plains. The sand-surcharged summer wind gets obstructed at the plantation site along the upwind margin of the road and deposits its load. Subsequently a sand heap grows there and advances through creep and saltation processes, especially through gaps between the tree/bush rows. Since grass cover is difficult to maintain under the open grazing system, and the rows of trees create a number of faster wind corridors, the sand begins to accumulate on the road much faster. Sometimes the tree rows and the fences, even on the downwind end of the roads, create new problems where none existed before. The fences attract more sand during the summer and sand ridges are created along the roads due to secondary re-circulation of the sand-surcharged wind in the upwind of the fences (Kar, 1989). The best way to solve the problem is to allow the sand to move across the road. The hard surface of the road acts as a faster zone of grain transportation and hence, if the surroundings of the roads are cleared of structures, the rest of the problem could be easily solved by the road itself (Kar, 1992, 1994).

The agencies responsible for sand dune stabilization have their own limitations in carrying out the programme. For example, a number of reactivated hair pin parabolic dunes in the zone of very high wind erosion index in the desert are now being planted with trees and shrubs, especially *Acacia tortilis* (and sometimes *Prosopis juliflora*), at the nosal part of the dunes. Only this small part is within the jurisdiction of the Government Forest Department. The rest of the area of any such dune belongs to the farmers nearby, who practice dry farming along the dune slopes, often using tractors instead of country plough, and open grazing during rest of the year. The natural shrubs on the dunes provide fuel-wood to these villagers. Faced with such a scenario, forest department attempts to develop a green cover in the land within its own jurisdiction. However, the relative paucity of a ground flora in the area of development and the non-availability of the long upwind parts of the dunes for re-vegetation programme make sand control on the dunes a difficult proposition (Kar, 1996). With increasing cultivation and grazing pressures on the dunes, the sand starts moving at an accelerated pace from the dune slopes, accrete over to the nosal part of the dunes and then moves forward along narrow, artificially created corridors of faster wind flow through the belts of planted trees to avalanche downslope, thus helping the dunes to prograde (Kar, 1996). There are many instances of such creeping dunes in spite of a plantation programme in the Thob-Shergarh-Ujlan-Phalsund area of Jodhpur and Jaisalmer districts.

So far, sand dune stabilization programme has not become a popular people's programme. Villagers hardly accept it as a viable proposition. This is because, for many farmers the dune is the only land resource to thrive on. Compulsion to grow more from the limited land available makes it increasingly difficult for them to keep a part of the land fallow, even for a year. The earlier practice of keeping the upper slopes of the dunes marked for grazing, is difficult to follow now. Technological advances have made it easier to deep plough the middle and upper slopes of the dunes through tractor, and growing of crops. It makes a better economic sense for a farmer to deep plough the higher slopes of the dunes and reap a harvest, than to keep the land as long fallow. The farmers are, therefore, willing to accept some degradation and consequent dune movement, rather than to leave the land for fencing and for less lucrative proposals. Therefore, to get people's support for sand dune stabilization programme it is necessary to introduce economically viable perennial plants like fruit trees or forest trees that can give good remuneration. *Acacia tortilis*, which was formerly disliked by farmers because of its slow growth rate and poor economic returns, has now been made to exude high-value gum regularly in appreciable quantity through the use of a gum-inducing chemical. In the northern part of the Thar farmers have started introducing horticultural crops like Ber (*Ziziphus nummularia*) and Kinno (a local variety of orange) on the sand dunes, although gram cultivation during winter, using conserved soil moisture, is still a popular rotational crop after pearl millet in summer. Several such viable alternatives to cropping on the dune slopes have been found through research interventions (Venkateswarlu, 1993).

Since plantation of trees and tall shrubs on the dunes has been taken up on a large scale, it is worthwhile to find out the optimum plant density on a given dune type under a given climatic situation. Although it is known that the dune sand has a better moisture holding capacity, it is yet to be calculated how many trees and shrubs can be held by the different types of old dunes on a sustainable basis. One estimation is that the old dunes can sustain 400 trees per ha, but this high value is, perhaps, related to the dunes where the loose sand apron is quite thick. If the loose sand apron becomes thinner, especially through deflation after the trees attain some maturity, or through the process of sand stabilization, then the moisture holding capacity will be quite different than in the loose sand apron. Consequently, the sustainability of such high number of trees becomes doubtful. Moreover, as we have discussed, different dune types have different patterns of sand transport. Therefore, under the trees and tall shrubs a sand-passing form, like a linear dune, will carry the loose sand away downwind at a higher speed than under a sand-accumulating form and will deflate a long stretch of dune surface. Once the loose sand is blown out, the moisture regime will change. We have discussed the results of plantation at the nosal part of hairpin parabolic dunes. In the case of the short-arm parabolic dunes also, the plantation at the nosal part provides a somewhat deceptive view of sand control. The trees at the crestal part and on the leeside of the dune act as barriers to the wind, but the incoming sand at the location creep and avalanche down the nose to increase the loose sand apron there. By contrast, the upwind arms will release more sand downwind, if the surface further upwind has natural stability, and will have an exposed *kankar* layer. Surely, the sustainable quantity of trees and shrubs on the old dunes needs to be modelled on the basis of dune form, wind regime, moisture depletion pattern, etc.

It is sometimes suggested that sand control measures should be initiated in the 'source areas'. Such suggestions are utopian, because most of the development agencies, and especially the agencies whose infrastructures are affected (e.g. the railways and the road maintenance departments), have very less land at their disposal and much less resources and expertise to provide a long term solution to the menace of sand deposition. What is needed, therefore, is a set of cheap and acceptable solutions for both the private landholders and the development agencies. Given the complexities of situation in which sand movement takes place in the dunefields the solutions will be more location-specific.

It is also increasingly realised by the participants in sand stabilization programmes that small shrubs and grasses are better sand binders than the trees. However, in most cases retention of a grass cover is a problem, especially because of uncontrolled grazing on the dunes and sandy plains. Bio-fencing, using locally non-palatable species, is a cheaper and more effective form of barrier. Unless it is demonstrated that biomass production from the shrub and grass cover on the dunes provides a better financial return to the owners of such lands, as compared to the crop cultivation, and unless the traditional mind-set of the villagers is tuned to such a message, there is less likelihood of shrubs and grass covers on dunes becoming a successful venture. At present there is no clear land use policy to enforce the conservation-related practices.

REFERENCES

Bagnold, R.A. 1941. *The Physics of Blown Sand and Desert Dunes.* Methuen, London, 265p.

Buckley, R. 1987. The effect of sparse vegetation on the transport of dune sand by wind. *Nature*, 325 (6103): 426-428.

Dhir, R.P., Kolarkar, A.S. and Singh, H.P. 1992. Soil resources of the Thar. In, *Perspectives on the Thar and the Karakum* (Eds., A. Kar, R.K. Abichandani, K. Anantharam and D.C. Joshi). Department of Science and Technology, Govt. of India, New Delhi, pp. 60-85.

Goudie, A., Allchin, B. and Hegde, K.T.M. 1973. The former extensions of the Great Indian Sand Desert. *Geographical Journal*, 139: 243-257.

Harsh, L.N., Tewari, J.C. and Kumar, S. 1995. Sand dune fixation and afforestation in arid regions. In, *Land Degradation and Desertification in Asia and the Pacific Region* (Eds., A.K. Sen and A. Kar). Scientific Publishers, Jodhpur, pp. 171-178.

Kar, A. 1989. Evaluation of present-day geomorphic processes in the Indian desert for assessing land degradation and natural hazards. In, Geomorphology and Environment (Eds., S. Singh and R.C. Tewari). Allahabad Geographical Society, Allahabad, pp. 643-654.

Kar, A. 1992. Morphological processes, human influences and land degradation in the Indian desert. In, *New Dimensions in Agricultural Geography*, Vol. 2 (Ed., N. Mohammad). Concept Publishing Co., New Delhi, pp. 123-150.

Kar, A. 1993. Aeolian processes and bedforms in the Thar desert. *Journal of Arid Environments* 25: 83-96.

Kar, A. 1994. Sand dunes and their mobility in Jaisalmer district. In, *India: Geomorphological Diversity* (Eds., K.R. Dikshit, V.S. Kale and M.N. Kaul). Rawat Publishers, Jaipur, pp. 395-418.

Kar, A. 1996. Morphology and evolution of sand dunes in the Thar desert as key to sand control measures. *Indian Journal of Geomorphology* 1: 177-206.

Kar, A. 1998. Present-day mobile crescentic dunes in the Thar desert, India. In, *Quaternary Deserts and Climatic Change* (Eds., A.S. Alsharhan, K.W. Glennie, G.L. Whittle and C.G.St.C. Kendall). A.A. Balkema, Rotterdam (Netherlands) and Brookfield (USA); pp. 155-164.

Kar, A. 1999. Quaternary period in Rajasthan: Geomorphic process response to forcing and implications for terrain evolution. In, *Geology of Rajasthan: Status and Perspective* (Ed., P. Kataria). M.L. Sukhadia University, Udaipur; pp. 175-212.

Kar, A. 2002. Granulometry of aeolian bedforms in the extremely arid Jaisalmer region of Thar Desert. In, *Recent Advances in Geomorphology, Quaternary Geology and Environmental Geosciences: Indian Case Studies* (Eds., S.K. Tandon and B. Thakur). Manisha Publications, New Delhi, pp. 261-286.

Kar, A., Felix, C., Rajaguru, S.N. and Singhvi, A.K. 1998. Late Holocene growth and mobility of a transverse dune in the Thar desert. *Journal of Arid Environments*, 38 (2): 175-185.

Kar, A. and Joshi, D.C. 1995. Sand movement and control of aeolian hazard. In: *Land Degradation and Desertification in Asia and the Pacific Region* (Eds. A.K. Sen and Amal Kar). Scientific Publishers, Jodhpur, pp. 19-40.

Kar, A., Singh, N. and Kumar, S. 1994. *Wind Erosion Control Measures for Gas Pipeline between Gamnowala Tar and Ramgarh (Jaisalmer District).* A report for the Gas Authority of India, New Delhi. CAZRI, Jodhpur, 20 p. (mimeo).

Kar, A., Singhvi, A.K., Rajaguru, S.N., Juyal, N., Thomas, J.V., Banerjee, D. and Dhir, R.P. 2001. Reconstruction of late Quaternary environment of the lower Luni plains, Thar desert, India. *Journal of Quaternary Science*, 16: 61-68.

Kar, A., Tsunekawa, A. and Miyazaki, T. 1998c. Potentiality of global positioning system in sand dune measurement: A case study from the Thar desert, India. In, *Quaternary Deserts and Climatic Change* (Eds., A.S. Alsharhan, K.W. Glennie, G.L. Whittle and C.G.St.C. Kendall). A.A. Balkema, Rotterdam (Netherlands) and Brookfield (USA); pp. 433-438.

Muthana, K.D. 1982. A review of sand dune stabilization and afforestation. *Proceedings of the Workshop on the Problems of the Deserts in India.* Miscellaneous Publication, 49, Geological Survey of India, Calcutta, pp. 363-368.

Narain, P., Kar, A., Ram, B., Joshi, D.C. and Singh, R.S. 2000. *Wind Erosion in Western Rajasthan.* CAZRI Publication, Jodhpur.

Pandey, S., Singh, S. and Ghose, B. 1964. Orientation, distribution and origin of sand dunes in the central Luni basin. *Proceedings, Symposium on Problems of Indian Arid Zone*, CAZRI, Jodhpur, pp. 84-91.

Prakash, I. and Gupta, R.K. 1976. An ecological exploration of an extremely arid tract of Rajasthan. *Indian Journal of Ecology*, 3: 11-21.

Saxena, S.K. 1977. Vegetation and its succession in the Indian desert. In, *Desertification and Its Control* (Ed. P.L. Jaiswal). Indian Council of Agricultural Research, New Delhi, pp. 176-192.

Singh, S. 1982. Types and formation of sand dunes in Rajasthan desert. In, *Perspectives in Geomorphology* (Ed., H.S. Sharma). Concept, New Delhi, pp. 165-183.

Singhvi, A.K. and Kar, A. (Eds.). 1992. *Thar Desert in Rajasthan.* Geological Society of India, Bangalore, 191p.

Skidmore, E.L. 1988. Wind erosion. In, *Soil Erosion Research Methods* (Ed., R. Lal). Soil and Water Conservation Society of America, pp. 203-233.

Thomas, J.V., Kar, A., Kailath, A.J., Juyal, N., Rajaguru, S.N. and Singhvi, A.K. 1999. Late Pleistocene-Holocene history of aeolian accumulation in the Thar desert, India. *Zeitschrift fur Geomorphologie* Supplement Band 116: 181-194.

Venkateswarlu, J. 1993. Problems and prospects in desertification control: role of Central Arid Zone Research Institute. In, *Desertification and Its Control in the Thar, Sahara and Sahel Regions* (Eds. A.K. Sen and A. Kar). Scientific Publishers, Jodhpur, pp. 249-267.

Verstappen, H. Th. 1968. On the origins of longitudinal (seif) dunes. *Zeitschrift fur Geomorphologie*, 12: 200-220.

Wadhawan, S.K. 1988. Evolution of Quaternary aeolian deposits of Jodhpur and Barmer districts, Rajasthan, India. *Proceedings, National Seminar on Recent Quaternary Studies in India* (Eds., M.P. Patel and N. Desai). M.S. University, Baroda, pp. 64-78.

Wadhawan, S.K. 1996. Textural attribute of recent aeolian stratigraphy in the Ghaggar basin of Thar desert, India. *Journal of Arid Environments*, 32: 59-74.

Wadhawan, S.K. 1998. Late Quaternary evolution of clustered parabolic megadunes in Thar

Desert, India. In, *Quaternary Deserts and Climatic Change* (Eds., A.S. Alsharhan, K.W. Glennie, G.L. Whittle and C.G.St.C. Kendall). A.A. Balkema, Rotterdam (Netherlands) and Brookfield (USA); pp. 185-195.

Wasson, R.J., Rajaguru, S.N., Misra, V.N., Agrawal, D.P., Dhir, R.P., Singhvi, A.K. and Rao, K.K. 1983. Geomorphology, late Quaternary stratigraphy and palaeoclimatology of the Thar dunefield. *Zeitschrift fur Geomorphologie, Supplementband* 45: 117-151.

Watson, A. 1990. The control of blowing sand and mobile desert dunes. In, *Techniques for Desert Reclamation* (Ed., A.S. Goudie). John Wiley, London, pp. 35-85.

Yallon, D.H. and Ganor, E. 1966. The climatic factor of wind erodibility and dust blowing in Israel. *Israel Journal of Earth Science*, 15: 27-32.

2

Meandering and Cut-Off of the River Bhagirathi

S. R. BASU, ANASUYA GHOSH AND SUNIL KUMAR DE

Introduction

The channel of the River Bhagirathi (the lower course of the river Ganga) has been so much characterized by conspicuous winding of its course, that Major Hirst (1916) regarded them as abnormal features, not really explicable by the usual riverine processes. In fact, this feature has been so much prevalent in the Ganga delta, that it has prompted in the naming of many streams, such as *Atharbanka* (eighteen bends) in North 24-Parganas, West Bengal and again in Khulna district of Bangladesh (Reaks, 1919).

The river Bhagirathi has a natural tendency to shape its course according to the varying discharges received from the Farakka Barrage and its western tributaries. In order to tackle the continuous excess in volume of water during the monsoon, the river has an inclination to increase its length by virtue of meandering. When the curvature of the river reaches a certain stage, the tendency for short-circuiting steps in and a good length of the loop is by-passed through the process of cut-off. Such a continual change in the course of the river, shifts the balance in the river regime also.

In the past, when the discharge was heavier, being aided by the larger supply from the parent river Ganga, the river increased its length by meandering so as to reduce the gradient of the thalweg. But today, as the river is subjected to a reduced volume of discharge through the feeder canal of the Farakka Barrage, a tendency to straighten its course has taken place so as to increase its gradient, which enables it to maintain a perennial flow throughout the course in spite of its reduced discharge. But again increase in gradient of river tends to increase the velocity of the flowing water and as such the process of erosion is heightened, endangering the nearby settlements (Fig. 2.1). These eroded materials will ultimately increase the suspension load of the river water, which will eventually be unable to carry them downstream, and will drop them on its own bed to form sandy shoals, which will ultimately help in the winding course of the river (Basu,1976).

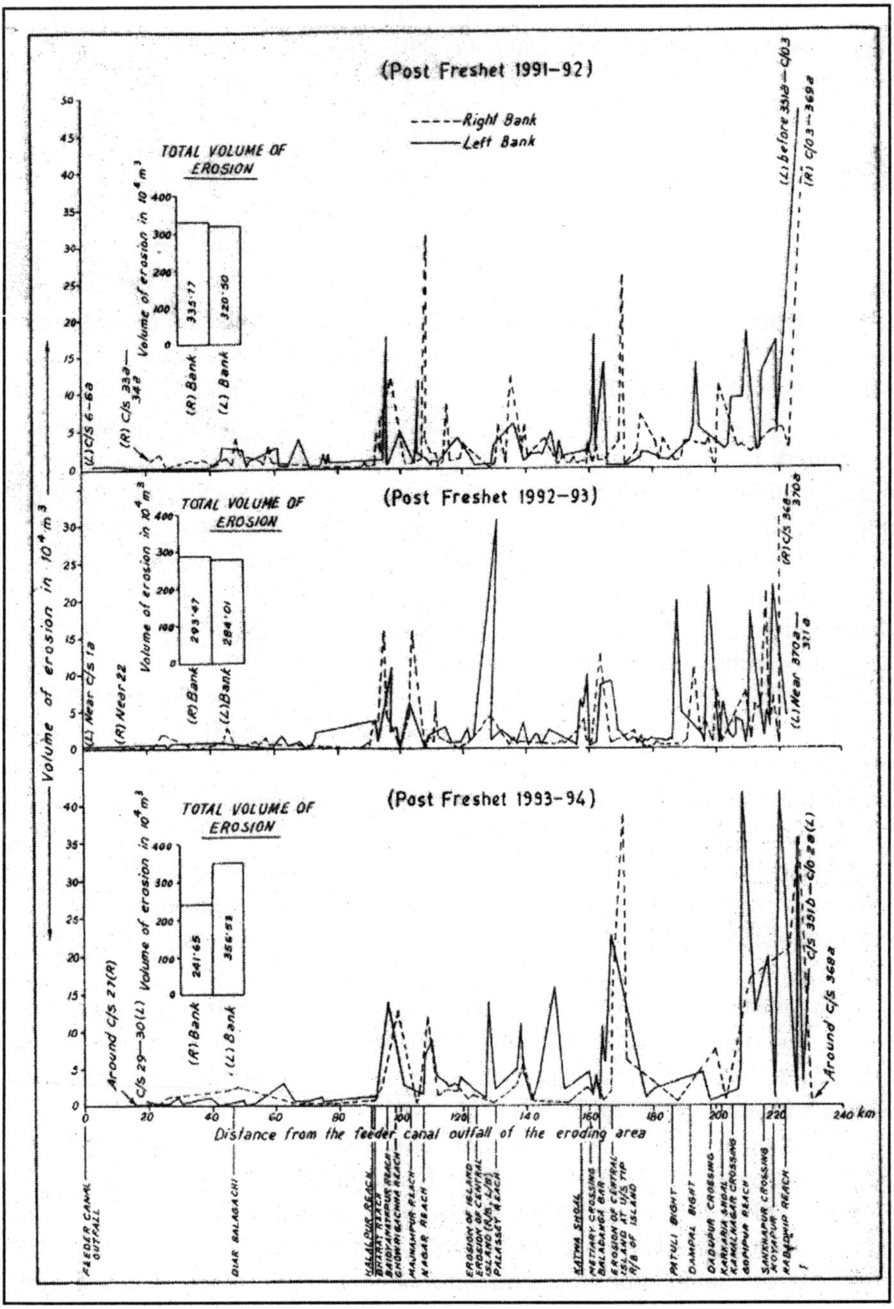

Fig. 2.1 : Volume of Erosion along the Banks of the River Bhagirathi from Feeder Canal Outfall to Nabadwip

To have a clear concept of the river curve and their forms of the river Bhagirathi, studies like the sinuosity, its relation with variance have been made. An attempt has also been made to predict the breaking point of a meandering river which ultimately leads to the formation of a cut-off.

Study of the Sinuosity

The study of sinuosity has been done for the three reaches of the river Bhagirathi for the year 1993-94.

In the upper reach (Jangipur Barrage to Narayanpur) the sinuosity ranges from 1.09 to 5.33 (Fig. 2.1, Table. 2.1). In this reach, the river has a winding as well as a straight course. The sinuosity index is mainly within 2, in most part of the river. This is due to the less sinuous course of the river, mainly from Hazarduary to Baidyanathpur. In this part of the river the sinuosity index is minimum about 1.09. Wherever the river has less sinuous course or the sinuosity index is within 2 the river is in between the straight and the regular stage.

In some places like at Ramchandrapur–Gadikhal area the river has a winding course and here the sinuosity index has increased and is within 2 to 2.44. At Maheshpur too, the river is sinuous and the index is 2.10. As we go downstream, the river becomes sinuous at Baidyanathpur where a cut-off has taken place in the year 1984. But the sinuosity index at Baidyanathpur is about 1.82. This is because the river has straightened its course after the cut-off. Thus, it is seen from Fig. 2.2 that with the increase of sinuosity index slightly above 2, the river in these areas is in between the regular and irregular stage.

At Diar-Balagachi, a conspicuous meander has been formed with a wide loop. The river has formed a sinuous curve and the sinuosity index here is 5.33 (Fig. 2.2) which is the maximum in the entire river course. Due to such a wide loop, a cut-off is expected in near future. This part of the river is in the tortuous stage.

The lower reach, that is from Katwa to Nabadwip, the river has a winding course, due to the flat gradient on which it flows and also due to the amount of sediment it is carrying. In this part of the river, the sinuosity index ranges from 1.27 to 3.32 (Fig. 2.3a, Table 2.2).

The sinuosity index is the minimum at Karkaria and the index is 1.27. So the river here falls in between the straight and the transitional stage.

From Patuli till Dampal Bight the river has a distinct winding course and thus the sinuosity index has increased to 2.34 in this area. Similar situation is observed from Sankhapur till Nabadwip where the index is 2.43, which is quite high. This region falls in between the regular and the irregular stage as is evident from the Fig. 2.3a.

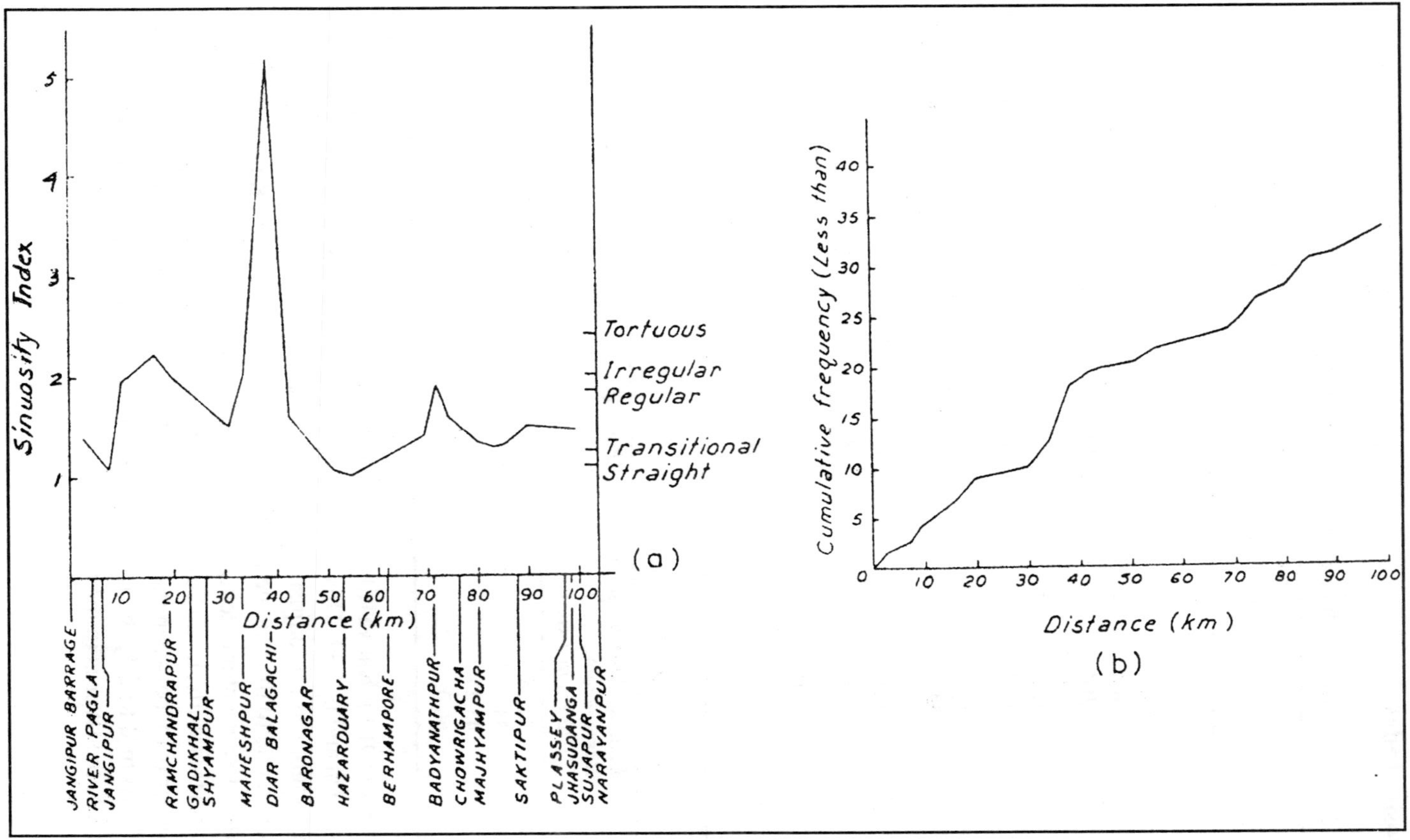

Fig. 2.2 : Sinuosity Character of the River Bhagirathi from Jangipur Barrage to Narayanpur (Post Freshet 1993-94)

Sinuosity Index

Table 2.1 : From Jangipur Barrage to Narayanpur Reach

Sl. No.	*Distance from Jangipur Barrage to Narayanpur (2 cm to 1 km.)*	*Thalweg Distance*	*Airline Distance*	*Sinuosity Index*
1.	2.5	9	6.5	1.38
2.	7.5	10	8.2	1.22
3.	10.0	16	8.6	1.86
4.	16.5	20	8.2	2.44
5.	20.0	18	9.0	2.00
6.	31.0	21	14.0	1.50
7.	34.0	13	6.2	2.10
8.	38.0	16	3.0	5.33
9.	42.5	15	9.2	1.63
10.	51.5	6	5.3	1.13
11.	54.5	12	11.0	1.09
12.	69.5	11	8.0	1.38
13.	72.0	10	5.5	1.82
14.	74.0	8	5.0	1.60
15.	80.5	8	6.0	1.33
16.	83.0	12	9.4	1.28
17.	85.5	13	10.0	1.30
18.	90.0	15	10.0	1.50
19.	100.0	13	8.8	1.48

Table 2.2 : From Katwa to Nabadwip

Sl. No.	*Distance from Jangipur Barrage to Narayanpur (2 cm to 1 km.)*	*Thalweg Distance*	*Airline Distance*	*Sinuosity Index*
1.	22.25	44.5	13.4	3.32
2.	34.5	25	10.7	2.34
3.	43.25	18	14.1	1.27
4.	60.0	32.5	13.4	2.43

The area from Katwa till Char-Chakundi has the maximum sinuosity index in this reach and is 3.32. This part of the river falls under the tortuous stage. The river is sinuous as it is adjusting to the additional discharge from the river Ajay. Moreover, due to the change in gradient of the river bed from Katwa, the river meanders to a greater extent, thus causing the sinuosity index to increase.

From this study, it can be concluded that, when the sinuosity index is within 2, the river is in between the straight to the regular stage. When the sinuosity index is from 2 to 2.5, the river is in the regular to tortuous stage. When it is above 2.5, the river is in the tortuous stage.

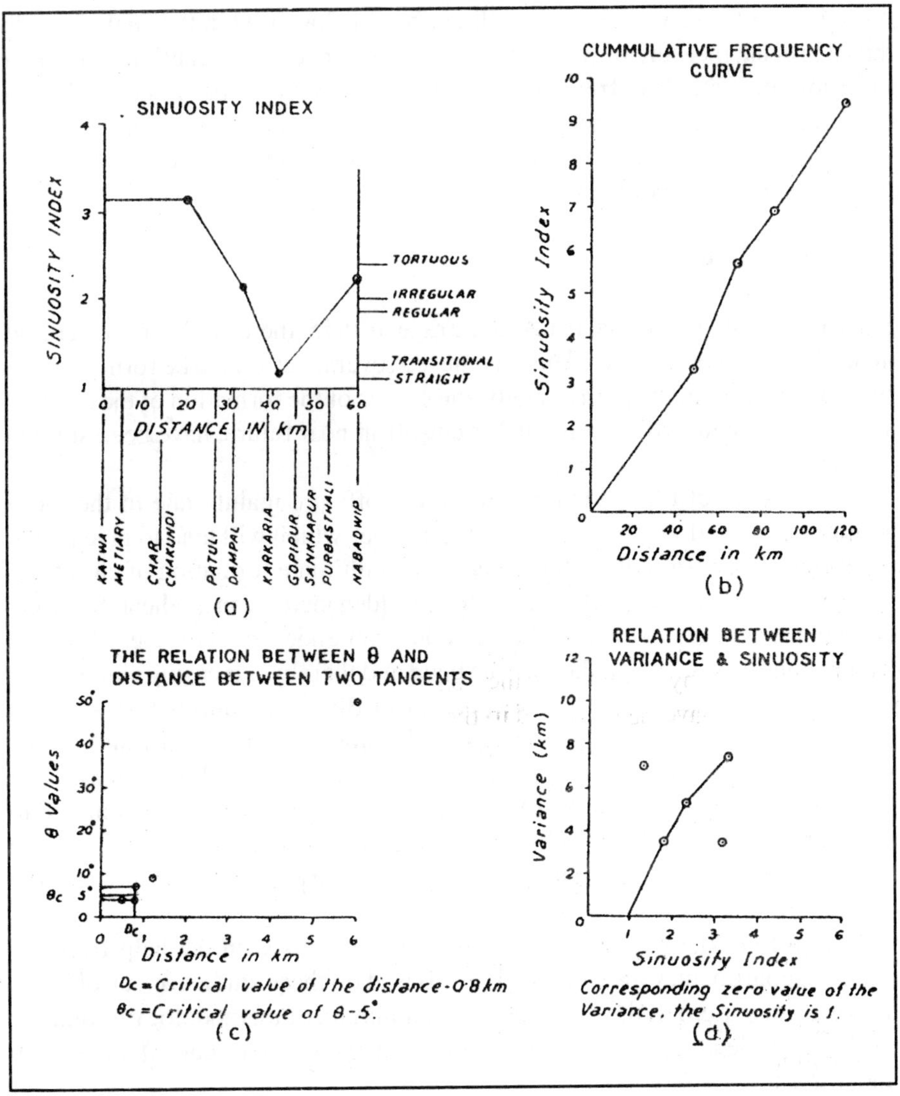

Fig. 2.3 : Sinuosity Character of the River Bhagirathi

Relation between Variance and Sinuosity

The relation between variance and sinuosity is shown in Fig. 2.3d. This study has also been done in the lower reach only since the river in this reach is very sinuous. Speight (1965) points out that, "the use of variance as a sinuosity measure carries the obvious advantage that the sinuosity component due to given wavelength of meander can be isolated. The area under spectral curve is proportional to the variance of the directions of flow used in the analysis. It is thus a precise measure of the sinuosity of the channel, if precautions are taken to eliminate very low frequency oscillations likely to be associated with topographic obstacles rather than with the phenomenon of meandering."

It appears from Fig. 2.3d, that the corresponding zero value of the variance, the sinuosity is 1 (extrapolated).

Study of Cut-Offs

Since the installation of the Farakka Barrage in 1975, there has been an increase in the discharge of the river. This has caused several cut-offs to be formed in the river Bhagirathi. In the present study, the causes of the formation of the cut-offs and whether there will be any further cut-off in near future, have been studied in details.

The causes of the formation of the cut-offs are mainly due to the loose texture of soil and varying discharge of the river which will tend to give rise to progressive extension of meander bends with ultimate formation of cut-offs.

The formation of cut-offs has led to the sudden increase in gradient. Scouring takes place actively in the river bed and the river endeavours to re-establish its former conditions, by lengthening the bend and thereby increasing sinuosity. A group of cut-offs have been formed in the river Bhagirathi during post-Farakka days reducing the length of the river by about 23 km and thus almost straightening the channel (Figs. 2.4 and 2.5). This invariably increases the scouring action throughout the river. But the scouring action is not confined to the river bed alone and the banks are also vigorously attacked bringing in more silt load to the river and thereby reducing the velocity of water with consequent depositions.

Cut-offs in the river have been studied in details with the help from the Survey Department, Calcutta Port Trust. Based on these studies of cut-offs and meandering pattern of the river, further formulation of methodology to find out the situation leading to the breaking of the meanders into cut-offs, has been undertaken.

For the study of cut-offs an angle is taken between the two meander curves. The angle is said to be "θ". The value "θ" is defined to be the angle made between two tangents in the outward direction and the tangents being drawn from such two points on the curvature of the meanders so that the angle between the tangents just shifts from inward to outwards direction (Fig. 2.6). It may appear that there will be angles of various low magnitudes in each case, but it is observed that there is a consistency in the value of angles when meander reaches the breaking point (Basu, 1967).

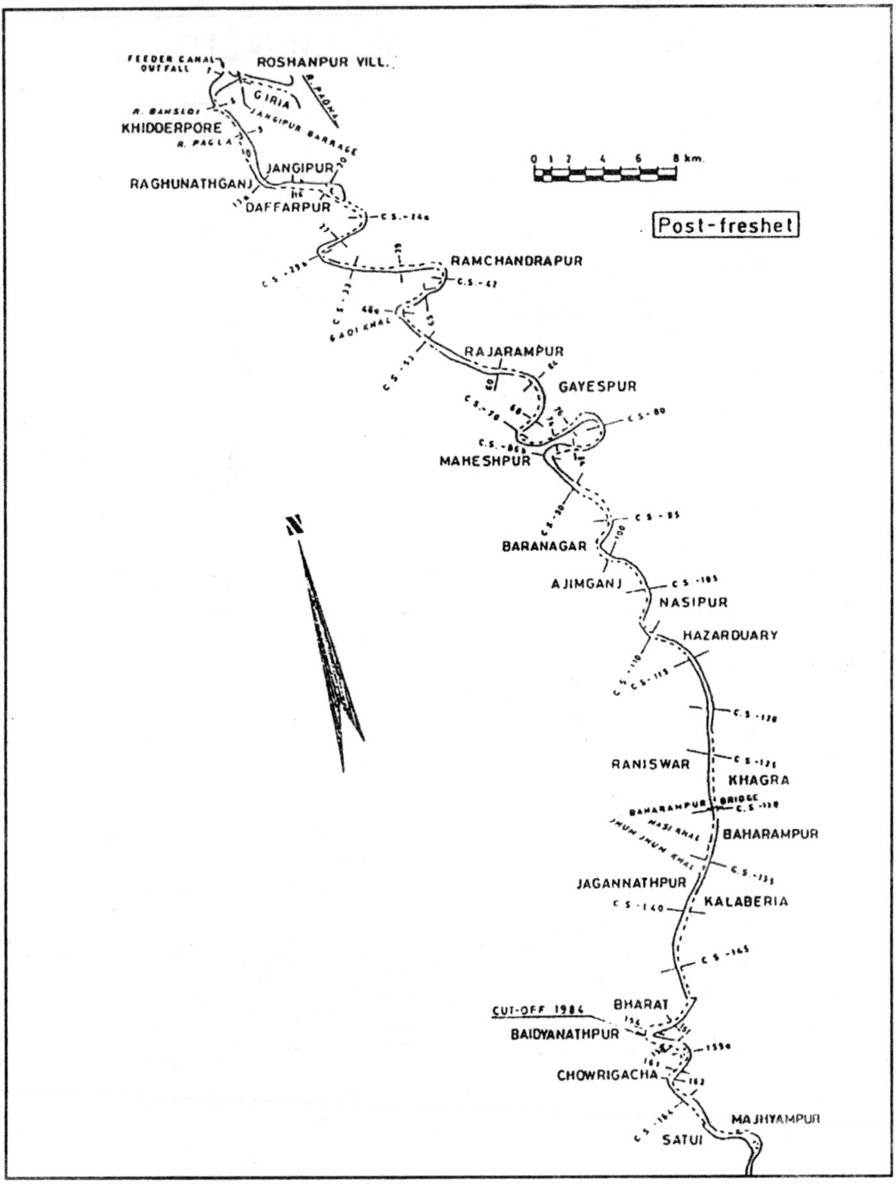

Fig. 2.4 : Key Plan of the River Bhagirathi Showing the Cross-Section Points

In developing stages, the angles are not possible on outward direction between the tangents drawn on any point from the curvature. The tangents are either parallel or divergent towards the outward direction.

The value "θ" of the meander curve has been obtained by tangential procedure. The relation between "θ" and the distance (km) is shown in

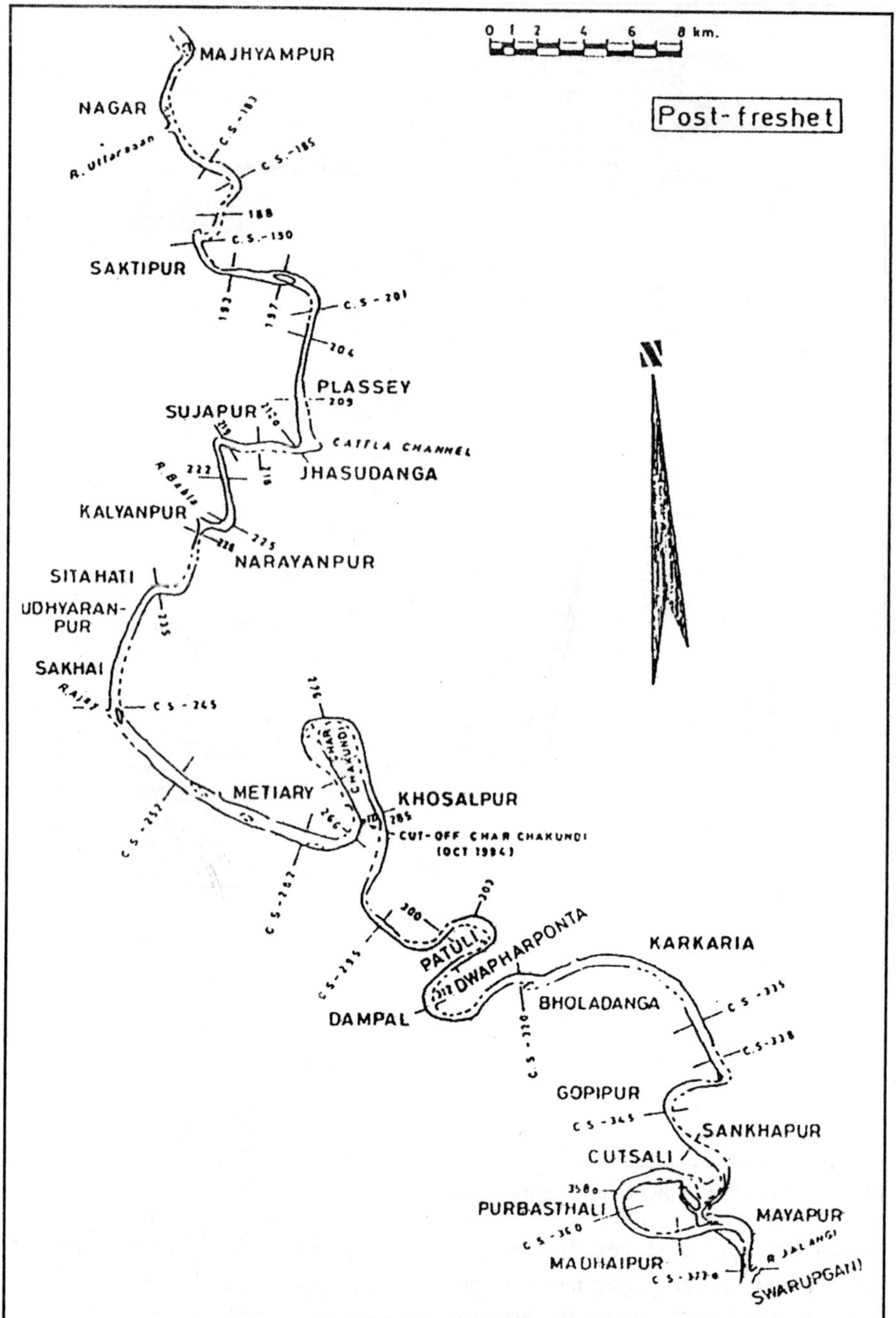

Fig. 2.5 : Key Plan of the River Bhagirathi showing the Cross-Section Points

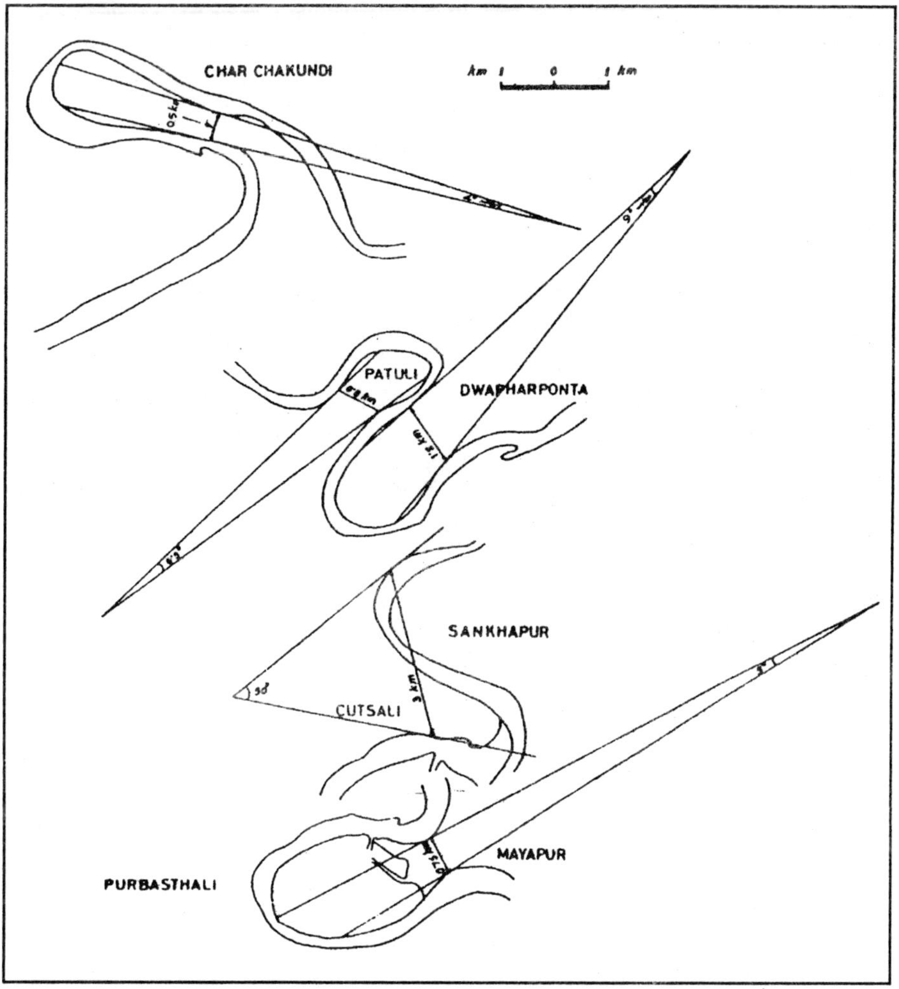

Fig. 2.6 : The Meander Curves of the River Bhagirathi with Angles Made by the Tangents

Fig. 2.3c. Since the river has a meandering course in the lower reach, so this study has been mainly confined to this reach. When the θ value reaches 5° or less and the distance between two tangents are 1 km or less, the meander reaches the breaking point (Basu, 1976). Hence this "θ" may be called "θ_c" (c stands for critical value). Here the critical value of the distance is 0.8 km and critical value of θ is 5°.

With this idea, the present cut-offs and the cut-offs which will occur in near future have been studied.

Meander curves which have occurred at Char-Chakundi, Patuli, Dwarphonta, Gopipur-Cutsali Reach and at Purbhasthali-Mayapur Reach have been studied.

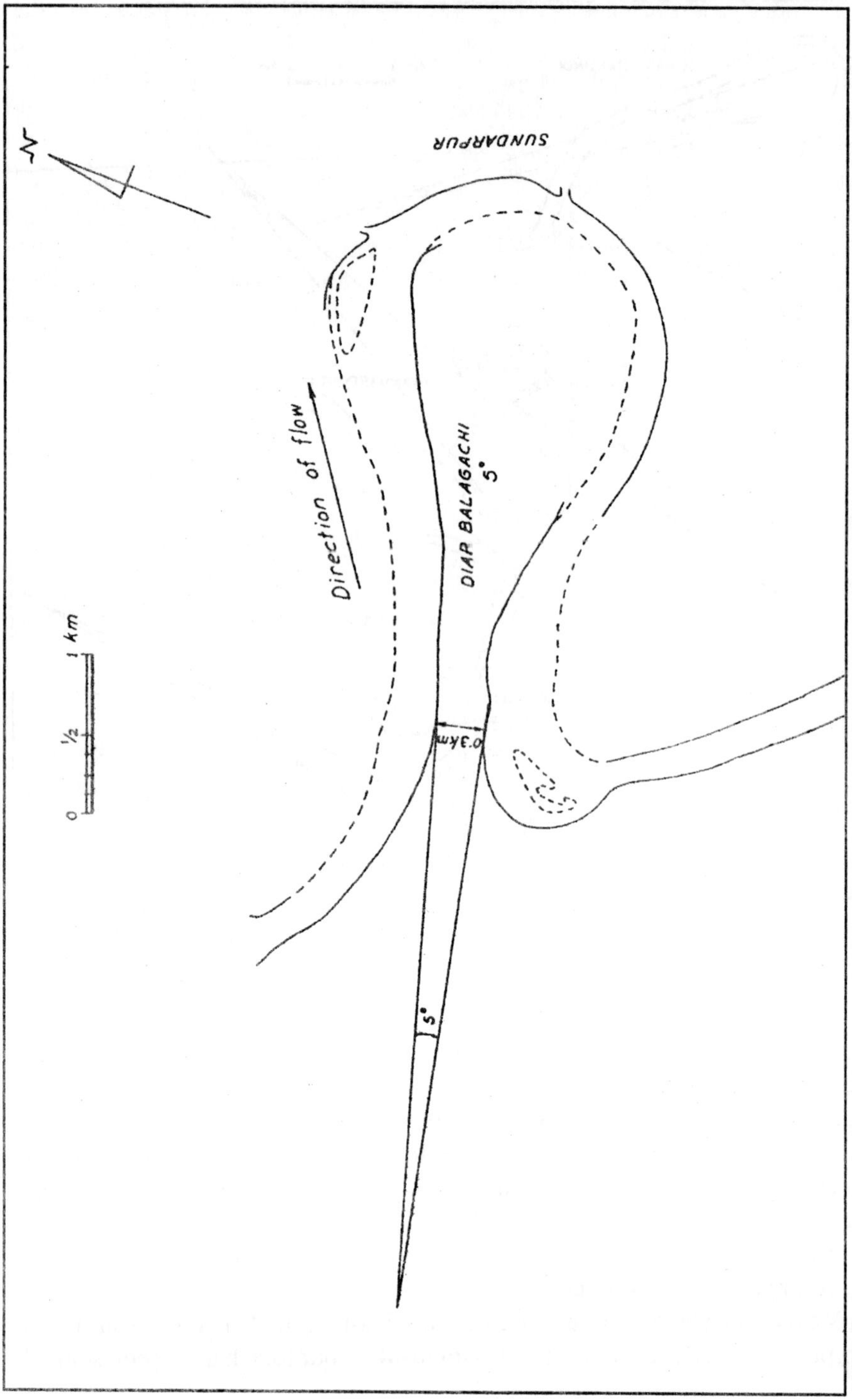

Fig. 2.7 : The Meander Curve of the River Bhagirathi at Diar Balagachi with Angles Made by the Tangents

Tangents have been drawn and angles have been measured to find out whether they have reached the critical value or not. At Char-Chakundi and at Purbasthali-Mayapur reaches it is seen that the angles made between the two tangents are 4° and 5° and the distance between these two meander curves is 0.5 km and 0.75 km (Fig. 2.6), respectively. Thus, it is seen that the meander curves have reached their breaking points and this is evident from the fact that in these two places cut-offs have already occurred. At Baidyanathpur too, it is found that the angle between the two tangents drawn along the meander curve formed at this place is less than 5° and the distance between the two tangents is about 0.6 km and has been studied in details at the later part of this paper. At the other meander curves, it is seen that the angle made between the two tangents of the meander curves at Patuli, Dwarphonta, Gopipur-Cutsali Reach are, 6.5°, 9°, 50° and the distance between the meander curves are also 0.8 km, 1.2 km and 3 km, respectively. From the study of these meander curves it is seen that though meanders have been formed in these places but they will not form a cut-off in near future.

A separate map has been drawn of Diar-Balagachi (Fig.2.7) which is the only conspicuous meander to be formed in the upper course. At Diar-Balagachi it is seen that the angle between the two tangents is 5° and the distance between the two tangents is 0.3 km, so there is a chance of a cut-off in near future. Diar-Balagachi loop which is 43 km downstream from the Feeder Canal outfall. The erosion near the neck of the loop during the year 1993-94 has been observed to be continuing though in slow pace and is one sided (i.e. the erosion occurring at the neck of the upstream side of the right bank only, thus reducing the land width at the neck, approximately 245 m, as measured in November 1993. The neck width measured in December 1992 was about 255 m (Report CPT, 1992-93). It is anticipated from the trend of the river channel, mode of erosion and quality of the soil which is clay-rich, that the probable cut-off is not imminent within a few years.

The following study is related with the three cut-offs which have already occurred in this river during the post–Farakka days. These cut-offs are found at Baidyanathpur, Bishnupur-Char Chakundi and at Sankhapur-Mayapur and have reduced the length of the river by about 23 km.

Cut-off at Baidyanathpur

The cut-off at Baidyanathpur (Fig. 2.8) took place during the freshets of 1984, which is situated about 95 km downstream of the Feeder Canal outfall.

The river at Baidyanathpur formed a wide loop with a very narrow neck. Severe erosion took place at the neck of the loop, causing the cut-off to take place, thereby straightening the loop and reducing the river length by 2 km

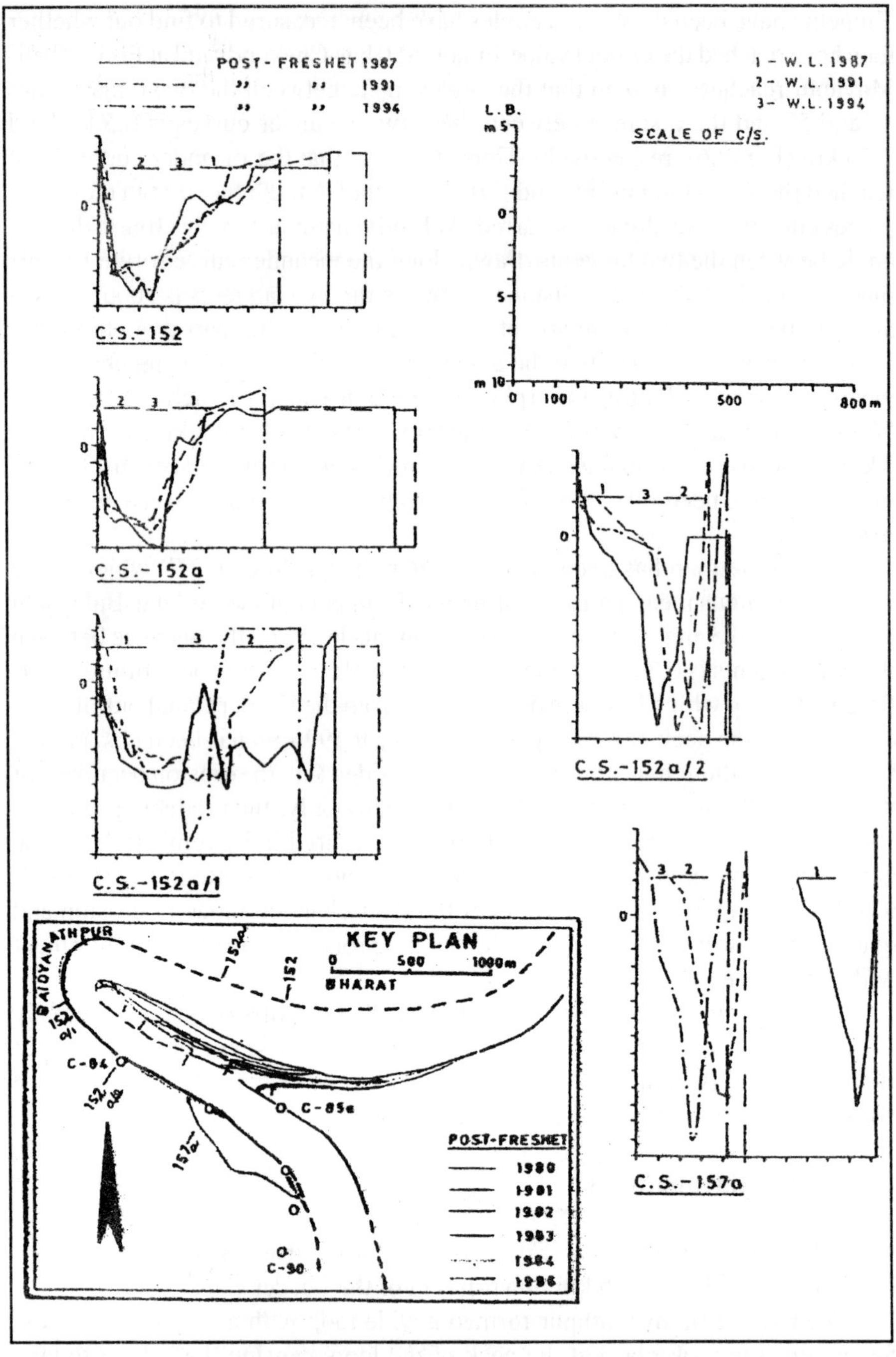

Fig. 2.8 : Cut-Off at Baidyanathpur Reach showing its present Configuration

(Report CPT – 1992-93). Due to the reduction in length of the river, the hydraulic gradient increases. The stepping in the gradient increases the velocity of the flowing water, resulting in the erosion of the banks. From the survey during the years of 1987, 1991 and 1994, that is after the cut-off which took place in the year 1984, it is seen that the river is still adjusting to the resultant high velocities associated with heavy erosion of concave banks and deposition in the convex banks (Report CPT, 1993-94). Before the formation of the cut-off the depth of water here was about 15 m. Thus, the velocity was very high and the same volume of water flowing through the narrow neck at Baidyanathpur reached the breaking point and produced a cut-off. After the cut-off is produced, the depth of the water at this reach automatically reduces to about 14 m with the widening of the channel.

Cut-off in between Bishnupur and Char - Chakundi

Another cut-off took place in between Bishnupur and Char-Chakundi during the post-freshet of 1994. It is 156 km downstream from the Feeder Canal outfall. The active zone for "cut-off" lies in the neck of the loop. During the year 1993, the eroding zone was limited between upstream right bank and downstream right bank. In the year 1994, the eroding zone shifted and thereby reducing the neck width to approximately 175 m as measured in December, 1993 (Report CPT, 1993-94). Further erosion near the neck led to the cut-off of the meander and the length of the cut-off loop is about 11.85 km, which means that the length of the river is shortened by another 11.85 km.

The present configuration of the cut-off at Char-Chakundi has been shown in Fig. 2.9. From the cross-profiles it is seen that at the neck of the loop the depth of the water is about 12m. The depth of the water from the left bank has a tendency to move to the right bank. Since the velocity of the water has been increased as it has to pass through the narrow loop at Char-Chakundi, vigorous erosion produced a cut-off. Thus, the length of the river has been shortened by about 12 km, thereby increasing the hydraulic gradient and velocity of the river.

Cut-off at Sankhapur-Mayapur Reach

This cut-off took place at the Sankhapur-Mayapur loop which is 218 km downstream from the feeder canal outfall. Noticeable bank erosion has been observed since 1982 with the possibility of a cut-off of the meander. During the freshets of 1989, the left bank breached and the river joined the lower portion of the small canal (taking off near left bank of the river) and finally, joined the main river course, thereby reducing the river length by 9 km. This cut-off is

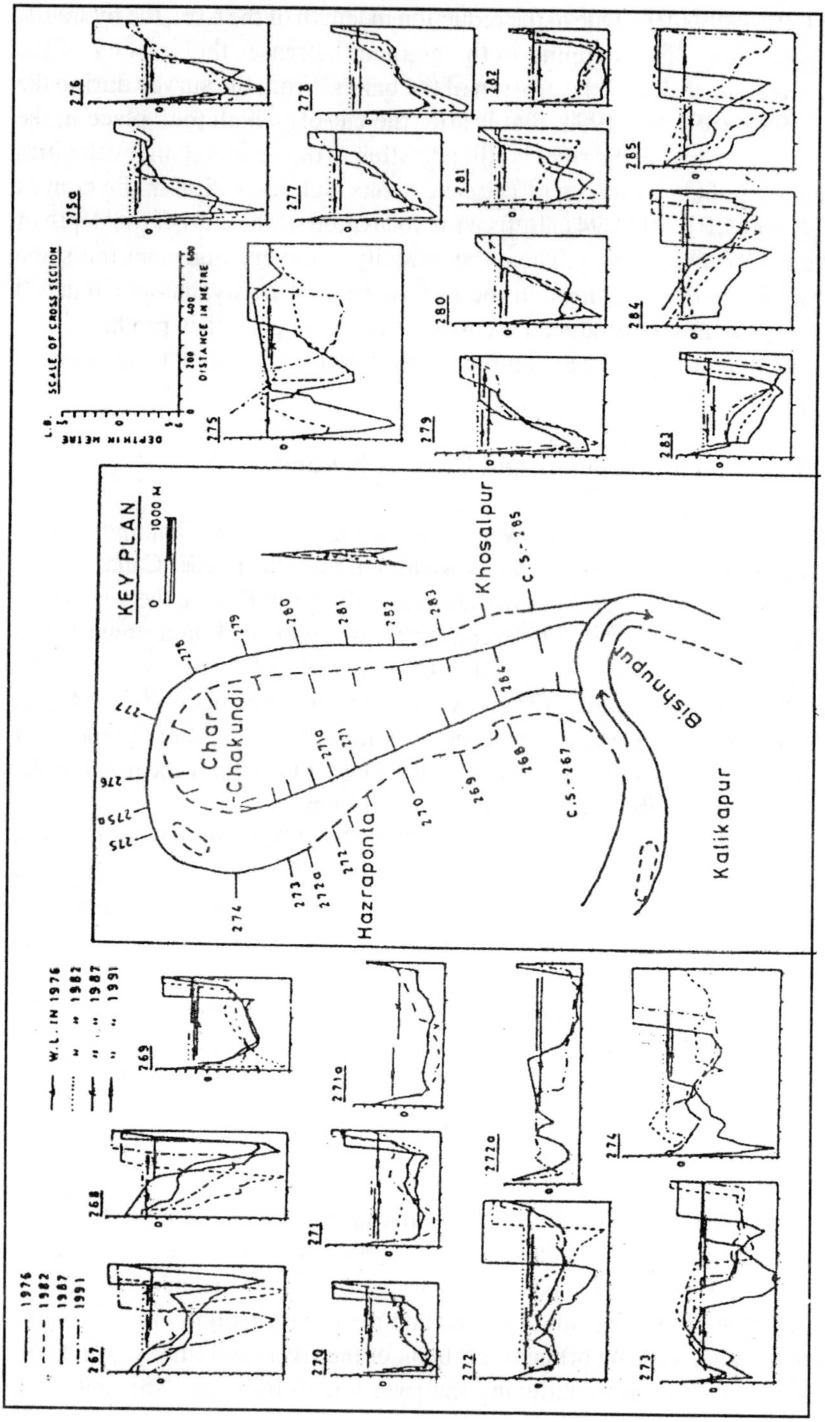

Fig. 2.9 : Cut-Off at Char-Chakundi showing its present configuration

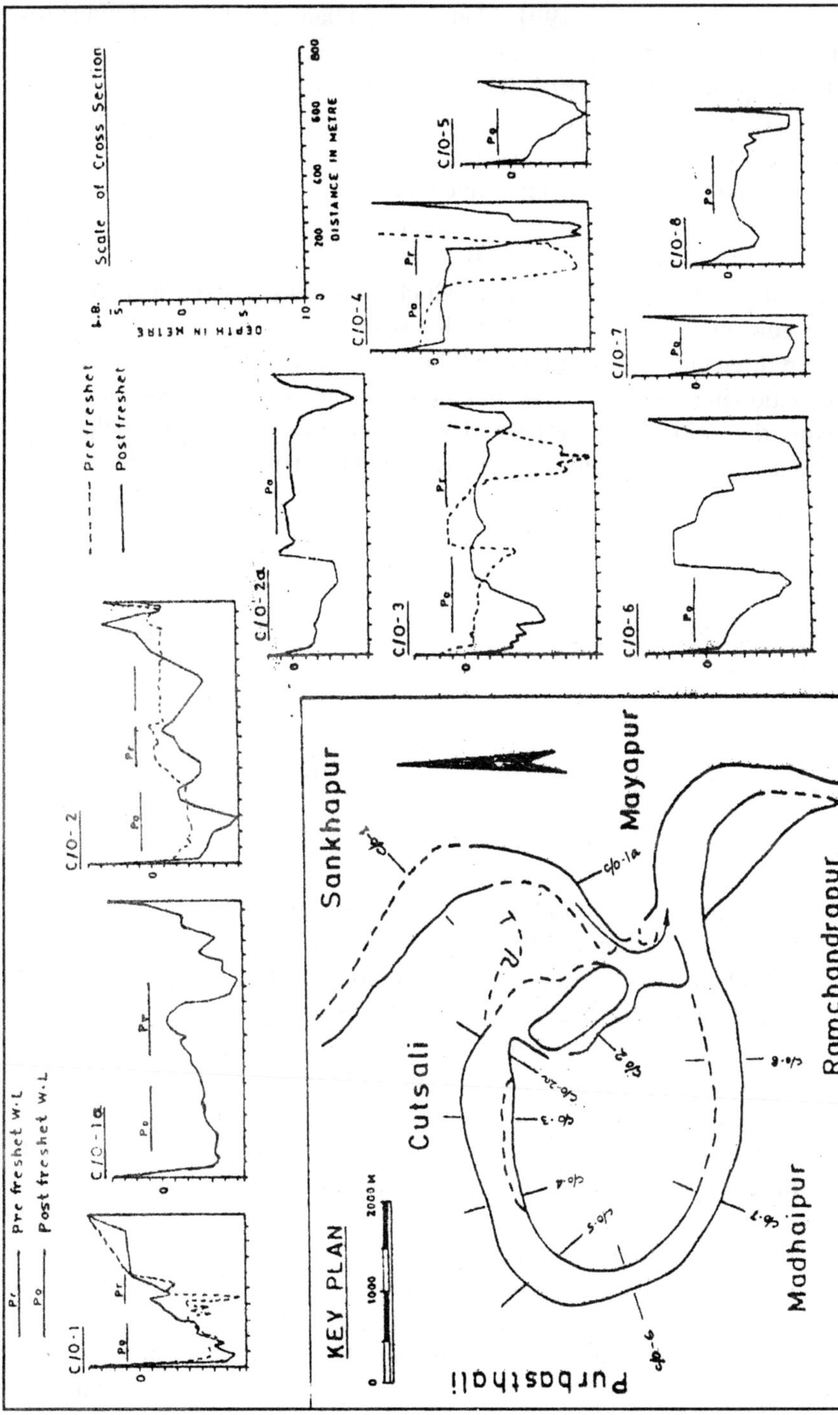

Fig. 2.10 : Cut-Off at Sankhapur-Mayapur showing its present configuration

causing severe erosion since 1991-92 at the right bank (Nawadwip Township). The effect of the cut-off is still continuing. The severe erosion is taking place at both the banks (Report CPT, 1992-93). In the Fig. 2.10, the present configuration of the cut-off at Sankhapur-Mayapur is shown. Cross-profiles have been drawn, showing the pre and post-freshet conditions of the river. The depths of the water in both the conditions are the same. Since this portion of the loop is totally cut-off from the main river, it practically remains dry during the pre-freshet or the dry season. Only it has water during the freshets, with a depth of about 5 to 11 m, practically in all the cross-sections which are surveyed and drawn just after the rainy season or during the post-freshet season.

These three cut-offs have reduced the length of the river by about 23 km. This reduction in length of the river has increased the hydraulic gradient, thereby increasing the velocity of the river. This increase in velocity has helped the river Bhagirathi to remove the sediments brought from upstream and the western tributaries to the river Hugli.

Conclusions

1. From the study of sinuosity, it is seen that the river is more sinuous in the lower reach than in the upper reach. The river has been divided into various stages on the basis of sinuosity index. The various stages are the straight, transitional, regular, irregular and tortuous stages.
2. Wherever, the river has a straight course or the sinuosity index is within 2, the river is in between the straight and the regular stage. As for example from Hazarduary to Baidyanathpur in the upper reach and around Karkaria in the lower reach.
3. When the sinuosity index is within 2 to 3, the river is in between the regular and irregular stage. As for example in certain places like Ramchandrapur, Gadikhal and also from Patuli to Dampal Bight and Sankhapur to Nawadwip if falls under this stage.
4. When the river is very sinuous, that is if the sinuosity index is above 3, the river is said to have reached the tortuous stage and can cause "cut-offs". This is seen at Diar-Balagachi where the sinuosity index is 5.33, and from Katwa to Char-Chakundi where the index is 3.32. These are the areas where the sinuosity is the maximum in the entire river.
5. From the relation between variance and sinuosity it is seen that with the corresponding 0 value of the variance, the sinuosity is 1.
6. A cut-off in the river can only form when the angle between the two tangents drawn along the meander curves is 5° or less and the distance between the tangents is less than 1 km. This has been found to be true in case of the "cut-offs" which have already taken place at Baidyanathpur, Char-Chakundi and at Sankhapur-Mayapur.

7. At Diar-Balagachio, a wide narrow loop has been formed. Here the angle between the two tangents is 5° and the distance between them is less than 1 km, so there is a possibility of a cut-off in near future. But due to the stickiness of the clay soil found here, the probable cut-off is not so imminent in near future.

REFERENCES

1. Basu, Subhashranjan (1967): Fundamental Problems of Meander Formation with Special Reference to the Bhagirathi River; *Indian Journal of Power & River Valley Development,* Vol. XVII, No. 2, pp. 15-23.
2. Basu, Subhashranjan (1976): Some Fundamental Problems of Meander Formation with Special Reference to the Bhagirathi River; *Geographical Review of India,* Vol. 38, no.1, pp 54-67.
3. Das, B.C. and Mukherjee, B.N. (1966): *Intermediate Dynamics*; eleventh edition, Calcutta; Ch. XII, pp. 202-206.
4. Fergusson, J. (1863): Delta of the Ganges; Geological Society of London, *Quart. Journal*; Vol. XIX, pp. 321-354.
5. Friedkin, F.C. (1945): A Laboratory Study of the Meandering of the Alluvial Rivers; *U.S. Waterways engr. Exper. Sta.*; p.40
6. Hirst, F.C. (1916): *Report of the Nadia rivers*; the Bengal Secretariat Book Depot; Ch. VII, p. 21.
7. Leopold, L.B., Wolman, M.G. and Miller, J.P. (1963): *Fluvial Processes in Geomorphology*; W.H. Freeman and Company, San Francisco & London; Ch. VII, p. 302.
8. Matthes, G.H. (1941); Basic Aspects of Streams Meanders; Am. Geophys. Union Trans., Vol. XXII, pp. 632-636.
9. Mitra, A.N. (1944) : Training of Rivers; *the Journal of the Institution of Engineers* (India); Vol. XXIV, No. 3, p. 37.
10. Mitra, B. (1958): *Annual Report of the River Research Institute, West Bengal*; W.B. Govt. Press, Alipore; Pub. No. 14 & 15.
11. Miller, J.P. (1959): Geomorphology in North America; *Polish Geographical Review,* Vol. XXXI, No. 3-4, p. 574.
12. Prus-Chacinski, T.M. (1954): Patterns of motion in open channel bends; *Assoc. International d'Hydrologic*, Pub. 38, Vol. 3, pp. 311-318.
13. Reaks, H.G. (1919): *Report on the physical and Hydraulic Characterisitics of the rivers of Delta; Report of the Hooghly River and its Headwaters*, the Bengal Secretariat Book Depot, Calcutta; Vol. I, App.II, Ch. II, Sec. 54, pp.29-48.
14. Report on the River Bhagirathi (July 1992–June 1993), Vol. I, Hydraulic Study Department, Calcutta Port Trust, Berhampore (West Bengal).
15. Report on the River Bhagirathi (July 1993–June 1994), Vol. I, Hydraulic Study Department, Calcutta Port Trust, Berhampore (West Bengal).
16. Report on the River Bhagirathi (July 1994–June 1995), Vol. I, Hydraulic Study Department, Calcutta Port Trust, Berhampore (West Bengal).
17. Speight, J.G. (1965): Meander Spectra of the Angabunga River; *Journal of Hydrology,* Vol. 3, pp.1-15.

3

Combating Desertification in Dry Agro-Ecological Region of Cold Desert using RS & FIS

A Case Study of Fast and Micro Watershed, Ladakh District, J&K*

P.S. DHINWA, AJAI AND MAYANK BHAVSAR

Introduction

Desertification means land degradation in arid, semi-arid and dry sub-humid areas resulting from various factors including climatic variations and human activities (Earth Summit, Rio, 1992). The term embraces loss of biological or economic productivity and complexity in agricultural, pastoral and forest land. The essential causes of desertification have now been identified as climatic fluctuations and non-sustainable forms of land use. The most frequently cited forms of unsustainable forms of land use are overexploitation, overgrazing, logging and inappropriate irrigation practices (Horstmann, 2002). In India about 175 m ha of land area is suffering from various kinds of degradation problems and 20.17 per cent of area is lying under wastelands (Wastelands Atlas of India, 2000). For meeting the need of food, additional area has to be brought under cultivation which may cause environmental imbalance since the area under forestry has already been declining at the rate of 1 mha per year due to deforestation. The other option lies only in increasing the productivity per unit area through systematic appraisal of our soils and climatic resources for optimizing land use on sustainable basis. Large scale irrigation, coupled with extensive use of fertilizers, poor drainage and inappropriate agricultural practices has led to undesirable increase in soil salinity, reducing previously fertile land to an impoverished and unproductive state (Williams, 2001). In the last decade it has become evident that the processes of desertification have been very much enhanced mainly through rapid increase of human and animal population (Sharma, 1993). India has large area under arid and semi-arid zones. According

* Paper presented by ISRS sympoosium on Watershed Management from 9-12 December, 2003, Thiruvanthapuram.

to National Bureau of Soil Survey and Land Use Planning, Nagpur, the arid, semi-arid and dry sub-humid regions constituting the 'drylands' (as per Thornthwaite Classification, 1955) cover 228.3 m ha of the total area (328m ha) of the country (Anon., 2001).

Study Area

The Fastan micro-watershed (1 F3A6) lies in the northern part of Ladakh district (Anon.,1990). This micro-watershed is drained by Fastan Lungpo, which is a tributary of Shyok River and a part of the Indus River system. Geographically, it is bounded by latitude 34° 50′ to 35° 00′ North and longitude 77° 04′ to 77° 20′ East. The total area of the watershed is 185 sq. km. This watershed lies in Western Himalayas Cold Arid Ecoregion and Ladakh Plateau Cold, hyper arid Ecosub-region (A13Eh1). The subregion constitutes the leeward flanks of Western Himalayas facing Tibetan Plateau. The mean annual temperature (MAT) is 11 degree C and ranges from 9° to 14°C. The mean annual precipitation (MAP) ranges from 100 to 115 mm. In view of its location in the rain shadow side, the area receives erratic and scanty precipitation from westerly depression mainly as snow during winter months (November to March). The precipitated snow which account for 37 per cent of the total freezes and forms thermal pan for crop growth during winter season. The moisture index (MI) varying from between 85 and 87 per cent suggests that the area is under hyper aridic condition. The length of crop growing period (LGP) is less than 60 days (Velayutham *et al.,* 1999). The Ladakh plateau is rugged mountainous terrain interspersed with narrow to nearly broad valleys. The elevated ranges remain under perpetual snow cover and the gently to moderately sloping and nearly level intermountain valleys have mostly skeletal soil cover. The moderately to gently sloping and nearly level intermountain valley comprise soilscapes of Cryothents and Cryorthids great groups (Sehgal *et al.,* 1989). The soils in the area are in general shallow to moderately deep, coarse textured, mostly skeletal, calcareous and strongly alkaline represented by Ladakh series. The Ladakh soils are poorly developed (A-C profile) moderately deep, coarse loamy Typic Cryorthents. The occurrence of gravel's and pebbles is a common feature in subsoil. The clay content of the surface soil is around 10 per cent with low organic matter (0.1%). The clay and organic carbon content decrease with depth while the soil reaction is moderately to strongly alkaline though the pedon reflecting poor soil development, inadequate leaching and accumulation of soluble salts in the soil. The charge characteristics is low elite constitutes the dominant clay mineral in the exchange complex (Sehgal, 1973). The plant available moisture capacity (PAWC) is very low.

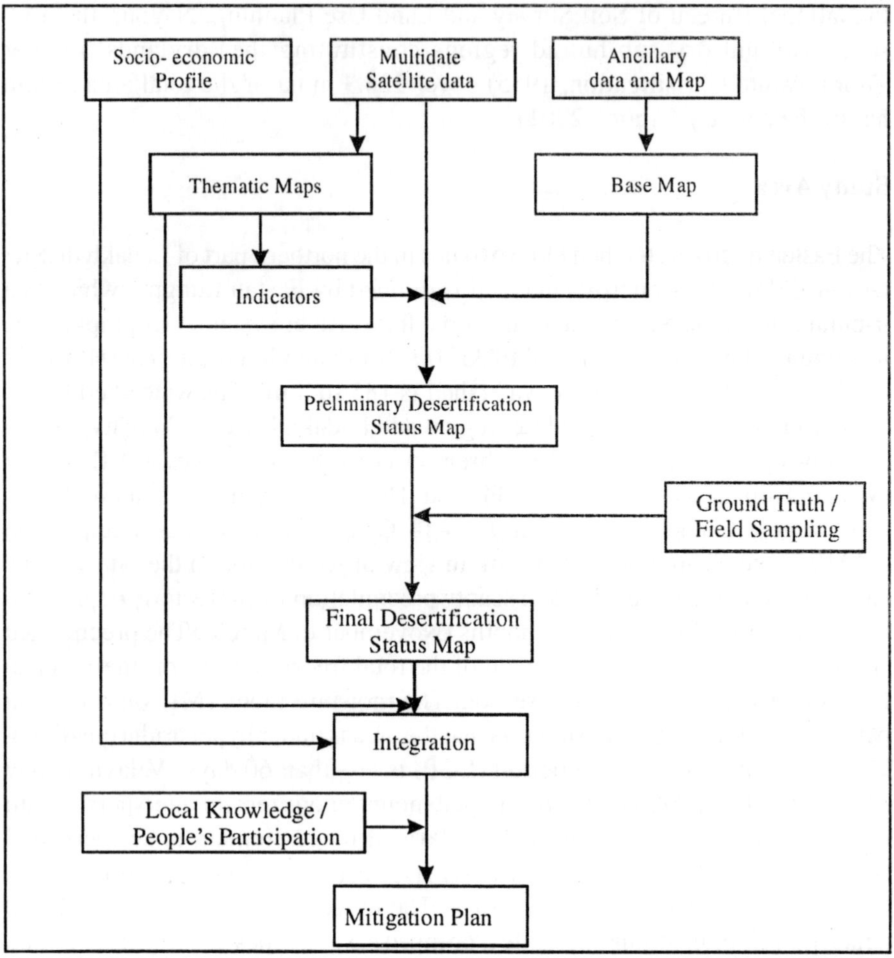

Fig. 3.1 : Methodology for Desertification Status Mapping/Mitigation Plan using Satellite data

Data Used

The multi-temporal satellite data of October 1999 and May 2000 of IRS-1D, LISS-3 and ancillary information has been used for desertification status mapping.

Methodology

The methodology and classification system is finalised after brain storming discussion with various collaborating agencies and academia (Fig.3.1). By

Table 3.1 : Desertification Status Mapping: Parameters amenable to Remote Sensing and Conventional Survey

Process of Degradation	*Desertification Indicators*	*Degree of Severity*		
		Slight	*Moderate*	*Severe*
Vegetal Degradation	% of Crown Cover*	40 -20	20 7 10	< 10
	Plant Community @	Clima X	Long Lasting Secondary	Ephemeral Secondary
	% of climax species @	> 75	75-25	< 25
	Decrease of total plant cover, % @	< 25	25-75	> 75
	Loss of forage (grasses) % @	< 25	25-75	> 75
Salinization	Crops Supported*	Two Crops	One Crop (Kharif)	No Crops
	pH@	< 8.5	8.5	> 8.5
	Soil Salinity Ecenin ds/m @	4-8	8-30	> 30
	Salinity of ground water g/litre @	3-6	6-10	10-30
	Loss of yield of main crop @	< 25	25 - 50	< 50
	Soil residue, % @	0. 20-0.40	0.40-0.60	> 0.60
	Salinity of Irrigated water g/litre @	0.5-1.0	1.0 -1.5	> 1.5

* Amenable to Remote Sensing
@ Amenable to Conventional Survey

Table 3.2 : Area under various Land Use/Land Cover Classes/Processes of degradation/ Degree of severity

Land Use/ Land cover	*Process of Degradation*	*Symbol*	*Degree of severity*	*Area (sq. km)*	*% total area*
Forest	Vegetal	NA	Not Apparent	10.50	06.56
		Fv1	Slight	41.00	25.63
		Fv2	Moderate	32.50	20.31
		Fv3	Severe	15.00	09.38
Agriculture	Salinization	Is1	Slight	08.50	05.31
Irrigated		Is2	Moderate	03.50	02.18
Land with Scrub	Vegetal	Sv1	Slight	22.00	13.76
		Sv2	Moderate	15.00	09.38
		Sv3	Severe	15.00	09.38
			Total	160.00	100.00

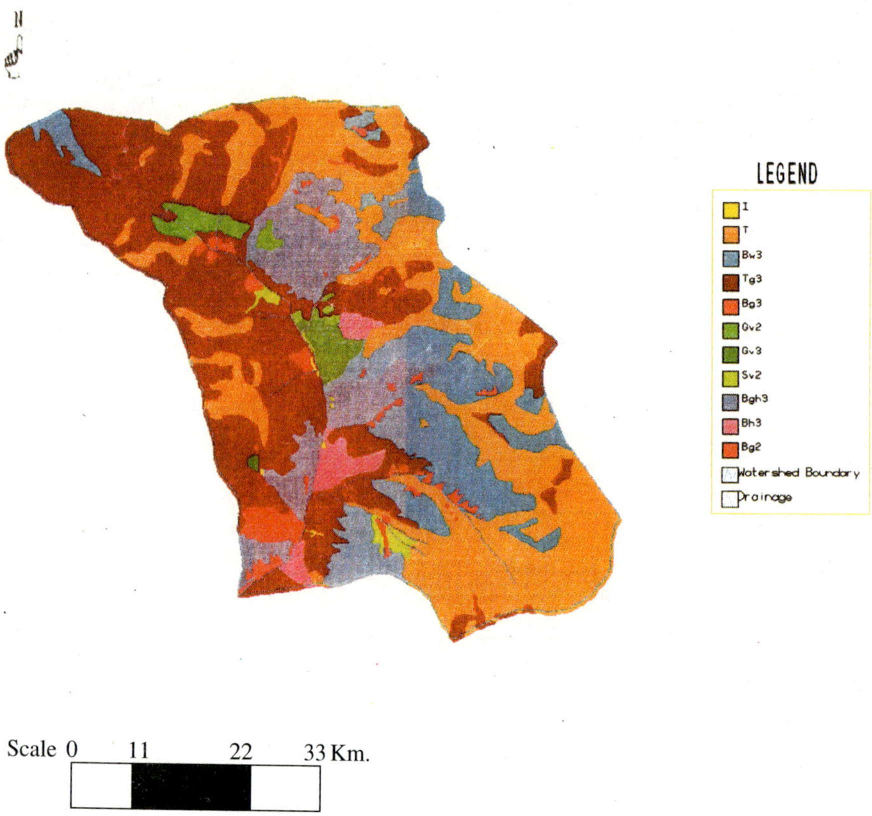

Classification System

Level I *Land use/cover*	*Level II* *Degradation process*	*Level III* *Severity of Degradation*
Agriculture unirrigated -D	Vegatation degradation-v	Slight-1
Agriculture irrigated-I	Water erosion-w	Moderate-2
Forest-F	Wind erosion-e	Servere-3
Grossland and/Grazing land-G	Water logging -I	
Land with scrub-S	Salinization/Alk-S	
Barren/Rocky area-B/Sc	Man made-m	
Dune/Sandy area-E	(Mining/Quarrying, Brick kiln	
Water body/Drainage-W	Industrial effluent, city waste	
Other-I	Urban Agg. etc)	
(Urban, Goacial, Peri-Glocial,	Mass movement-g	
Man-made etc.)	Weathering-h	

Fig. 3.2 : Desertification Status Map Fastan Micro Watershed, Ladakh District, J&K Cold Desert

adopting the above mentioned classification system and employing the visual analysis and interpretation technique and by using multi-temporal satellite data a desertification status map is prepared for Pastan micro-watershed, Ladakh district J&K, (Annexure I). Various categories of land use/land cover, processes of degradation and its degree of severity are delineated on the map. Slight, moderate and Severe categories of vegetal degradation are delineated on the basis of percentage of crown cover, climax of plant community, decrease of total plant cover and loss of forage (grasses). Slight, moderate and severe categories of mass wasting and weathering are delineated based on texture, association, size of boulders, slope and aspect of the mountain (Table 3.1). Spatial database is created for various thematic layers, viz., Land use/land cover, slope, ground water prospect and land capability using ARC/INFO GIS software. The analysis of thematic layers is carried out in GIS environment and based on some criterian analysis, the mitigation plan is generated(Fig.3.2).

Results and Discussion

The results obtained through visual analysis reveal that mass wasting process is very dominant in valleys and on the piedmont slopes facing the valley while weathering is prominent both on mountain tops and on piedmont slopes facing the valley. Sand blasing is also observed at many places (Table 3.2 & Fig. 3.2).

Conclusion

The multi-temporal satellite data reveals that anthropogenic factors and vegetal degradation are very prominent for desertification in this region and 86 per cent of area is affected by vegetal degradation and 8 per cent of the region is reeling under salinity.

Acknowledgements

The authors are grateful to Dr. K. Shankara, Director SAC, and Dr. V. Jayaraman, Director EOS, Bangalore for their guidance and to Dr. S. K. Pathan, Head, LPPDIFLPG for his valuable comments and critical evaluation of paper.

REFERENCES

1. Anon, 1990. *Watershed Atlas of India. All India Soil and Land Use Survey (AISlUS),* Ministry of Agriculture Cooperation, Government of India, New Delhi.
2. Anon, 1992. *Managing Fragile Ecosystems: Combating Desertification and Drought,* Agenda 21, United Nations Conference on Environment and Development, Rio de Janeiro, Brazil, June 3-14, p.131-150.
3. Anon, 2000. *Wastelands Atlas of India.* Ministry of Rural Development, Department of Land

Resources, Government of India, New Delhi-110 001 and National Remote Sensing Agency, Department of Space, Government of India, Balanagar, Hyderabad-500037.

4. Anon, 2001. INDIA, National Action Programme to Combat Desrtification in the context of United Nations Convention to Combat Desertification (UNCCD) Status of Desertification. Vol. 1 Ministry of Environment and Forests, New Delhi, p. 29.
5. Dhinwa, P.S. and Ajai (2002). *Desertification status mapping in cold desert—A case study of Kathua micro watershed, Kathua District (J&K)*, Proc. of International Seminar at Hyderabad during December 21-26, 2002.
6. Dregne, H.E., 1986. *Magnitude and spread of the desertification process—Arid land Development and the combat against desertification—An Integrated Approach.* USSR Commission on UNEP, Moscow.
7. Horstmann, B., 2002. *Desertification a world problem. Agriculture + Rural Development,* 9(1), DLG-erlags-GmbH, Eschborner LandstareBe122, 60489 Frankfurt am Main, Germany.
8. Rao, U.R., 1991. *Space and agricultural management.* In 'Space and Agricultural Management,' Special Current Evening session, International Aeronautical Federation, 421AF Congress, Montreal Canada, pp.1-10.
9. Sehgal, J., Mandal, D.K., Mandal, Cd Vadivelu, S., 1992. Agro-Ecological Regions of India. 2nd Edition, *Tech. Bull., Pub. 24*, p.130, NBSS&LUP, Nagpur.
10. Sharma, H.S., 1993. Geomophological aspects of the processes of desertification in the northeastern Rajasthan (India). *Z. Geomorph.* N.F., Berlin, Germany.
11. Thornthwaite, C. W. and Mather, R.J., 1955. *The water balance publication in Climatology,* 8 (1) DIT, Laboratory of Climatology, Centerton, NJ, p.104.
12. Velayutham, M., Mandal, D.K., Mandal, C., Sehgal, J., 1999. *Agro-ecological Subregions of India for Planning and Development.* NBSS Pub. 35, NBSS&LUP (ICAR), Nagpur.
13. Williams, J., 2001. *GIS Processing of Geocoded Satellite Data.* Praxis Publishing Ltd., Chichester, UK, p. 2398.

ANNEXURE-I

CLASSIFICATION SYSTEM FOR DESERTIFICATION STATUS MAPPING

A three level hierarchical system of classification has been followed for the desertification status mapping. This is explained as bellow:

LEVEL 1: Land Usel Land Cover

The following categories have been identified as below:

Agriculture—Unirrigated
Agriculture—Irrigated
Forest
Grass land/Grazing land
Land with Scrub
Barren I Rocky Area
Dunel Sandy Area
Water body/Drainage
Others (Urban, Glacial, Peri-glacial, Man-made etc.)

LEVEL 2: Processes of Degradation

Type of morphological processes resulting in degradation.
Vegetation Degradation
Water Erosion
Wind Erosion
Water Logging
Salinization/Alkalinization
Man made (Mining/Quarrying, Brick Kiln, Industrial Effluents, City Waste, Urban Agg. etc. Mass Movement.

LEVEL 3: Severity of Degradation

This level represents the degree and severity of the degradation.
Slight
Moderate
Severe

4

Neotectonic Activity and Seismicity in and around Sambhar Lake, Rajasthan

J.K. SHARMA AND M.L. SHARMA

Introduction

The post-Neogene reactivation of the major faults like great boundary fault, Sardarshar fault, Dausa-Jaipur-Sambhar-Lake fault etc. occurring along the eastern and western margins of the Aravalli range have recently been highlighted. Vaidyanadhan (1977), Sen (1983), Sinha-Roy (1986), Dassarma (1986), Singh (1998), Prakash (2000), Vaidiya and Rajagopalan (2000) etc. In recent years Neotectonic activity is mostly associated with seismicity which exemplifies in the area of study. Neotectonism have been described on the basis of a variety of morphotectonic features like knick-points, lineaments, fault generated scraps, multitiered hanging valleys, sag ponds, subsequent nature of drainage with strong preferred orientation along fractures, seismic level difference in earlier erosional surfaces in the contiguous blocks etc. This paper presents the records of some such evidences along with their control over the patterns of the epicentres in the area of study.

The study area of Sambhar Lake and its environs is located in a gap of the Aravalli range about 60 kms west of Jaipur in Rajasthan.

It is an intermontane basin with endorheic drainage. It holds a unique distinction amongst the saline basins of Rajasthan, due to maximum thickness of quaternary deposits (2m to 150m) and being the principal source of salt having 234 square kilometres of area.

The geology of the area around the lake has been described by Hacket (1880), Heron (1971), Sen and Ramalingam (1976), Sing (1974) etc. Neither any information is available on the relationship of neotectonism and seismicity in the region nor any work has so far been carried out to determine the linkage between recent tectonic movements and seismocity. Therefore, here an attempt has been made to establish the relationship between the two.

Material and Methods

The present study is carried out over an area of about 6000 sq. kms in the catchment area of Sambhar salt source in the Aravalli hills around the Mendha, Rupnagar, Khari, Khandel, Turatmati internal ephemeral rivers. The area is bounded by E longitude 74° 35′ 5″ to 75° 55′ 10″ and N lati -tude 26° 30′ 5″ to 27° 30′ 55″ (Fig.4.1). The study include quaternary geological and geomorphological mapping with the help of aerial photographs (1:50,000 scale). Periodic Photographs (1:25000 scale) around the critical patches of neotectonic activity corroborated with a large scale surveying and field checks. Moreover, survey of India toposheets (1:50,000 scale) of the basin area have also been consulted to strengthen the text of the article. Field survey and field checks have also been carried out in different times and years in the area.

Geology and Geomorphology

The Delhi Supergroup of the rocks are overlain by fluvial aeolian and playa deposits of quaternary period. The topography of the area is generally undulating which exhibits a number of different geomporphic units like NNE-SSW trending discontinuous denudational hills like Korsina, Nawa, etc. isolated hill Rocks, Inselbergs—Lal pir Ki Dungari, Shakambari Mata Dungari etc. Sand dunes, sand sheet, lakes, playa etc. Landforms of structural and erosional origin (Fig. 4.2) include hogbacks and denuded hummocky mounds and pediments carved out of the Proterozoic rocks. The axial zone of the Aravalli hills is an important tectonic zone described as the South Delhi fold belt (Sinha-Roy, 1984) and as the Delhi synclinorideum (Heron, 1953). There are two levels of fluvial terraces within elevation from the mean lake level. The palaco channels tend east-west as low lying linear zones. The aeolian landforms which generally cover the lacustrine and fluvial landforms are present in the form of longitudinal, obstacle and barchan shaped dunes as well as sand sheet and undulatory sand mounds. The present lake surface is at an elevation of 360m above m.s.l. and present 'self-rising' or 'Puffy ground' appearance described by Rosen (1991). Indications of a paleobore line at an height of 1 to 1.5 m above the present level are locally present. Swain, Kutzbath and Hastenrath (1983) postulated a maximum lake level upto 381m contour indicating a 21m. rise during the quaternary period based on evidences of topographic charts and land set imagery. Figs.4.3 and 4.3A shows the major old faults and lineament within the basin. The Aravalli sequence is separated from the banded gneissic complex (BGC) by a thinnest and a durtile shear zone. The South Delhi Thrust (SDT) dipping steeply towards N.W., defines the contact of the Delhi sequence with the Aravalli and the BGC rocks. This thrust was responsible for the initial uplift of the Aravalli hill axis. The Sambhar Jaipur-Dausa fault (SJDF) is a right lateral strike slip fault linked with the (SDT) (Sinha-Roy, 1984) and South Delhi Thrust. These faults have a bearing on the neotectonic activities in Sambhar basin. Lineaments as we, know are the linear features on land surface

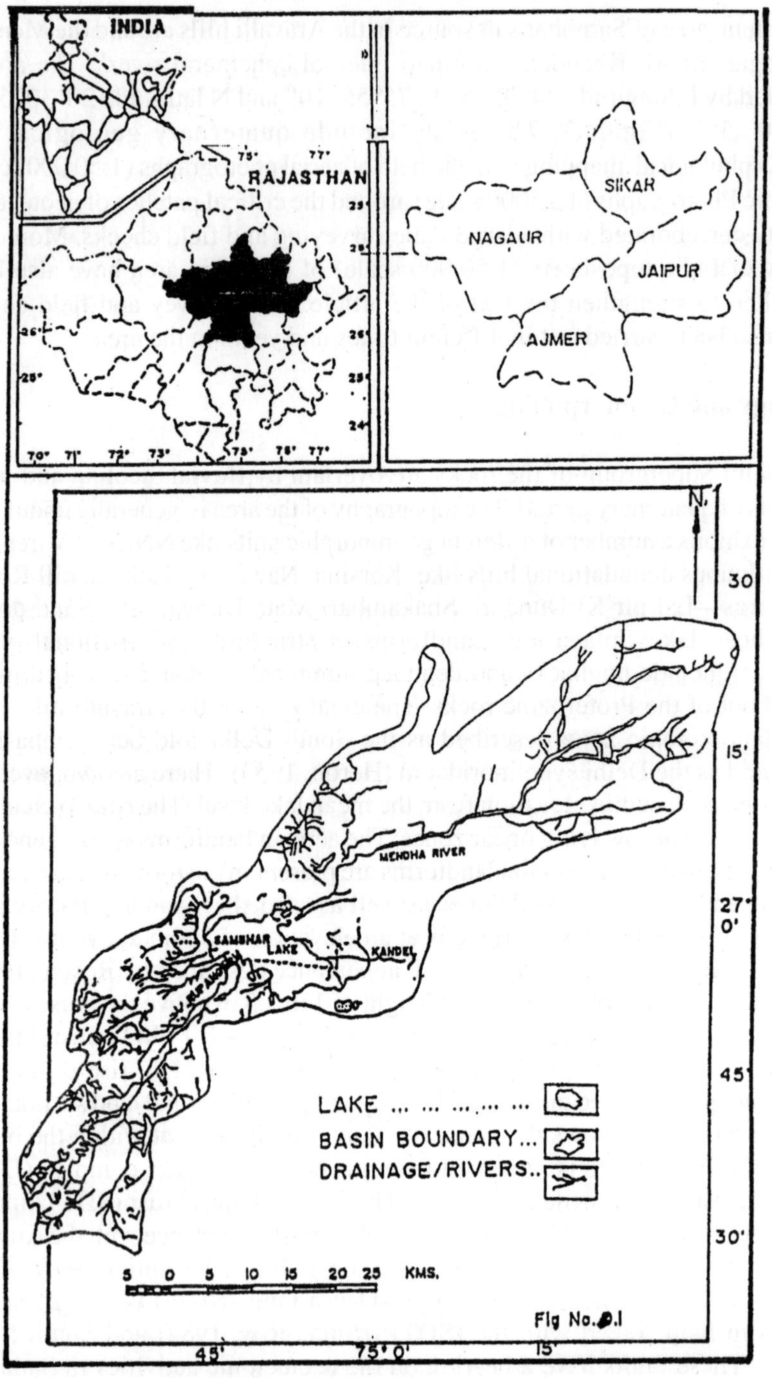

Fig. 4.1 : Location Map of Sambhar-Lake Drainage Basin and Inland Drainage Basin

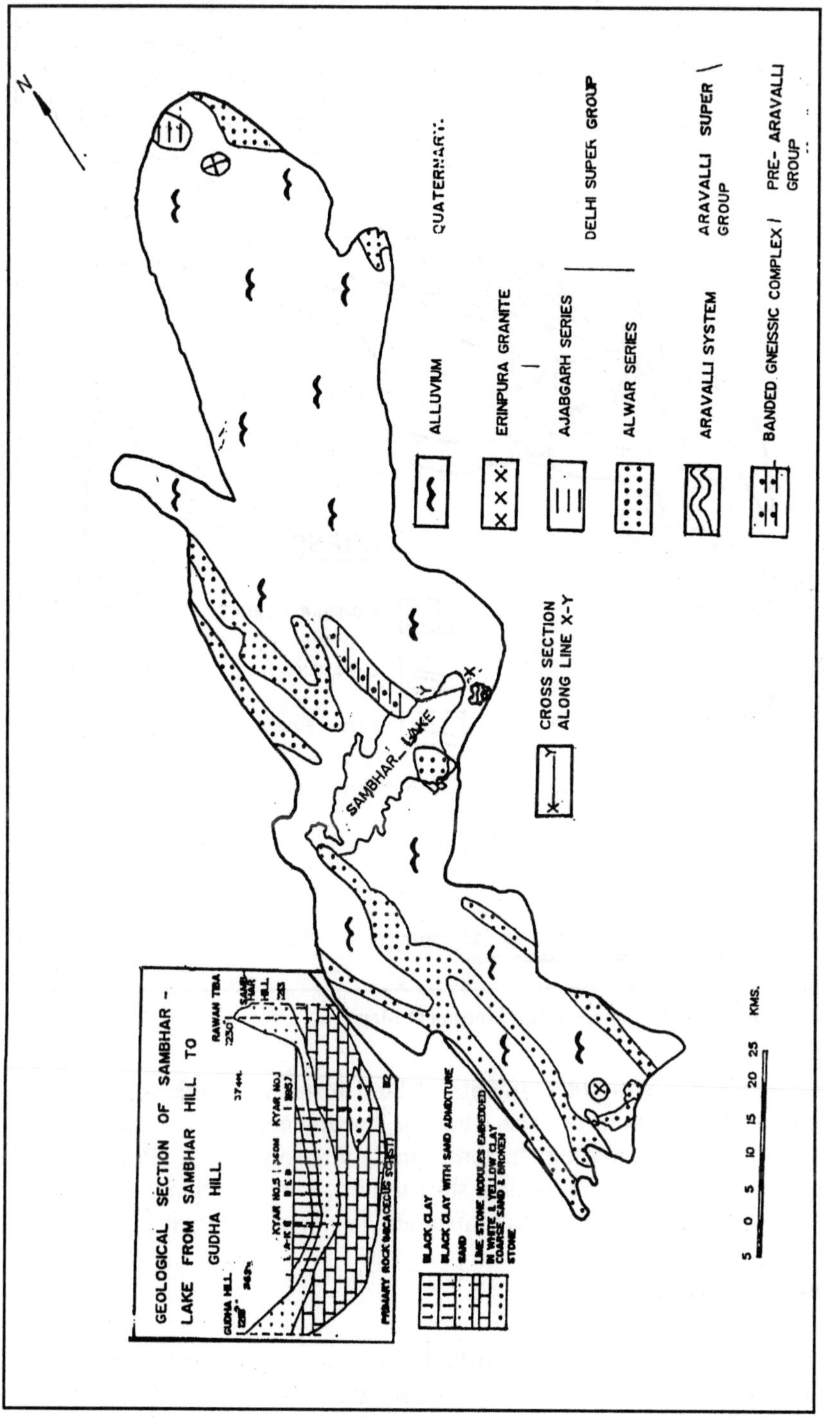

Fig. 4.2 : Sambhar Inland Drainage Basin Geology

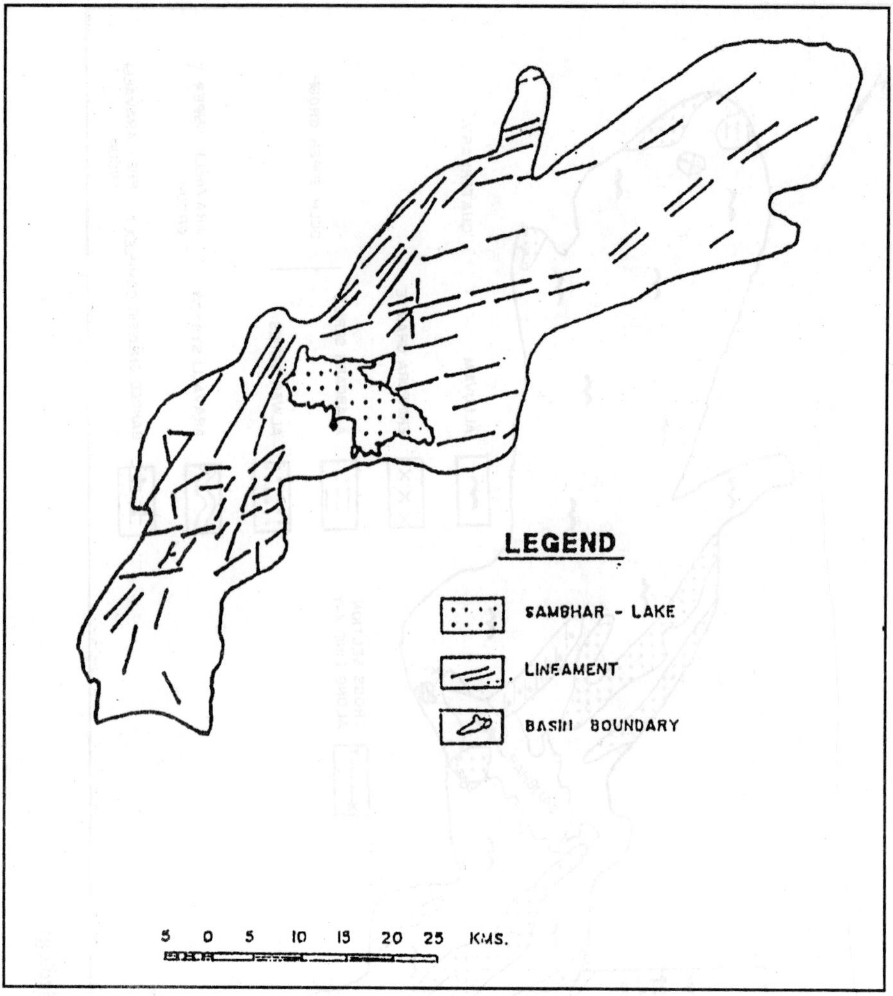

Fig. 4.3 : Sambhar Inland Drainage Basin Lineament Map

corresponding to the straight segments of land forms and generally reflects sub-surface plainer structures. The study of lineaments is very useful for investigations in geomorphology tectonics including seismology (Powar *et. al.* 1978, 1979, 1980, 1982, 1983). A map showing lineaments in Sambhar basin is important study material for neotectonism and seismosity (Fig. 4.3).

Result and Discussion

Neotectonic activity is discernible through the following features in Sambhar lake basin which is ultimately responsible for the seismosity and earthquake epicentres.

Tectonic Sag

The intersection of Lineaments (Fig. 4.3) where recent crustal movements are observed shows development of tectonic sag ponds. The development of quaternary depositional basin at the top of Precambrian rocks indicates that the region under study might have experienced tectonic movement at the beginning of quaternary sedimentation. It is further evidenced by variation in thickness of quaternary sediment from 2.5 to 40m in the Rupangarh and from 30m to 70m in the Mendha River basin. This also indicates existence of graben in between the Rupangarh and Mendha river basins within which the Sambhar salt lake formed (Sural and Dassarma, 1987) Fig. 4.4.

The presence of 100m to 130 m thick quaternary alluvivum was also reported in the adjoining part to the north-west of the Sambhar lake thereby indicating existence of subsiding basin attributed to tectonism, which is further supported by the shifting attitude of Mendha and Rupangarh rivers which shows a lateral shift towards north-west in the Sambhar salt lake area living behind many channels (Fig.4.2) Fluvial facies occure mainly along the Mendha river as terraces and present flood plains respectively. Facies as referred to here and as defined by Smoot (1991) are characterized by distinctive litho-assemblages are geomorphic expressions and denote depositional environments. The quaternary sedimentation depicting the subtle changes in climatic conditions is considered to be controlled by neotectonism which is manifested in geomorphological, structural and stratigraphic anomalies such as drainage rehersals, unpaired terraces, structure controlled drainage, scarp and scarplets on hogbacks, residual hills etc. (Fig. 4.5) (Sundarama and Pareek, 1995).

Sedimentation and Palaeo-channels

In Sambhar and its environs, three generations of drainage network have been identified on the basis of fluvial facies (Fig. 4.5A). The three distinct aeolian phases coeval with three lacustrine or Pluvial phases interspersed by the wet phases fluvial sediments are discernible in the area.

Fluvial facies occure along Mendha and Rupnagar streams. The F-3 fluvial facies is marked by a cap of blown loamy soil (Fig. 4.5B). The distribution pattern of F-3 facies suggest deposition by ephemeral or braided rivers. These facies were marked by migration of rivers and shifting of locales of deposition. The fluvial action was more dominant in the north of Sambhar lake as evidence by extensive terracing and delta occurrence. At Kuni and Khatwari Khurd villages and in-between Tyod and Navrangpura near Sambhar lake palaeo-channels were also inferred (Fig. 4.5). Couple of Nalas dissecting the aeolian country and meeting the Mendha river in upstream direction is certainly quite significant as drainage reversals.

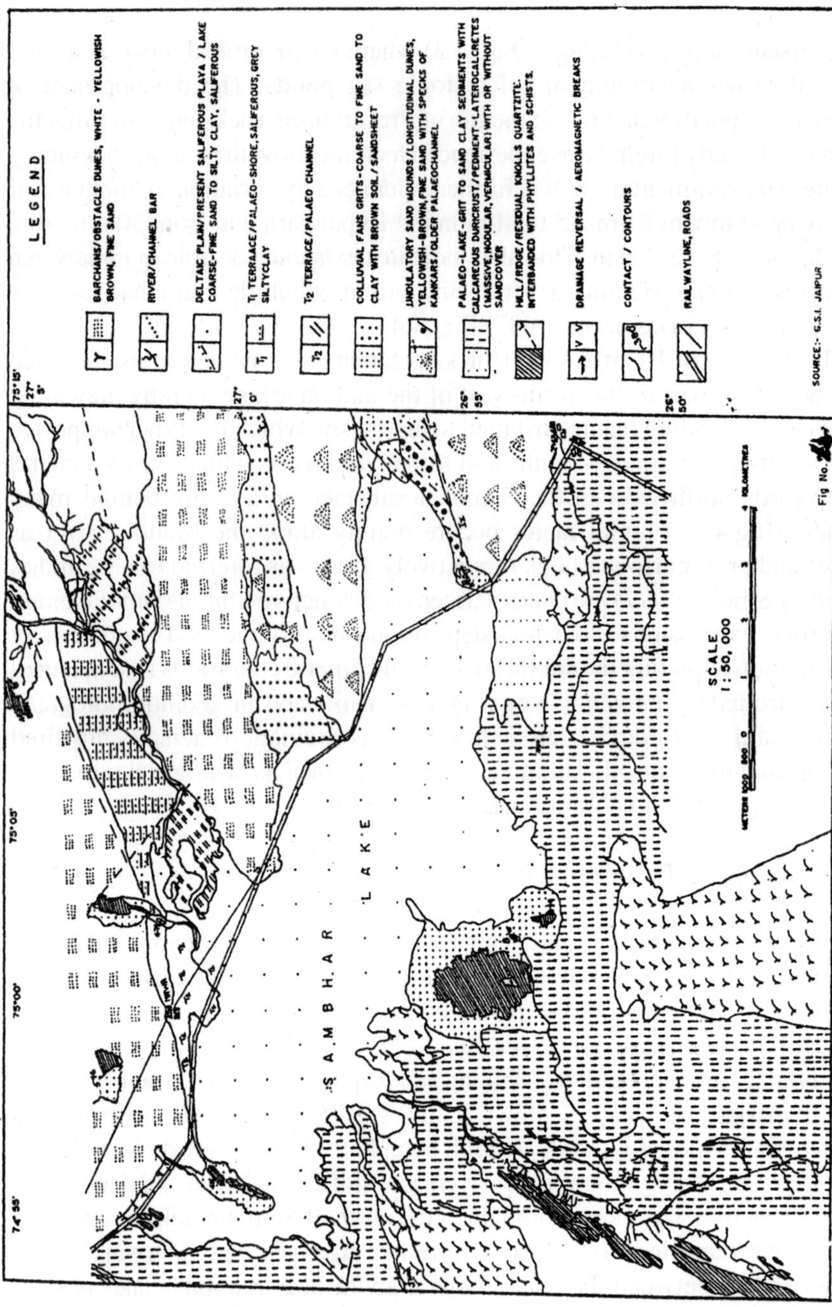

Fig. 4.4 : Quaternary Geology of Sambhar Lake, Rajasthan

The oldest fluvial deposits are the reworked rock fragments laterocal creates and other clastics as colluvial fills by ephemeral rivers. In the southern and western parts of the lake author has observed occurrence of denuded hills of pre-cambrian rocks and extensive pediment tracks with latosols. Ephemeral drainage of Rupangarh river dissecting the Pediment country can be seen as similar to dominance of Mendha river basin in the northern part where extensive terracing and deltaic deposition are seen. Features of neotectonic evidences were reported by Dassarma (1986). The Palaeochennels are linearly parallel stretches indicating sedimentation in stepped graben. It all denotes the neotectonic activity in the basin (Fig. 4.5.B).

Rejuvenation of Fault and River Entrenchment

The ephemeral streams of Sambhar basin have a tendency to form point bar deposit on convex side and lateral cutting on concave side which is a important indication of initial stage of entrenchment. Where the tributaries have not yet attained the equilibrium due to changing base-level. Rejuvenation of fault plain along the valleys and subsidence in the lake or playa area are suggested for the changing base level. River entenchment due to subsidence is also indicated by the formation of gullics in obstacle due field area near Nawa on the western side of the hills which is drained by east flowing tributaries against the general slope.

Tilt and Warping

About 6° to 8° of tilt towards north-west and warping were recorded in the sediments of first generation palaeo drainage in the upper reaches of the Turatmati. North-east of Khandela.This tilt holds the view of neotectonism in the basin.

Fault and Dislocation

In and around Sambhar lake, there is the presence of a faulted Kankar horizon which shows a vertical dislocation of about 1 metre was observed within one of the abandoned playa deposits in the central part of Mendha river basin. Offsetting of first generation of Palaeo drainage sediments against the second generation Palaeo drainage sediments was also observed North-west of Kandela. Kankar Pan horizons are also traceable in the lake enroute to Korsina and Mata Pahari, Rupangarh river low reaches are full of Kankar horizons thereby depicting the changes of climate in the quaternary era and dislocation in Sambhar lake basin.

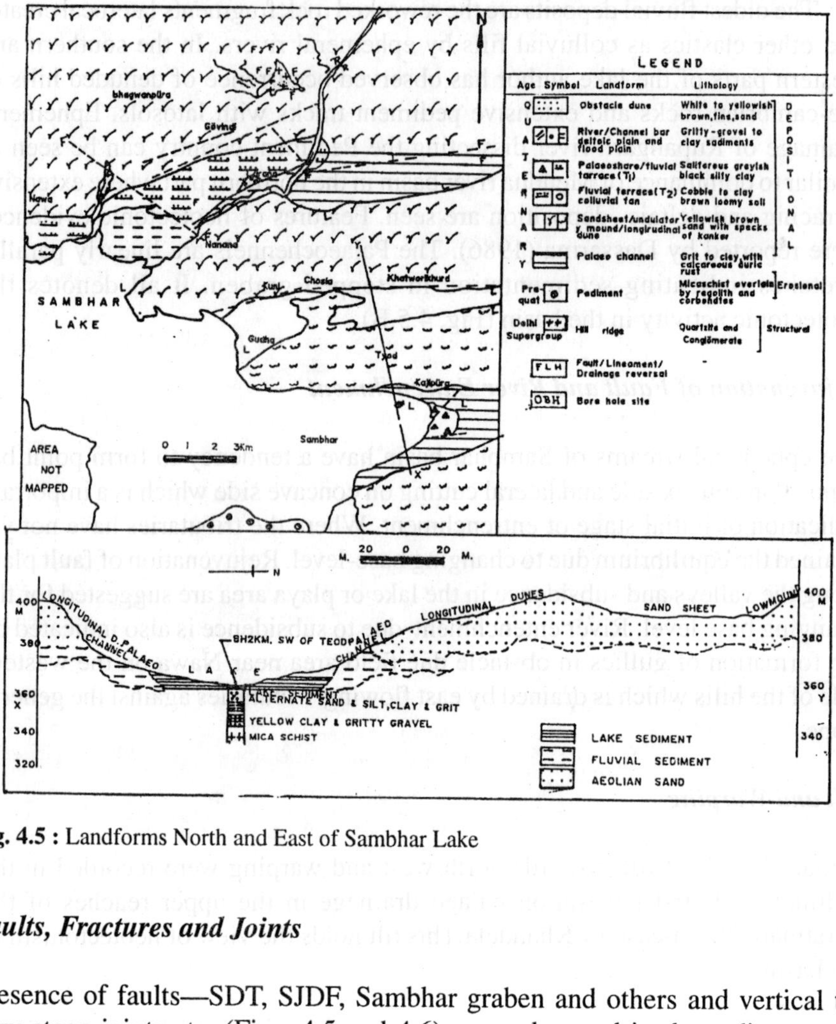

Fig. 4.5 : Landforms North and East of Sambhar Lake

Faults, Fractures and Joints

Presence of faults—SDT, SJDF, Sambhar graben and others and vertical to very steep joints etc. (Figs. 4.5 and 4.6) were observed in the sediments of second generation of palaeo drainage north-east of Khandela. Some of these fractures are filled up by the precipitation of calcium carbonate. Such fractures and joints widely traceable in the basin are again the testimony of neotectonism. These facults, fractures and joints shows the weaker zone on the earth surface and thus liable to reactivisation and seismic activity.

River Terrace

In upper reaches of Rupangarh river two sets of fluvial cut and fill terrace and rock cut terrace have been identified whereas only one set of fluvial cut terrace

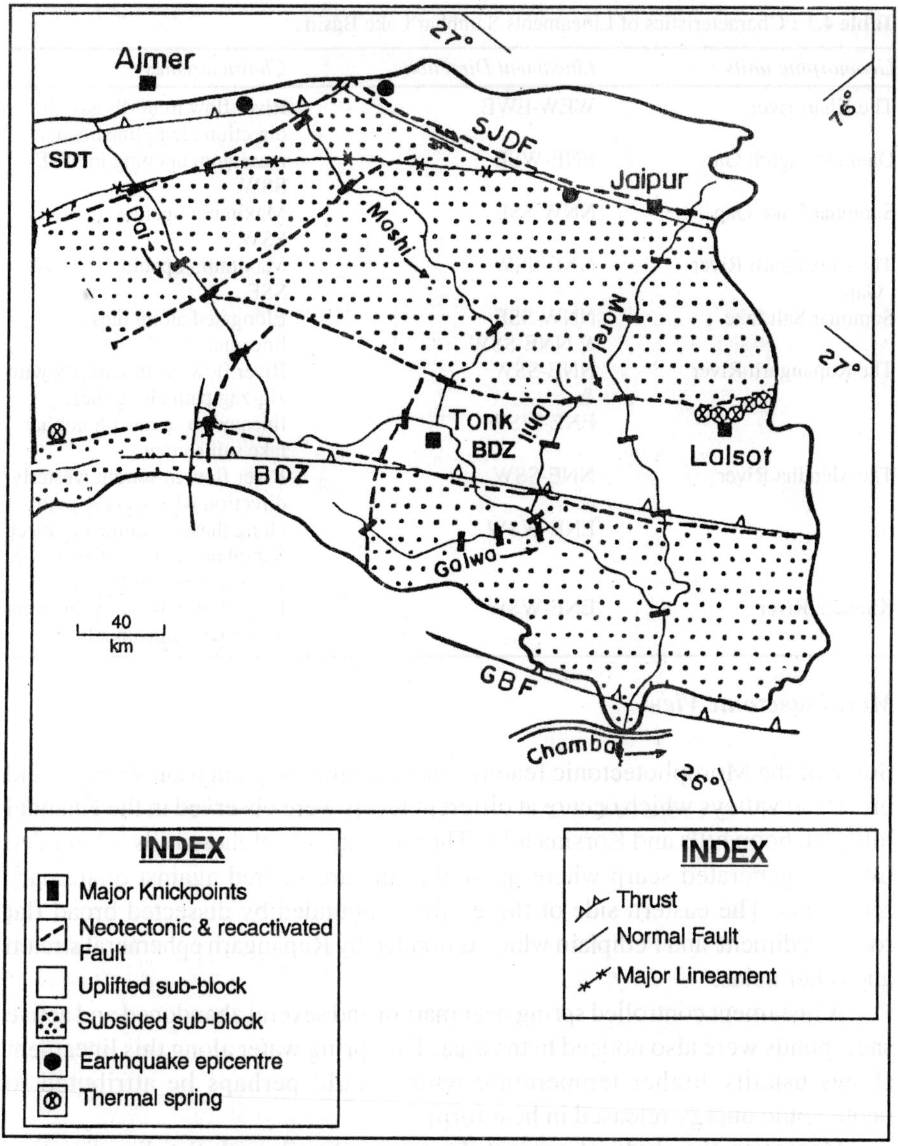

Fig. 4.6 : Faults and Fractures Around Sambhar Lake Basin

were observed in the upper reaches of the Mendha river. A level difference of about 5m in the present terrace of second fluvial phase is observed in the upper reaches of Mendha river and Rupangarh river which is attributed to neotectonic activity in Sambhar basin. It is again a symbol of seismic activity.

Table 4.1 : Characteristics of Lineaments Sambhar Lake Basin

Geomorphic units	*Lineament Direction*	*Characteristics*
The Khari river	WEW-EWE	River flow in north-east do direction along lineaments
Danta Ramgarh Gap	ENE-WSE	maximum opening in NNE-SSW.
Sambhar Lake Gap	NNW-SSE	Maximum opening in NNE-SSW.
The Rupangarh River Gap	NNE-SSW	Maximum opening in NNW-SSE.
Sambhar Salt lake	NNW-SSE & NNE-SSW	Elongated along these lineaments.
The Rupangarh River	NNE-SSW & ENE-WSW	River flow north-easterly with zig-zag path along these lineaments to meet Sambhar lake salt source.
The Mendha River	NNE-SSW & ENE-WSW	River flow in sought westerly direction with zig-zag path along these lineament to meet Sambhar salt lake. This is the master stream of the lake.
Khandel River	ENE-WSW	River flow in westerly direction along these lineaments.

Morphotectonic View

Some of the Morophotectonic features like multitiered plantation surfaces and broad flat valleys which occure at different levels were observed in the Khandel hills, Bichoon hills and Korsina hills. The western side of these hills is bounded by fault generated scarp where quartzite hills are faulted against quaternary sediments. The eastern side of these hills is bounded by dissected broad flat rocky Pediment and Pediplain which is drained by Rupangarh ephemeral stream and other nalas.

A lineament controlled spring near maroth and several abandoned and active sage ponds were also noticed in the area. The spring water along this lineament shows usually higher temperature which could perhaps be attributed to neotectonic energy released in heat form.

Moreover, the study of present drainage network with the dip, direction and alignment of channel in preferred orientation suggest that the major control of these courses is structural. Although the overall drainage represent a coarse dendritic pattern modified by structures yet the structural control over the drainage lines of the area is obvious.

These sets of lineaments identified on space imagery as well as on aerial photographs corroborated by toposheets which the stream courses are aligned (Table 4.1). These directions are NNW-SSE, ENE-WSE and NNE-SSW

(Fig. 4.3). The influence of the directions of these lineaments is also noted in the development of hill gaps as well as Sambhar salt lake where longer axis parallel to these orientation.

The inter-section of NNE-SSW and NNW-SSE and ENE-WSW trending lineament where evidence of recent crustal movements can be discerned shows development of sag ponds. The Mendha river master stream appears to be flowing in a younger graben which is bounded on both sides by uplifted blocks occupied by Khandel hills and Rupangarh river basin in the south. The upliftment has also created a divide along eastern margin of this graben which have divided the flow of neighbouring Bandi river towards south paralleling the divide which causes Bandi river to flow into Yamuna system as against the other rivers of the area which are internal in nature. The dominance of salinity, water logging, flooding and sand movements in the Mendha river basins and soil erosion in Rupangarh and Khari, Khandel Basin also indicate control of these features.

It is obvious on the basis of disposition of Palaeo-drainage as well as present drainage the NNW-SSE trending lineament is the youngest, Which has caused offset in the ENE-WSW and NNE-SSW trending lineament. The ENE -WSW trending lineament shows parallelism with Palaeo-drainage and abandoned playas of second generation indicating a clockwise rotation of the palaeo drainage from first generation to second generation.

The Sambhar lake basin itself is located adjacent to the Banas basin is an impressive geomorphic feature whose tectonic origin has been highlighted by many authors (Sinha Roy, 1986; Roy, 1999). Almost E-W trending long axis (22 kms) of the lake is parallel to a dextral strike slip fault zone. The fault geometry suggests that the lake depression has formed within the Naraina Bagru-Sambhar graben (Fig. 4.6) that probably opened as a pull apart basin due to movement along the strike slip fault zone (Fig. 4.6). The uplift of the Jaiput neotectonism sub-block might have caused stream blockage and ponding particularly Rupangarh river to the south-west. The right angle course deflection of Bandi river near Jobner from EW upstream to north-south down stream is due to the capture of a stream of the Mendha drainage system by the Bandi river of the Banas drainage system. The down stream segment of the beheaded stream in the west led to ponding in the Sambhar lake and formation of the wind gap in the Aravalli hill range.

Neotectonic Implications

The pre-cambrian Rajasthan cration with its Proterozoic flood-thurst belt has been neotectonically active in the quaternary and recent times. Many authors have reported evidence of neotectonism in Jaipur upland and Sambhar lake environs and have explained the geomorphic features such as graben, disorganized course of rivers, river capture, erosion surfaces, lost rivers, etc. by neo-tectonic movements in new and active fault zones and along re-activated old structures. Heron (1938, 1953) postulated a Jurassiz and younger age for the uplift and warping of the Aravalli hill range. Sen and Sen (1983) arrived at the similar conclusion and demonstrated that the knick points in the west flowing rivers and the pediments of the Mendha and Rupangarh basin are caused by neotectonic uplift of the Aravalli hill range. The deflection of the river course and the origin of the lakes in south-east Rajasthan have been explained by both vertical and strike slip movements along active faults (Sinha-Roy, 1986). The structural depressions in the form of lakes especially Sambhar lake have been interpreted as pull apart grabens and half-grabens linked with strike slip faults. These neotectonic features and terraines uplift are the result of lake tertiary quaternary to recent indentation tectonics caused by westward push of the Rajasthan cration by the Bundelkhand Massif in response to (India-Tibet) collision related deformation of the Indian shield. Many active faults and lineaments have been delineated from remote sensing studies (Bakiwal and Ramasamy, 1987) in Rajasthan. Recent differential uplift of the Aravalli hill range especially in the Udaipur region, postulated by Heron (1953) has been substantiated in his studies.

The active nature of faults bounding the sub-blocks, is indicated by moderate (M=4-5) seismicity and by the occurrences of thermal springs. (Fig.4.7). The earthquack epicentre located on the Sambhar Jaipur Dausa fault (SJDF) west of Jaipur and sought-east of Sambhar lake and on the (SDT) South Delhi thrust north of Ajmer. Shows the weaker zone and reactivation of neotectonics. Other minor earthquake epicentres can be mapped all along the lineaments as suggested by the Department of Geology, Sirohi and Udaipur on the unwanted event of Balotra earthquake in October 2004. Thermal springs are located on the north-western edge of the uplifted sub-block in Lalsot area, and Maroth area and on the trace of the (BDZ) Banas dislocation zone which shows the outlet of geothermal energy through neotectonism.

Conclusion

The study has demonstrated the relationship of neotectonic activity and seismocity which is deep-seated and complimentary to each other. The neotectonic block movements are responsible for earthquake origin. It also helps to demarcate the sub-blocks bounded by active faults in the drainage basin.

(i) The north-eastern block of the basin is a major neotectonic block in the Aravalli hill range and the adjoining area.

(ii) These major blocks contain several notectonic sub-blocks having horst-graben structures.
(iii) Some of the old fault sand lineaments have been reactivated.
(iv) The north-east sub-block river capture by Bandhi river happens due to subsidence on the one hand and upliftment process on the other.
(v) Sambhar lake is a tectonic depression and is the result of stream ponding caused by strike slip faulting and differential uplift of the neotectonic sub-blocks in the Aravalli hill range.
(vi) Moderate seismicity and occurrence of thermal springs indicate the active nature of some of the old and new faults in the area.
(vii) Lineaments are also responsible for the earthquake epicentres in the blocks and sub-blocks in the area.
(viii) Faults, fractures, joints, lineaments shows the weaker zone and thus liable to seismic activity.

Acknowledgements

Authors are thankful to GSI, western circle, survey of India, Jaipur circle, Department of Geography, Rajasthan University, Jaipur and President Shri Bhawani Niketan Shiksha Samiti, Jaipur for providing the basin materials, facilities and all possible assistance in preparing this article.

REFERENCES

Backiwal, P.C. and Ramasamy, S.M. (1987) *Lineament fabrics of Rajasthan and Gujarat.* India record, Geol. Survey of India, Vol. 113 (7) pp. 54-64.

Biswas, R.K., Chattopadhyay, G.S. and Subrata Singha (1982) *Some observations on the salinity problems of the inland lakes of Rajasthan,* records Geological Survey of India, Misc. Pub. 49.

Das Sarma, D.C. and Basudev Sural (1987) Aeolian stratigraphy and neotectonics of Rajasthan, *G.S.I. Annual Review,* 187 p.

Das Sarma, D.C. (1986) *Neotectonism in Rajasthan, its manifestations and effects.* Proc. Int. Symp. on Neotectonism in South Asia, Survey of India Dehradun, India. pp. 282-288.

Das Sarma, D.C. (1988) Post orogenic deformation of the pre-cambrian crust in north east Rajsthan. In A.B. Roy (Eds) *Pre-cambraian of Aravalli mountain range Rajasthan India,* Mem. Geol. Sor, India, No. 7 pp. 109-120.

Ghosh, T.K., Vishwanathana, S. and Sural, B. (1991) Neotectonic Analysis of Mendha river basin, Rajasthan, India, *Internation. Journ. Remote Sensing,* Vol. 12, pp. 2585-95.

Hacket, C.A. (1880) *Salt in Rajputana,* Ro. Geol. Surv. India, 13 p. 197.

Heron, A.M. (1938) *The physiography of Rajputana.* Proc 25th Indian Sci. Congress, pt 2 pp. 199-132.

Heron, A.M. (1953) *The geology of central Rajputana.* Mem. Geol. Surv. India, Vol. 79 pp. 389.

Heron, A.M. (1917) *The Geology of north-eastern Rajputana,* Mem. Geol. Surv. India, Vol. 45, 128 p.

Powar, K.B., Sukhtankar, R.K., Patil, D.M. (1979) *Geomophology and tectonics of the west coast of India between Palghar and Revas,* Maharshtra technical report No. 2 Department of Geology, University of Poona, Pune pp. 60.

Powar, K.B., Sukhtankar, R.K., Patil D., Miand Sawant, P.T. (1978, 1979, 1980, 1982, 1983) *Geomorphology and tectonics of the west coast of India between Revas and Srivrdhan Colaba district, Maharashtra.* "Technical Report No. 1, Department of Geology, Univeristy of Poona, Pune, pp. 64.

Powar, K.B. and Patil, D.M. (1982) *The tectonic frame work of structural evolution of parts of deccan trap province of western Maharashtra.* India, A study based on remote sensing techniques in engineering geosciences, Prof. R.S. Mittal Commemo Vol. (Ed. B.B.S Singhal) Sarita Prakashan, New Delhi, pp. 20-33.

Powar, K.B. and Patil, D.M. (1983) *Lineaments, their Mapping and Analysis,* NAGI annuals Vol. II, pp.1-17.

Prakash, B., Kumar, S., Rao, M.S., Giri, S.C., Kumar, S., Gupta, S. and Srivastava, P. (2000) Holocene tectonic movement and stress field in the western Gangetic plain, *Curr. Sci.* V. 79 (4) pp. 438-449.

Roy, A.B. (1999) Evolution of saline lakes in Rajasthan, *curs. sci.* 76(3) pp. 290-95.

Raghav, K.S. (1991) Neotectonic Activities and quaternary sedimentation in parts of Jaipur upland; *Annals of arid zone,* 30(4) 301-308.

Rosen, M.R. (1999) Sedimentologic and Geo chemical constraint on the evaluation of bristol dry lake basin, California Palaeogeog, Palaeoclim, Palaeoecol. Vol. 84, pp. 229-257.

Sen, A.K. and Ramlingam (1976) Report on the geology of the area around Sambhar lake, Jaipur and Nagaur district, Rajasthan, progress report, G.S.I. (unpub.).

Sen, D and Sen, S. (1983) Post neogene tectonism along the Aravalli range, Rajasthan, India Tectonophysics, Vol. 93 pp. 75-93.

Sinha, R.S. (1986) Himalayan collision and indentation of the AraValli orogen by Bundelkhand wedge. Implications for neotectonics in Rajasthan. In proc. International symposium on Neotectonics in South Asia Wadia Inst. Himalayan Geol., Dehradun, pp. 13-214.

Sinha, R.S. (2001) Neotectonically controlled catchment capture, an example from the Banas and Chambal drainage basins, Rajasthan, *Curri. Sci.,* 80(2) pp. 293-298.

Sinha, R.S., Saxena, R.K. and Sinha R.K. (1996) Geothermal fields and regional tectonics in Rajasthan and Gujarat. In U.L. Pithale and R.N. Padhi (Eds) *Geothermal Energy of India,* Geol, Surv. India spec. pub. No.45 pp. 391-93.

Sinha, R. and Ray, Mahashay, B.C. (2000) Salinity in proterozoic fold belt development in Rajasthan India minerals Vol. 43 (3 &4) pp. 226-40. Model from two shallow cores at Sambhar salt lake Rajasthan, *Jour Geol. Soc.* India 4.56 pp. 213-217.

Sinha-Roys Neotectonic significance of Longitudinal river profile, An example from the Banas drainage Basin, Rajasthan *Jour. Geol Soc.* India Vol. 58 Annl. 2001 pp. 143-156.

Smoot, J.P. (1991) Sedimentary facies and depositional environments of early mesozoic. Newark supergroup basins, Eastern north America Paleogeogh, Palaeoclim, Paleoecol, Vol. 84 pp. 369-423.

Sundaram, R.M. and Pareek, Suresh (1995) Quaternary facies and Palaoenvironment in north and east of Sambhar lake, Rajasthan. *Jour. Geol. Soc.* India Vol. 46 pp. 385-92.

Sural, B. and Dassarma, D.C. (1981) Report on the studies in neotectonics in Parts of Pali, Nagaur and Sikar Districts, Rajasthan, unpub. G.S.I. report F.S.P. 1986-87, pp.1-9.

Swain, A.M., Kutzbath, J.E. and Hastenrath (1983) Estimate of Holocene precipitation for Rajasthan India, based on pollen and lake level data, quaternary research, Vol. 19 pp. 1-17.

Vaidyanadhan, R. (1977) Recent Advances in Geomorphic studies in Peninsular India, India *Soc. Earth Sci.* S. Roy volume pp. 13-35.

Vaidiya, K.S. and Rajagopalan, G. (2000) Large Palaeolakes in Kaveri basin in Mysore plateau; Late quaternary fault reactivation, *curr. sci.,* V. 78(9) pp. 1138-1142.

Wadhawan, S.K. and Sural, B. (1992) Quaternary aeolian stratigraphy and neotectonism in Rajasthan and Gujarat. Rec. Geol. Surv. India 4, 125(7) pp. 18-19.

5

Flood and Water Logging Problem in Hanumangarh District, Rajasthan

S.C. Kalwar, H.V. Sharma and Avinash

Flood is a natural climatic phenomenon and occurs in almost all countries particularly in tropical and sub-tropical areas. Floods may be expressed as high stream flow which overtops natural or artificial banks of corresponding stream while causes of water logging are seepage from canal, over irrigation by cultivators, absence of natural drainage and outfalls in the area and continuous ponding of water in depressions.

The Ghaggar floods in Hanumangarh district of Rajasthan are quite different than floods experienced in other parts of the country. Here floods are mostly desirable and necessary in the larger interest of the people. The development of water logging problem around depressions and other low lying areas has however complicated the problem of floods due to which villages after villages are being evacuated and shifted to other areas besides vast damages of agricultural fields in the region.

The study area, Hanumangarh district is situated between 28°44′ to 29°57′ north latitude and 73°47′ to 75°31′ east longitude (9656km^2) in northern part of the Rajasthan state having alluviam and sandy plain in arid climatic conditions.

The Ghaggar river is originated from the village Dagshai near Simla in Himachal Pradesh. It flows from Shivalik hills at an elevation of 1927 metres above mean sea level and then in the plains of Punjab. It passes through Ambala and Patiala districts in Punjab and Hissar district in Haryana. Its main tributaries are Markanda, Saraswati, Tangri and Chotang river (Fig. 5.1). The Ghaggar river enters into Rajasthan near Talwara jheel in Hanumangarh district and flows through Pilibangan, Suratgarh, Vijainagar, Jetsar and Anoopgarh in Hanumangarh district and reach Fort Abbas in Pakistan after crossing the international border occasionally when abnormal rainfall occurs in the catchment area. Hanumangarh, Pilibangan, Jetsar and Vijainagar are situated on the banks whereas Suratgarh and Anoopgarh town along with a number of small villages are situated in the nali bed and flood plains of Ghaggar river.

Although the Ghaggar river and all tributaries rises from the foothills of Himalayas but does not have perennial flow. It flows only during the monsoon

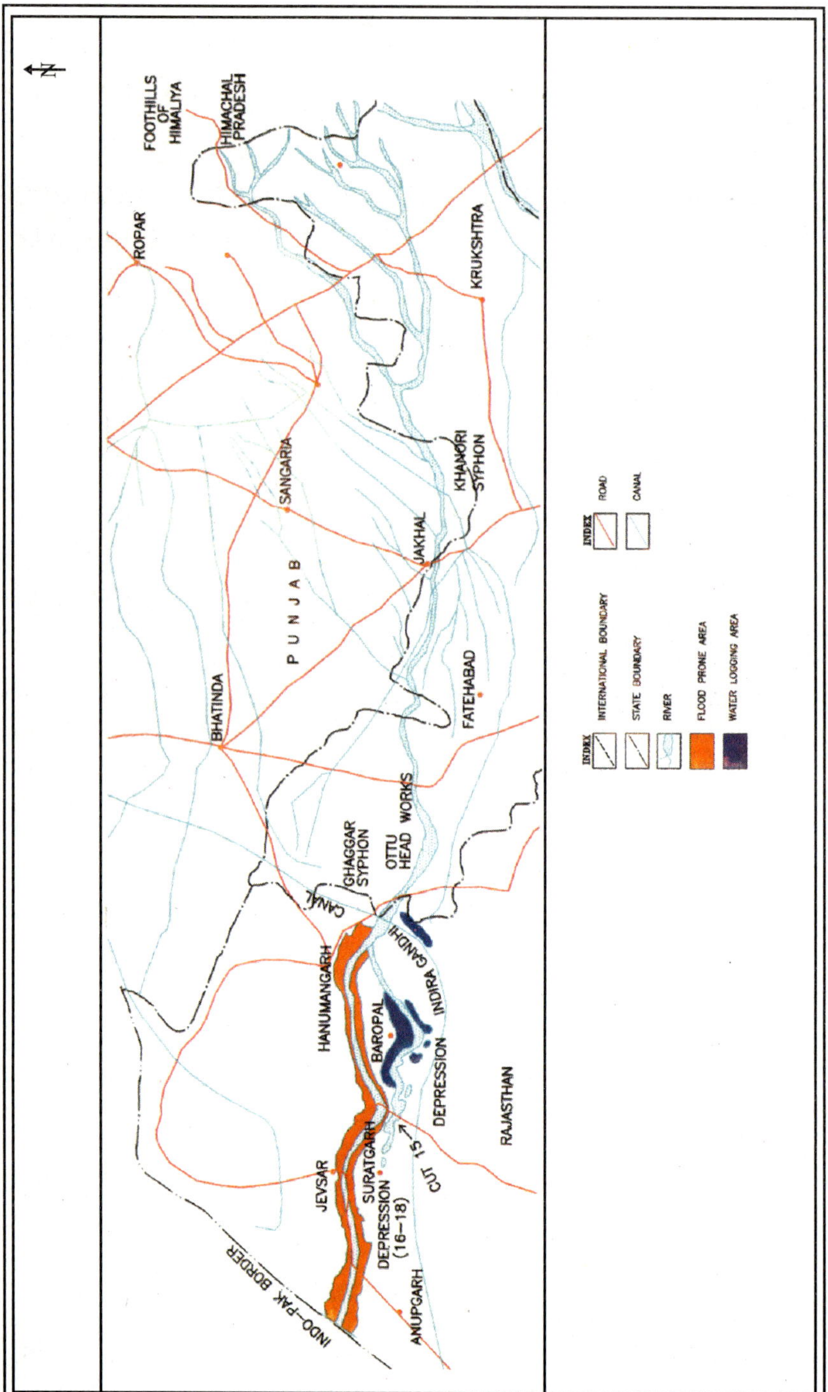

Fig. 5.1 : Ghaggar River Basin

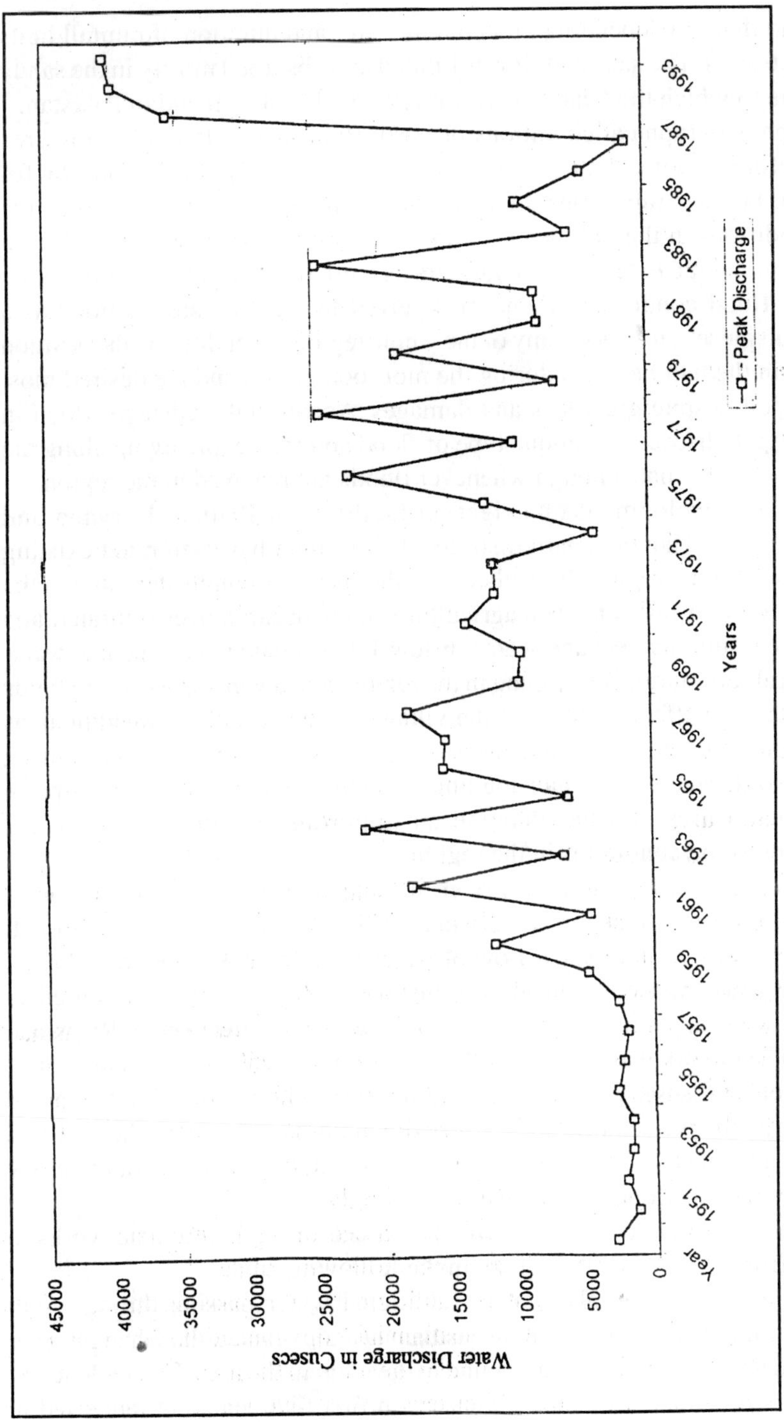

Fig. 5.2 : Ghaggar Flood Water-Peak Discharge at Ottu Headwork During 1950-1995

period and that too depending upon the intersity and duration of rainfall in its catchment areas. The seasonal flow is limited and absorbed mostly in the sandy region due to which the Ghaggar river rarely reaches its tail end in Pakistan.

The physiographical conditions of north Rajasthan belong to sandy area and therefore, whole of the riverbed consists of mostly fertile soils fit for producing bumper rabi crops after the monsoon season. In other words, the flood conditions in the nali bed are supposed to be most beneficial and most desirable to a large extent for the farmers as well as other people living in the area. The flood conditions whenever received in the area are regarded as a "boon in disguise" since economy of the whole region depends upon the position of floods in the nali bed area during the monsoon period and are desired most in the region despite great loss and damages. Accordingly, Ghaggar flood in Hanumangarh district is a unique type of flood and the people living along the nali bed feels fortunate enough whenever floods are received in the region.

The fast development of irrigation facilities in Punjab, Haryana and Ganganagar and Hanumangarh districts of Rajasthan have saturated existing layers of soil in the region. It resulted into the rise of groundwater table in the whole region at a fast rate. The agriculture fields became over saturated and started accumulating seepage water in low lying areas. Consequently water logging problem started developing in the region due to which agriculture fields after fields were affected. Most of the villages situated in the low-lying areas were affected first and it became necessary to shift to other safer areas during the last two decades. The water-logging problem has now been increasing in the command area and has become an important factor responsible for intensifying flood conditions in the region.

The Ghaggar river was a dead river in Rajasthan after the construction of Ottu pick-up weir (Near Sirsa, Haryana) 1896-97 but remained alive in nali shape afterwards with seasonal flow of water from the catchment area below Ottu during the monsoon period. The history of floods, however, started in Rajasthan when an additional runoff of 2972 cusecs was received in Rajasthan from Ottu headwork was passed for the first time in 1950 due to which a peak discharge of 2972 cusecs was received in Rajasthan. This additional water passed through the old river course created flood conditons all over in the nali bed areas in Ganganager and Hanumangarh districts received flood waters every year and treated as flood prone areas accordingly.

The peak discharge data attained and passed through Ottu head works to Rajathan from 1950 to 1995 is given in the following table.

The above table reveals that the additional water passing through Ottu headwork from 1950 onward into Rajasthan has rejuvenated the Ghaggar river due to which flood conditions are bound to develop in the area. The tendency of floodwaters remained around 2000 cusecs in first five years but increased in

Table 5.1 : Peak Flood Discharge through Ottu Head Work (Haryana) to Rajasthan 1950-95

Year	*Peak Discharge (Cusecs)*	*Year*	*Peak Discharge (Cusecs)*
1950	2972	1971	13800
1951	1356	1972	11584
1952	2174	1973	11560
1953	1735	1974	4046
1954	1713	1975	1225
1955	2755	1976	22408
1956	2321	1977	999
1957	1991	1978	24461
1958	2652	1979	6902
1959	4897	1980	18725
1960	11867	1981	8092
1961	4657	1982	8330
1962	18000	1983	24583
1963	6547	1984	5766
1964	21436	1985	9553
1965	6120	1986	4760
1966	15589	1987	1290
1967	15434	1988	35677
1968	18188	1993	39863
1969	9758	1995	40413
1970	9607		

Source : Hydrology, Irrigation Department, Govt. of Rajasthan, Jaipur.

subsequent years. It was 2755 cusecs in 1955, 11867 cusec in 1960, 18000 cusec in 1962 and 21436 cusec in 1964. Consequently the floodwaters in Ghaggar river had to flow further on the old course of Ghaggar river accordingly. It reached Pilibangan in 1955 and flow about 30 kilometres away from Hanumangarh in 1958. The Ghaggar river water flow 65 kilometres in 1959 and thereafter 100 kilometres in 1960. It touched Pakistan border in 1962 by flowing about 150 kilometres from Hanumangarh and crossed the border in 1964 when the peak discharge of 21436 cusec was received from Ottu pick-up weir.

The tendency of increasing discharges in the Ghaggar river form Ottu caused regular supplies of additional water and consequent development of flood conditions in the districts of Ganganagar and Hanumangarh. The Ghaggar floods water crossed international boundary with Pakisthan in 1964 and caused a great worry to the people of the region as well as the state government. The Government of India was also perturbed and decided to treat this problem as a national problem.

The carrying capacity of Ghaggar syphon was only 18000 cusecs. The flood peak of 18725 cusecs was observed in 1980 but it rose to 24583. cusecs in 1983 (Fig. 5.2) and created more flood problem in Ganganager district both in the nali bed as well as in GDC. The maximum flood peak of 35677 cusecs was observed on 2nd October, 1988 at Ottu pick-up weir. It was the worst flood observed in the living memory since the floodwater started flowing from 6th July 1988 and continued till 10th October, 1988 with a break of only 23 days.

In Rajasthan state the peak discharge data of Ghaggar flood waters passing in nali bed through the railway bridge at Suratgarh railway crossing from 1976 to 1988 are mentioned in the following table.

Table 5.2 : Peak Discharge of Ghaggar Flood at Suratgarh

S.No.	*Year*	*Period*	*Flood Volume Cusecs/per day)*	*(Peak Discharge (cusecs)*	*Date*
1.	1976	27 Aug. to 11 Sept.	71805	7408	31.08.76 02.09.76
2.	1977	28 July to 22 Aug.	79761	4972	11.08.77 12.08.77
3.	1978	8 Sept. to 26 Sept.	93742	8000	19.07.78
4.	1979	15 July to 25 July	9666	1091	19.07.79
5.	1980	16 July to 1 Aug.	45097	5066	19.07.79
6.	1981	30 July to 18 Aug.	38380	4204	19.07.80
7.	1982	30 July to 26 Aug.	61522	3302	12.08.81
8.	1983	25 Aug. 18 Sept.	97880	4399	13.08.82 14.08.82
9.	1984	4 Sept. to 14 Sept.	10480	1248	11.09.83
10.	1985	8 Aug. to 23 Aug.	21044	2016	07.09.84
11.	1986	No flood			14.08.85
12.	1987	No flood			
13.	1988	27 Aug. to 19 Sept.	90644	9349	

Source : Hydrology, Irrigation Department, Govt. of Rajasthan, Jaipur.

The above table shows that the Bhaggar floodwater exceeded 8000 cusecs only in two years, i.e. 1978 and 1988 in nali bed after the construction of GDC. It means that floodwater passed in such a quantity once in ten years in nali bed. It may be because of diverting more floodwater in Ghaggar diversion channel in order to avoid damages in residential areas situated in the flood plains.

The problem of Ghaggar floods shall be clear when details of some menorable flood events are also described. The following table shows flood volume passed through important locations with peak discharge data.

The above table reveals that the flood event occurred in 1988 was the severe most and remained for three months. Other flood events occurred in 1993 and 1995 although recorded more peak discharge data but both the flood events remained for short duration. The floodwater started receding subsequently after few hours due to which the thrust of floodwater did not affect much and absorbed in the riverbed and GDC in due course accordingly. It is also evident that the intensity of floods in Ghaggar river has been increasing and all flood protection works carried out by the Government during 1964-68 as well as after 1988 floods were not found sufficient and durable.

Table 5.3 : Ghaggar Flood volumes at important locations

Flood years ≥ important locations	*1988*		*1993*		*1995*	
	Volume MAF	*Peak Cusecs*	*Volume MAF*	*Peak Cusecs*	*Volume MAF*	*Peak Cusecs*
Ottu headwork	2.52	35677	0.787	39863	1.100	40413
Ghaggar syphon	2.36	32000	0.522	32128 31872	0.848	33132
Nalibed	1.005	17803	0.281	17244	0.362	17050
GDC	1.24	13802	0.241	14160	0.493	16075

Source : Hydrology, Irrigation Department, Govt. of Rajasthan, Jaipur.

Depressions

Natural depressions of sand dune existing around Suratgarh were selected for collecting surplus floodwaters. There are 18 depressions out of which 15 depressions are in the upstream side of cut No. 15 and remaining three depressions exist in the downstream side. Two depressions are joined by providing cuts at suitable sites. Bunds are provided for filling up gaps between two depressions so that water can be stored in natural depressions of sand dunes. The following table shows capacity of depressions at different reduced levels.

Table 5.4 : Capacity of 18 Depressions around Suratgarh in M.A.F.

Reduce level Metres	*From I to XV M.A.F.*	*From XVI to XVIII M.A.F.*	*Total M.A.F.*
580	0.01	0.10	0.11
582	0.02	0.12	0.14
584	0.03	0.14	0.17
586	0.01	0.15	0.19
588	0.05	0.17	0.22
589	0.06	0.17	0.23
590	0.11	0.18	0.29
592	0.14	0.20	0.37
594	0.17	0.22	0.39
596	0.21	0.23	0.44
598	0.26	0.25	0.51
600	0.32	0.27	0.59
602	0.37	0.29	0.66
604	0.42	0.31	0.73

Source : Hydrology, Irrigation Department, Govt. of Rajasthan, Jaipur.

The above table shows that at 580 R.L. the capacity of depressions in only 0.11 MAF and total capacity of water at maximum water level of 604.00 metres was estimated 0.73 MAF. The Ghaggar diversion channel (GDC) was completed in 1967-68 and floodwater was passed for trial purposes. The GDC received damages at several places.

It was though that the floodwaters stored in the depressions would either be absorbed or percolate into the bed of sand dunes or evaporate because of high temperatures in the region in an usual manner. The farmers may utilize these dry beds of depressions for agriculture and producing rabi crops as per practices adopted in the nali bed area.

Unfortunately, the proposed concept could not be successful since the floodwater remained standing in the depressions for a long period. The floodwater did not percolate into the bed of depressions due to deep and strong bed of gypsum beneath all depressions. The floodwater remained standing and started seeping through the sides of the saddle dams and sand dunes by lateral pressure resulting in accumulation of seepage water in low lying areas and creating water-logging problem around saddle dams. The water-logging problem intensified further and affected nearby villages situated around depressions. Manaktheri and Baropal villages near Suratgarh were affected in the first instance because of standing water in depression number VI (Fig.5.3). Both these villages and nearby fields were badly damaged in subsequent years and had to be got evacuated after paying huge compensation.

Detailed surveys around all depressions tell similar stories of other villages and deteriorating conditions of agriculture fields. It is sufficient to understand for the present that the standing water in the depressions after its collection during floods caused waterlogging problem and damaged agriculture fields to a great extent in subsequent years. The floodwater should have been passed towards other areas subsequently after the monsoon period in order to provide dry bed of depression to farmer for cultivation in rabi season cultivation. Bunds are constructed wherever necessary at their costs to drain out all the floodwaters at an earliest in order to prepare their fields for cultivation after drying the soil. The increasing tendency of floods and floodwaters into nali bed areas has, however, become a problem when the duration of floods is more and flood plains do not dry up for rabi cultivation in proper time. It is just a unique type of flood problem since it is required in the larger interest of the farmers but the floodwater should not continue for longer period in the region and attempts are aimed at in passing the floodwaters and drying the flood plains at an earliest.

The intensive irrigation in upper areas of Punjab and Haryana has saturated most of the agriculture fields. It has reduced flood absorption capacity in the region and developed waterlogging problem in the low lying areas. This factor is also an important cause in intersifying flood conditions in Ghaggar river. However, some other factors have also developed in the recent past and have aggravated the problem of floods in the region.

Causes of Ghaggar Flood

The Ghaggar floods are very much desirable in the Hanumangarh district since the floods plains become most fertile and are made available to the concerned farmers of the region for cultivation in rabi season after the cessation of floods. Almost all the farmers have been managing their agriculture farms situated in the flood plain areas and can use for cultivation. Bunds are constructed wherever necessary at their costs to drain out all the floodwaters at an earliest in order to prepare their fields for cultivation after drying the soil. The increasing tendency of floods and floodwaters into nali bed areas has, however, become a problem when the duration of floods is more and flood plains do not dry up for rabi cultivation in proper time. It is just a unique type of flood problem since it is required in the larger interest of the farmers but the floodwater should not continue for longer period in the region and attempts are aimed at in passing the floodwaters and drying the flood plains at an earliest.

The intensive irrigation in upper areas of Punjab and Haryana has saturated most of the agriculture fields. It has reduced flood absorption capacity in the region and developed water logging problem in the low lying areas. This factor is also an important cause in intensifying flood conditions in Ghaggar river. However, some other factors have also developed in the recent past and have aggravated the problem of floods in the region.

It would, therefore, be better if surplus water of Punjab, Haryana and northern Rajasthan may be utilized and diverted to the scarcity areas of adjoining districts in the larger interest of the people of state and the country as whole. Costs of bringing the surplus floodwaters to scarcity areas should not be a problem under circumstances in the present day era of technological advancement.

Suggestions

Investigations may also identify pocket holes of pervious layers of soil beneath the area so that these pockets can be utilized for filling up floodwaters down below the deepest zones and raise levels of underground water table. The wells/ tanks shall have to be dug deep to the deepest level of groundwater so that the floodwaters may reach down below while flowing through GDC or nali bed as usual. This will help in seeping the floodwaters into the ground and meet with the groundwater for raising and maintaining the level of underground water table in the region for future utilisation or for utilizing elsewhere. These open wells shall have to be joined with pipes of suitable design for passing silt free water into the wells. It would enable in storing the floodwaters as per original plan of diverting floodwaters into depressions and to utilize in nearby areas after the monsoon period as a regular feature. The owners of agriculture fields in the region may also be encouraged and provided assistance for making open wells in their fields at the lowest portion and utilize the water of the well with the help of diesel engine or electric tube wells as and when required in the fields.

It would be better if the present riverbed or existing dimensions of the water course be defined as such and provided strong embankments on both sides. Punjab and Haryana are utilizing the Ghaggar river in diverting floodwaters of all other drains and rivers as well as diverting seepage water of surplus zones and passing the same to Rajasthan without causing any damages in upper areas.

Various aqueducts and syphons are visible in the region which could have been avoided after detailed surveys and revising the whole canal system in providing water from any other channel and utilize all available water in more areas of Ganganagar and Hanumangarh district. Detailed investigations are necessary before the present network of canal systems in the whole region, i.e. around nali bed and GDC in the Hanumangarh and Ganganagar districts. The whole canal system requires drastic change for better management and utilization of available water resources to the optimum level. Detailed surveys are also required in the south of IGNP and GDC and proposals can be prepared for providing another canal parallel to both IGNP and DGC. This canal may be proposed abut 50-60 kilometres below the alignment of GDC right from Ottu headworks and avoiding Ghaggar syphon as well as other syphons and aqueducts in the region.

The flood waters of depression no VI passed through sides of saddle dams and developed water logging problem and damaging villages after villages. This could have been avoided by idea of emptying the depressions at the earliest.

It has been confirmed that the water in the reservoir is fit for irrigation but gets contaminated after seeping through sides. It is, therefore, better to transport the floodwaters from reservoir direct to other adjoining needy areas as early as possible for further utilization. Few pockets of gypsum free soils shall have to be identified out of hard pan of gypsum belt so that these pockets can be utilize for digging open well and diverting the flood waters into wells to raise ground water table.

Water is a scarce commodity for Rajasthan state. Therefore "each and every drop of water" may be rain water or flood water or seepage water or otherwise any surplus water" should be utilized to the optimum level at all costs. Rajasthan has still been having a vast virgin land and the same may turn into a green belt like Punjab and Haryana even if a small portion of this flood water is transported to this vast area.

REFERENCES

Bakliwal, P.C. and Sharma, S.B. (1981), Migration of the River Yamuna. *Journal of Geological Society of India*, Volume 121, Sept. 1981, pp (461-463).

Bose, N.K. (1960), Flood control : A Human Problem. *Indian Journal of Power and River Valley Development,* Volume X.

Krishan, A. and Kushwaha, R.S. (1972), Probability of Period of Rainfall deficiency in Arid and Semi-Arid Zones of Rajasthan, *Indian Journal of Agricultural Science*, Volume 42.

Meinser, O.E. (1942) : *Hydrology*, McGraw-Hill Book Company, New York.

Sharma, H.S. and Padmaja Rao, G. (1997), *Quantitative Geogmorphic Characterstics of Streams of Morel basin*, Geographical Review of India, Vol. 39, Dec. 1997.

Sharma, S.N., (1986), *Twenty Five Monsoons in Rajasthan* (1961-85), Hydrology, Irrigation Department, Govt. of Rajasthan, Jaipur.

6

Spatial Pattern of Land Degradation in the Northern Highlands of Orissa

PUNYATOYA PATRA AND BALESWAR THAKUR

Introduction

Land is the most basic of all resources and the primary medium for food production. It is also fixed resource and non-renewable over human time scale. However, the land is degraded day by day all over the world due to unusual/overpressure of population on land. Thus, there is causal link between population increase, land shortage, poverty and land degradation. To fulfil the demand of increasing population, it is necessary to examine the different types of degraded lands in terms of their quality, quantity and causes underlying degradation on the basis of which they can be regenerated and developed in a sustainable manner. However, such study is relatively rare in India, where two-thirds of the total population depend directly upon land for their livelihood. Therefore, a study of this kind is necessary for that part of India, where forces of degradation are quite active.

Literature Review

From the literature survey, it has been found that the studies directly related to land degradation problem are scant and disperse. The studies are varied dealing with different topics. Therefore, an attempt has been made to systematize the studies on the proposed research problem. Johnson and Lewis[1] examined critically the concept and different dimensions of land degradation and provide insight both into the causes of land degradation and the reasons thereof. The interaction of physical systems such as geomorphic, climatological and biotic conditions; and cultural systems such as level of technological development, types of government and interests of local people contributes to land degradation in both historical and contemporary contexts . Barrow's[2] perspective was global, but his approach was empirical rather than theoretical. He emphasized on the technical issues and stressed the value of integrating environmental and human elements on land degradation. The approach of Wong, Wong and Baker[3] was broader who were concerned with the major issues involved in the process of land remediations in different parts of the world. Here, the main emphasis is to review the extent of land degradation, ecological requirements for reclamation or restoration of derelict and disturbed lands and their beneficial end-users. Eden and Parry[4] tackled the issue of land degradation in the tropics. The environments analysed, range from rainforests through tropical and dry forests and wetlands to cities, and examples are drawn from all the principal cultural regions of developing countries. They also analysed environmental change due to land degradation and evaluate policies of the respective countries. The approaches of all these books discussed above are global, i.e. citing examples from various parts of the world under different physical and socio-economic conditions. The authors stress concept, nature and issues involved in land degradation.

A number of studies have focused on nature and extent of land degradation in India. Bhardwaj, Narain and Singh[5] discussed the extent and distribution of degraded lands in India. The physical, chemical and biological causes of land degradation are outlined and a classification is presented. Soil survey criteria useful in mapping eroded and saline soils and areas prone to landslides are discussed. Recommendations for the management of eroded land are prescribed for several regions of India in terms of mechanical conservation measures, suitable cropping practice and landuse. Abrol and Sehgal[6] assessed the nature and extent of land degradation problems in India with their impact of degraded land on the environment. The rehabilitation and conservation of resource base are described by them. Nagaraja, Narayan and Gautam[7] mapped different types of degraded lands in a number of districts in India using remote sensing data. Jadhav[8] prepared a map of degraded lands in Southern Haryana using aerial photographs and landsat imageries. The causes and effects of degradation are also discussed by him. Raina, Joshi and Kolarkar[9] mapped degraded land in arid regions of India using landsat Thematic Mapper Subscences. According to them, land is degrading in this area due to cultivation of marginal areas and overgrazing of pastures. Both accelerated wind erosion on sandy surface and water erosion on the shallow soils of piedmont areas are common in this region. Dwivedi and Deka[10] delineated degraded lands of parts of Nowgong and Mikir hills of Assam through visual interpretation of Landsat TM data on 1: 50,000 scale. The results demonstrated the potentiality of T.M. data for differentiating three categories of eroded lands, i.e. uneroded to slightly eroded, moderately eroded and severely to very severely eroded lands.

Nature of Problem

Land degradation is the result of complex interaction of physical and cultural

Table 6.1 : Image Interpretation Key of Degraded Lands

Waste lands	*Tone/Colour*	*Size*	*Shape*	*Texture*	*Pattern*	*Location*	*Association*	*Remarks*
Degraded forest land	Dark brown to light red; subject to canopy cover	Varying in size	Irregular	Coarse to mottled	Non-contiguous	Hills/ mountains	Different forest types/sub-types and plantations	Originates mostly from biotic disturbance caused to dense forest cover; degraded forest cover is less than 20 per cent of canopy.
Shifting cultivation area	Light yellow, subject to vegetation cover	Small	Regular to irregular	Fine to mottled	Scattered	Hill slopes	Forest cover	Jhuming is a cyclic process by which forests are slashed and burned to bring the cleared land under cultivation. Forest blanks are openings without any tree cover within forest areas and they occur in patches.
Land with or without scrub	Light yellow to brown to greenish blue, subject to moisture and surface cover	Varying in size	Irregular	Fine to mottled depending on vegetation cover	Contiguous and scattered	Terrain with varying lithology	Gentle relief with moderate slopes and surrounding agricultural lands	Thin soil cover on the top often supports scrub and grass or devoid of vegetation where surface erosion is dominant. It includes foothills, stony and gravelly wastes, sheet eroded land and long fallows.
Barren rocky/ stony outcrop	Greenish blue subject to varying rock types	Varying in size	Irregular	Medium to coarse	Linear to contiguous scattered	Hill slopes plateaux and plains	Barren and exposed rocky/ stony wastes	Rock exposures of varying lithology often barren and devoid of soil cover and vegetation. They occur within hill forests as openings or scattered as isolated exposures on plateau and plains.

(Contd.)

Table 6.1 *(Contd.)*

Waste lands	*Tone/Colour*	*Size*	*Shape*	*Texture*	*Pattern*	*Location*	*Association*	*Remarks*
Gullied land	Light yellow to bluish green tone subject to moisture and depth	Varying in size and depth	Irregular and broken	Medium to coarse	Linear to dendritic	Usually along the river and stream courses	Severely eroded areas in plains along stream coarse and on sloping grounds made of loose sediments	All gullies and ravines contribute considerably to soil erosion. Gullies with steep sides which are very deep and wide are known as ravines.
Sandy area	White to light yellow, subject to moisture content	Varying in size	Regular to Irregular	Fine	Contiguous to linear	Desert, river bed and coastal (onshore plains)	Shifting desert sands, sand dunes, coastal beach sands	Areas which have an accumulation of sands *in situ* or transported. In coastal or inland areas as desert sands, sand dunes, dune sands, levee sands etc.
Mining/ Industrial wastelands	Dark blue to light blue	Small	Irregular	Mottled	Scattered	Near mining areas	Large industrial complexes	Degradation due to mine dumps and large scale industrial effluents.

processes in which quality of land reduces to lower rank and is initiated by later (cultural processes) and intensified by former (physical processes). Therefore, the problem is unique in each socio-ecological condition. The present study focuses on the land degradation problem in tropical forest environment within the Indian socio-political condition. The Northern Highlands of Orissa is selected as the study area because it is one of the most ecologically fragile areas of the country. A little carelessness regarding land use affects the area itself as well as its adjacent areas. Largescale deforestation leads to soil erosion in the area itself and when these soil gets deposited in the river valleys of nearby coastal plain, flood havoc increases there.

The Northern Highlands of Orissa has attracted attention of the government during the beginning of last century, mainly due to its rich mineral resources. Since independence, the establishment of Rourkela Steel Plant, National Aluminium Company, Hirakud dam, Thermal Power Plants and paper mills have changed the natural landscapes of this region into cultural one. Recently, Bharat Aluminium Company is in the process of establishing its branch at Gondhamardan mountain in Baragarh district. However, very few environmentalists have realized the continuous degradation of land resulting due to combined action of largescale mining, industrialization, development of plantation agriculture, shifting cultivation, flash floods, droughts etc. A number of studies conducted on this problem have thrown light on the nature of land degradation in this region. Though the use of fertilizers and high yielding varieties of seeds has increased, the production of some agricultural commodities has decreased. This trend has shown that the study area, the richest region of Orissa with respect to natural resources, is degrading day by day. Again, the adjacent coastal plain is also affected when the eroded soil from this region is transported and gets deposited on the lower reach of the rivers. Both physical and cultural factors are responsible for this degradation.

Objectives

Keeping this in mind, the following aspects of land degradation are proposed to be studied : (a) to measure and describe the spatial pattern of distribution of different types of degraded land, and (b) to explain the causes and interlinkage underlying land degradation.

Data Source and Methodology

The study is based on both primary and secondary data. To study spatial pattern of degraded land in the study area, the IRS IB LISS I and II data of the year 1999 are used, the resolutions of which are 76.5 metre and 32.25 metre, respectively. To achieve this, first of all the base map on 1:2,50,000 scale showing administrative boundaries, location of important towns, villages, roads, railways, drainage etc. with their names is prepared. In addition to these general informations, the IRS–IB index map and the legend of degraded lands are shown on the base map. Then, the monoscopic visual interpretation of FCC images of IRS-IB (hard copies) are done. The image characteristics such as colour, texture, pattern, size, shape, location and association have helped to identify and delineate degraded lands. The analysis of work can be broadly grouped under three phases of activity, viz., preliminary analysis, ground truth and final interpretation. Preliminary analysis includes procurement of IRS-IB LISS I and II data, establishing criteria and standardization of degraded land classification system, interpretation of different types of degraded lands and, lastly, identification of doubtful areas of the imagery. Ground truth includes physical verification of doubtful areas, as identified on the imagery and modification of boundaries of degraded lands. Final interpretation includes the development of visual interpretation key on the basis of image characteristics (Table 6.1), delineation of different types of degraded lands on imageries and, then, transformation of these informations on base map and, finally, reduction of base map to required scale. To study the causes and interlinkages underlying degradation, sample villages have been identified by using random sampling technique. Then, by observing degraded lands and interviewing local people, different inferences have been drawn.

The Study Area

The Northern Highlands of Orissa embraces an area of 56,776 sq. km and is roughly rectangular in shape. It is surrounded by the Chhotanagpur Plateau on the north, Madhya Pradesh Uplands on the west, the Lower Ganga Plain on the north-east, the Orissa Coastal Plain on the east and south-east and the southern Orissa Highlands on the south. The region comprises nine districts, namely, Sambalpur, Deogarh, Baragarh, Jharsuguda, Sundergarh, Keonjhar, Mayurbhanj, Dhenkanal and Anugul (Fig.6.1). Administratively, the region comprises of 51 tehsils, 158 police stations, 14,713 villages and 42 towns.

Geologically, the region is a part of the Gondwana landmass, the major part of which is composed of granite, gneiss, charnokite, Khondalite and Dharwarian rocks. However, the granites and gneisses dominate and are spread all over the region. A comparatively small patch of Pre-cambrian-Cuddapah formation is found towards the south of Hirakud dam. The coal bearing beds occur in the faulted trough of the Mahanadi river. These are mainly composed of glacial deposits of the Permo-Carboniferous (Lower Gondwana) period and occur in Sambalpur and Dhenkanal districts.

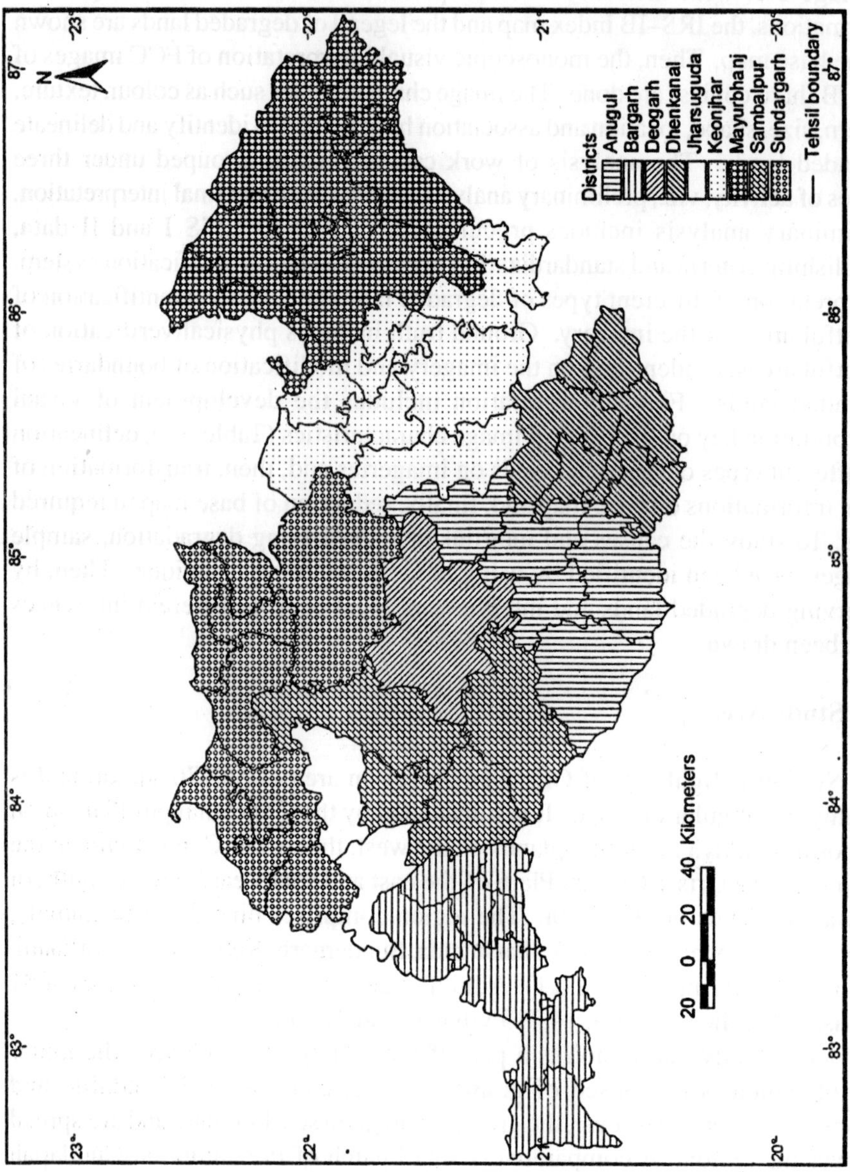

Fig. 6.1 : Northern Highlands of Orissa—Administrative Divisions

The Northern Highlands of Orissa can be broadly divided into four geomorphological regions, such as the Eastern Ghats, the Rolling Uplands, the Subdued Plateaus and the Major River Valleys and Flood Plains. The region is drained by four major river systems—the Mahanadi, the Brahmani, the Baitarani and the Subarnarekha. All these rivers contain a number of tributaries.

The soils of the region can be broadly divided into six categories, namely, lateritic soil, red soil, black soil, mixed red and yellow soil, mixed red and black and alluvial soil. However, red soil predominates all over the region. The lateritic soil is mainly found in the western and eastern part, while mixed red and yellow soil is found towards the north-western part. The mixed red and black soil is found in a small patch towards the south of Hirakud reservoir and along the Mahanadi valley. The Orissa Highland is the most thickly forested part of the state. Vegetation mainly comprises of tropical semi-evergreen, tropical moist deciduous and tropical dry deciduous. The mineral resources are rich and extensive in this region, the important being iron-ore, coal, manganese, chromite, dolomite, fireclay, graphite, bauxite, copper, lead, asbestos and mica.

The climate of the region is tropical, characterized by high temperature and medium to high rainfall in most part of the year. The maximum annual average temperature is 27° C, and the mean annual rainfall is 138 cm. The average humidity of the area varies from 45 per cent in the month of March-April to 75 per cent in the month of July-August.

The Orissa Highland is the home of about 117 lakh people with an average density of about 207 persons per sq. km. Of the total population, the Scheduled Castes and the Scheduled Tribes comprise 11.98 per cent and 36.98 per cent, respectively. Agriculture, forestry, animal rearing are the major economic activities among the local people and mining and industrial activities are the major economic activities of the migrants. Paddy is the staple cereal and is raised by the tribal people as cash crop also. The average yield is much below the national average. The average size of agricultural holding varies from 4 to 7.5 acres.

Spatial Distribution of Degraded Lands

Total eight types of degraded lands have been identified in Northern Highlands of Orissa which cover about 18 per cent of the total area. These degraded lands are degraded forest, forest blank/shifting cultivated land, land with/without scrub, sandy area, gullied land, rocky/stony outcrop, mining wasteland and industrial wasteland. The distribution of degraded lands is shown in Fig. 6.2 and Table 6.2. Out of the total degraded land, degraded forest covers 27.86 per cent area which is found in all the districts. Sambalpur has maximum area under degraded forest

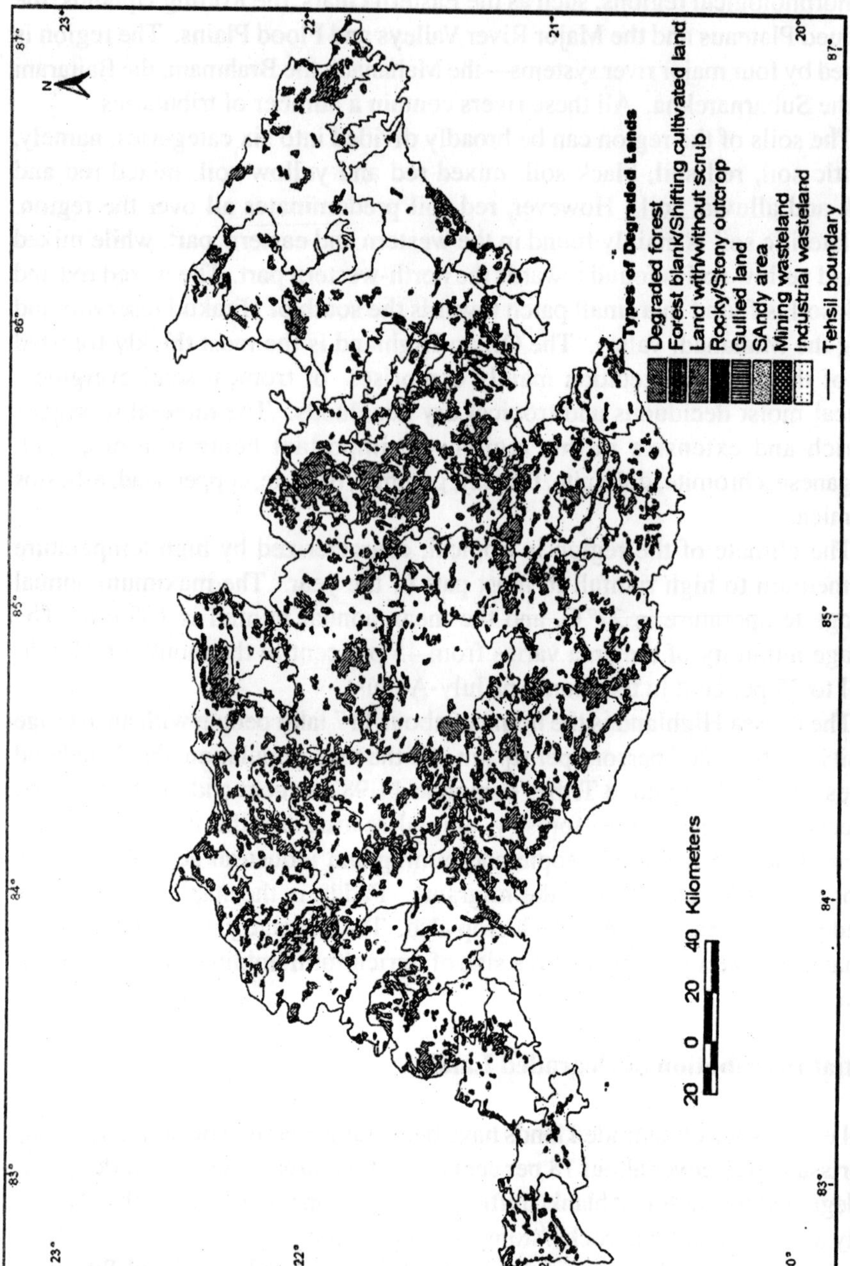

Fig. 6.2 : Northern Highlands of Orissa—Degraded Lands

Table 6.2 : Distribution of Degraded Land

Sl.No.	*Types of degraded land*	*Area covered (sq km)*	*Percentage to total degraded land*
1.	Degraded forest	2756.09	27.86
2.	Forest blank/Shifting cultivated area	26.27	0.33
3.	Land with/without scrub	6575.75	66.36
4.	Rocky/Stony outcrop	264.76	2.16
5.	Gullied land	34.77	0.43
6.	Sandy area	117.52	1.25
7.	Mining wasteland	143.14	1.51
8.	Industrial wasteland	3.11	0.10
	Total	9921.41	100.00

(25%), followed by Sundergarh (21%) and Deogarh (14%), respectively. Out of the total degraded land, forest blank covers 0.33 per cent area which is found in the districts of Keonjhar (31%),Sundergarh (30%), Deogarh (23%) and Jharsuguda (16%). Out of the total degraded land, land with/without scrub covers maximum area (66.36%), which is found in all the districts. Sundergarh has maximum area under degraded forest (25%), followed by Sambalpur (18%) and Deogarh (11%), respectively. Out of the total degraded land, rocky/stony outcrop covers 2.16 per cent area which is found in all the districts except Bargarh. Mayurbhanj has maximum area under rocky/stony outcrop (43%), followed by Keonjar (42%)and Sundergarh (5%), respectively. Out of the total degraded land, gullied land covers 0.43 per cent area which is found in the districts of Sambalpur (61%), Keonjhar (12%), Dhenkanal (10%), Sundergarh (9%) and Jharsuguda (8%).Sandy area covers 1.25 per cent of the total degraded land. Mayurbhanj (22%) has the maximum area under this type of land, followed by Anugul (15%), Sambalpur (12%) and Keonjhar (12%). Mining wasteland covers 1.51 per cent of the total degraded land, 62 per cent of which is covered by Keonjhar (41%) and Mayurbhanj (21%) districts. One patch of industrial wasteland is identified near Rourkela Steel Plant in Sundergarh district.

Causes of Development of Degraded Lands

The degraded forest land is the land where the quantity and/or quality of the original tree cover has decreased. Out of the total degraded land degraded, forest covers about 29 per cent in the study area. The main causes behind the development of degraded forest are demand of local people for fuel, fodder, timber trading, illegal cutting/theft of timber, existing socio-political condition, eucalyptus plantation etc. Timber trading is by far the most responsible factor for degradation of forest in the region, beginning from the first decade of twentieth century. Since 1990, timber logging has been restricted, on the one hand, and demand on timber has been increased due to increase in population, on the other hand. As a result, the gap between supply and demand has increased tremendously. If it is not planned to supply the demand, it would be difficult to stop illegal cutting or theft of timber. Illegal cutting of timber perhaps occurs at a larger scale than legal trade of it. Besides, the requirements of rich people to lead luxurious lives cannot be met by legal timber trade. Therefore, it is natural that organized looting of forests is carried out by an exclusive group of people, such as contractors, smugglers, politicians, bureaucrats etc. There is no estimate of how much of timber is lost in this way, but it is definitely done on a large-scale. Besides this, it is a common feature throughout the region that the party/organization who takes lease for a certain period goes on logging the trees after their lease period also. Like this, whenever an industry, mine or multi-purpose project is set up, thousands of trees are cut down in the demarcated areas. The traders and contractors also get chance to cut down the trees illegally from the periphery areas. During the large-scale project work, it is very difficult to catch illegal traders.

Forest blank is the area where the total vegetation on a particular patch of forest is cleared due to shifting cultivation or some other reason. Out of the total degraded land, forest blank covers 0.33 per cent of area. Forest blank is found mainly in Keonjhar and Sundergarh districts. Shifting cultivation, locally known as "Podu Chasha", was the dominant landuse system in the region uptill 1940s. The tribal people had evolved this mode of cultivation in response to the most difficult terrain and topography under inhospitable environment through centuries of struggle with nature. The problem began when the population increased, on the one hand, and land devoted for shifting cultivation decreased due to implementation of new forest policy, development of other economy like mining, industry etc., on the other hand. It resulted reduction in shifting cycle, onsite and offsite erosion, run-off, nutrient loss, loss of biodiversity and deterioration in watershed hydrology. The clearance of forests, burning of debris and use of land for cultivation proved to be a big shock to the ecosystem, leading to a change in vegetation and resulting in the preponderance of secondary growth of species, such as weeds and bamboos, which are different from those of the seasonal monsoon forest. It is known that todays natural forests are the end product of a long process of evolution over millennia. In the sub-tropical forests of the region, soil cover is extremely thin since the soil is continuously washed away by heavy rains. If these forests, with not much soil support are once cleared, they are likely to be lost for ever. Therefore, the inevitable consequence of shifting cultivation in such areas is complete barrenness of land.

Land covered mainly by dry grasses and scrubs intermingled by dry deciduous trees here and there, which gives semi-desert look, are termed as *land with/without scrub*. Ruthless cutting from degraded forest and natural growth of grasses and scrubs on abandoned mining and shifting cultivated area leads towards the development of this type of land. Sometimes, agricultural lands get spoiled due to the contamination of sewerage flowing and dust blowing from nearby mining areas; gradually, these lands give no return to farmers and farmers finally stop cultivating these lands. As a result, scrubs and unwanted plants grow there and good agricultural land is converted into scrub land. Another most important reason of conversion of this type of degraded land from agricultural land is that of lowering of water table in some parts of this region. Due to ruthless cutting of forests for mining, industries, construction of roads/ railways, water table in some pockets decreases. The upland agricultural areas do not get water for irrigation. On the other hand, rainwater drains out very quickly from these upland areas. As a result, crops do not sustain in these lands and lands are left as fallow. Gradually, these types of fallow lands become scrub as they grow there very fast. In the higher altitude, shrubs in these scrub lands are so dry and thorny that even the cattles do not go near these shrubs.

Rocky/stony outcropped land is at the last stage in the process of degradation of thick dense natural vegetation. Due to the continuous cutting of trees for different purposes, the thick dense forest gradually becomes open forest, then, the open forest gradually becomes land with or without scrub. If land with or without scrub is located on the hill top or on small hillocks with loose soil cover, and/or on steep slope, then, soil erosion takes place very fast. Gradually, the total soil cover erodes away, land becomes bare, and underlying rocks or stones are exposed. This type of degraded land is termed as "rocky/stony outcrop". When sun rays fall on this barren rocky land, the rock expands and at night it contracts. Due to alternate expansion and contraction, the rock starts breaking down to smaller pieces and these broken rocks start falling down on steep slopes and rolling down on gentle slopes. Therefore, the barren rocky areas are much prone to landslides and rockfalls. Rocky or stony lands are devoid of any type of soil and vegetation. Rarely, small trees are found on the flanks or cracks/joints of the rock which grow from the seeds dropped by the birds.

Land where a large number of gullies is developed which erodes away surface as well as sub-surface soil is called *gullied land*. Gullies are formed due to the practice of cultivation on steep slopes without following the accepted principles of terracing. Overgrazing and trampling of soil by cattle also aids in the formation of gullies. It has been seen that steeper the slope, greater is the hazard of gully formation. Very often, formation of a gully starts on the bank of natural waterways (stream or river) which has been eroded to a severe magnitude. They quickly extend back into drainage area and increase in depth as they cut upward along the slope. Branches of side gullies may be formed when these gullies cut across the natural drainage or depressions of waterfalls.

Thus, a network of gullies or ravines soon develops. Gullied land covers less area as few and small gullies develop on the hard terrain. Most of them are so small that which are not recognized by satellites.

As the name suggests, *sandy area* is covered by sands along with pebbles and *kankars*. They are very coarse, i.e. larger in grain size unlike the sands found in the coastal areas. Due to deforestation in catchment areas, soil erosion takes place and bedrocks get exposed. Then, bedrocks get weathered by the action of external forces and are broken intc smaller sizes and, finally, transported into the nearby rivers. During the rainy season when flood water crosses the river bank, these pebbles, *kankar*, coarse sands get deposited on the nearby land. This process continues year after year and these types of degraded lands come into existence. Gradually, the depth of sand increases. These areas are, generally, covered under water during rainy season. Gradually, the area is exposed when water in the river bed recedes after rainy season. Thus, the areas remain wastes throughout the year and are, generally, used by the nearby villagers for dumping solid wastes. Sandy area covers less area in comparison to other degraded lands as sands are deposited along river valleys and in the high land areas like Northern Highlands of Orissa, river valleys are very narrow.

Land which is degraded due to faulty methods of mining is called *mining waste land*. Most of the areas affected by mining operations had been either under agriculture or forests. In both the cases, agricultural activities or forestry have been automatically stopped due to mining operations which bring high economic return to the land for certain period of time, till the minerals get exhausted. After that, land remains abandoned and is, generally, not regenerated for economic use. In this way, the lushgreen vegetation and fertile agricultural land are converted into mining wasteland. Besides this, land degradation in and around mining areas is caused by unscientific methods of mining, both opencast and underground, ore processing, solid waste disposal and tailing etc. Though faulty method of mining is practiced in the region, mining waste land covers less area, because after short period of mining operation, shrubs and bushes grow on the abandoned mining area and the area is termed as land with/ without scrub.

All the industries in the study area were established by destroying large forest areas or vast agricultural land. It could not be avoided because mineral resources of the region had to be developed. However, the problem began when these industries started disposing untreated solid and liquid wastes to nearby agricultural fields, rivers etc. Moreover, the nature of land degradation due to one type of industrial set up differs from other type of industrial set up. *Industrial wasteland* covers least areas of all degraded lands, because the sizes of this type of degraded land are so small in comparison to the resolution of IRS 1B satellite imagery that most of them cannot be recognised.

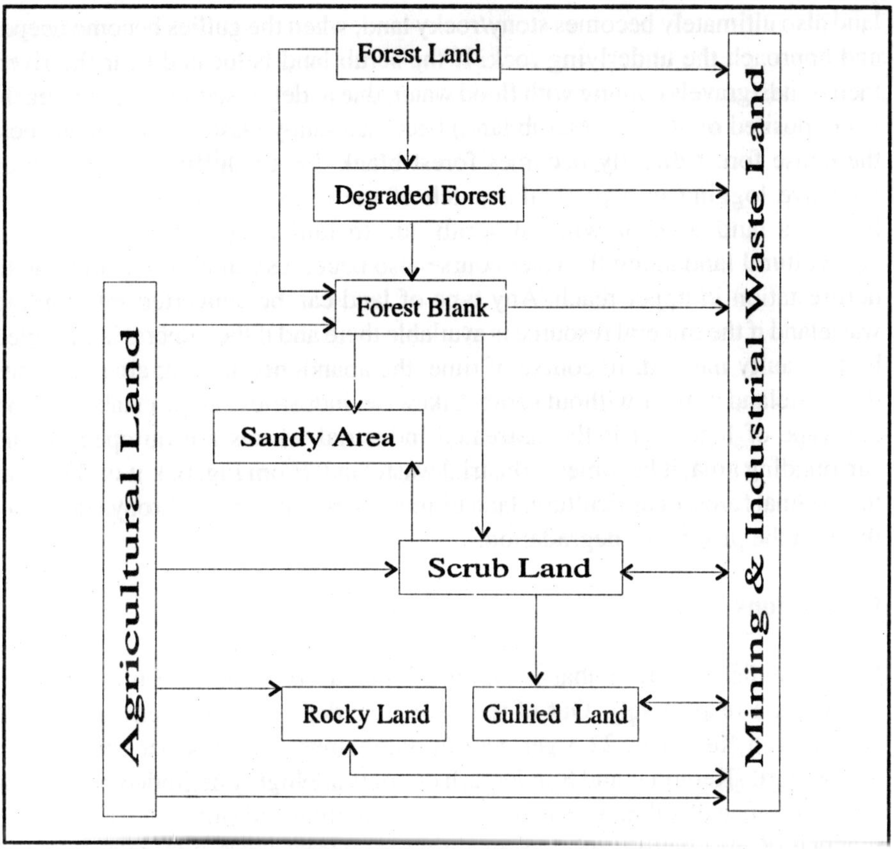

Fig. 6.3 : Stages of Land Degradation

Interlinkage Underlying Degradation

Fig. 6.3 explains the stages of land degradation which shows the interlinkage between each type of degraded land. The main cause behind land degradation in the study area is deforestation due to establishment of mining, industrial and multipurpose projects and logging of trees for various purposes. Due to regular cutting of trees, the natural dense forest becomes degraded forest; then, the degraded forest becomes forest blank; and then, the forest blank becomes land with and without scrub. Because of overgrazing, land with and without scrub becomes either rocky/stony wasteland or it becomes gullied land. If the rock underlying the land with and without scrub is hard enough, then, it becomes stony or rocky wasteland, when the total soil erodes away in the form of sheet erosion. If the rock underlying the land with or without scrub is not hard, then, rill erosion takes place leading to the formation of gullies. Over time, gully land also ultimately becomes stony/rocky land, when the gullies become deeper and approach the underlying rock. If the scrub land is located near the river, then, sands/gravels coming with flood water, due to deforestation in upper reach, get deposited on it and it (scrub land) becomes sandy waste land. Sometimes, the dense forest directly becomes forest blank due to shifting cultivation or intensive logging in a particular patch of the forest. Agricultural land also becomes land with or without scrub due to faulty agricultural practices. Agricultural land along the river course also becomes sandy wasteland due to deforestation in upper reach. Any type of land can be converted into mining wasteland if the mineral resource is available there and if the mineral is extracted by the faulty method. In course of time, the abandoned mining area becomes the wasteland with or without scrub. Likewise, industries can be established on any type of land and if the untreated industrial wastes are dumped on the surrounding area, it becomes industrial wasteland. From Fig. 6.3, it is clear that the original forest or agricultural land ultimately becomes rocky/stony wasteland through the process of degradation.

Conclusions

Of all the districts, Keonjhar is highly degraded, followed by Sambalpur and Sundergarh respectively. Out of the total area, degraded lands cover 27.5 per cent area in Keonjhar, 22.5 per cent in Sambalpur, and 19.4 per cent area in Sundergarh . Keonjhar and Sundergarh districts are highly degraded because of the occurrence of valuable minerals in its undulating and sloppy terrain and the practice of unscientific mining operation to extract these minerals. Besides, practice of shifting cultivation was a common feature up till last few years among tribal population which constitute more than 60 per cent of total population. Timber trading is also another reason which leads towards land degradation in these two districts. In Sambalpur the main reasons behind land degradation are illegal cutting, theft of timber and existing socio-political conditions. Of all the districts, Mayurbhanj is least degraded. Only 9.5 per cent of the total land is degraded in this region. Most of the area of Mayurbhanj district is covered by Similpahar reserved forest, where logging of trees, shifting cultivation, mining activities etc. which lead towards land degradation, are not prohibited.

The main cause behind land degradation in the study area is deforestation due to establishment of mining, industrial and multipurpose projects and logging of trees for various purposes. Each type of degraded land is interlinked with each other in such a way that it has shown different stages of land degradation. Most of the area in the Northern Highlands of Orissa goes through four stages of degradation, i.e. degraded forest (first), forest blank (second), land with/

without scrub (third) and rocky stony outcrop (fourth) respectively. Forest blank covers very less area as the duration of this stage, i.e. second stage of degradation is very less. Whenever degraded forest becomes forest blank, then after some days shrub and bushes appear on it and it is converted into scrub land. Out of the total degraded land, scrub land covers about 66 per cent area. In other words 66 per cent of the degraded land has reached third stage of degradation, which is alarming. If the proper regeneration measures are not taken then, scrub land will become rocky/stony land, the ultimate stage of degradation. This type of degraded land is called unculturable wasteland which cannot be regenerated. It may take thousands of years for soil formation on this land by natural process.

NOTES

1. Johnson, D.L. and Lewis, L.A. (1995) *Land Degradation : Creation and Destruction*, Cambridge: Blackwell.
2. Barrow, C.J.(1991) *Land Degradation : Development and Breakdown of Terrestrial Environments*, Cambridge : Cambridge University Press.
3. Wong, M.H., Wong, J.W.C. and Baker, A.J.M. (1999) *Remediation and Management of Degraded Lands*, New York: Lewis Publishers.
4. Eden, M.J. and Parry, J.I. (1996) *Land Degradation in the Tropics : Environmental and Policy Issues*, London: Pinter.
5. Bhardwaj, S.P., Narain and Singh, G. (1990) *Soil Surveys for Assessing Land Degradation with Special Reference to Water Erosion*, NBSS Publication, No.25, 64-81.
6. Abrol, I.P and Sehgal, J.L. (1992) "Degraded lands and their rehabilitation in India", *Soil Resilience and Sustainable Landuse : Proceedings of a Symposium held in Budapest*, 28th Sept. to 2nd Oct. 1992, also published in the 2nd Workshop on the Ecological Foundations of Sustainable Agriculture (WEFSA II), 1994, 129-144.
7. Nagaraja, R., Narayan, L.R.A. and Gautam, N.C. (1989), "Utility of landsat thematic mapper data for wasteland development", *The National Geographical Journal of India*, 35: 38-52.
8. Jadhav, A.S. (1990) "Impact of environment on land degradation in South Haryana: a geomorphological approach using remote sensing technique", *Proceedings of the Twenty-Third International Symposium on Remote Sensing of Environment*, 18-25 April, Bangkok, Vol.1, 517-526.
9. Raina, P., Joshi, D.C. and Kolarkar, A.S. (1991) "Land degradation mapping by remote sensing in the arid region of India", *Soil Use and Management*, 7(1): 47-52.
10. Dwivedi, R.S. and Deka, C.R. (1990) "Degraded lands in Northeast India as imaged by lands at T.M", *Asian-Pacific Remote Sensing Journal*, 3(1): 35-37.

7

Interpretation of Ravine Land in Sub-Watersheds of the Morel River, Rajasthan

R.D. Doi

Introduction

Of course, ravine land is an outcome of accelerated stream erosion. At the same time, ever increasing human intervention has intensified this kind of land degradation process. As conceived by the United Nations Environment Programme (UNEP) overcultivation, deforestation, overgrazing and unskilled irrigation are the main causes of farm land degradation (Tolba, 1982). It is widely held that 'catchment management is particularly difficult in sensitive semi-arid environments' (Richards, 1982, p.3). As the result the *'ravines'* are constantly posing a serious socio-economic and environmental problems by damaging especially marginal lands in the peripheral locations of such wastelands. In fact, 'increased rate of erosion affected by human activities through land use changes is defined as *accelerated erosion* (Singh and Dube, 2002 p.110). Hence, 'the damage to the agricultural land by network of ravines and later its conversion to badland topography is a serious concern to the scientific community since long back (Pani and Mohapatra, 2001).

In the absence of proper land conservation practices in the State like Rajasthan, the already ravinous land is becoming more wasteful. There is a fair agreement that "removal of vegetation is one of the major causes for soil erosion' (Aruchamy *et al.,* 2001, p.7). In real terms, formation of ravines result in the loss of land and consequent shortage of food, fuel and fodder production. 'The habitations of people are encroached by the advancing ravines which pose serious socio-economic and ecological problems' (Pandey, Saxena and Gaur, 1993, p.10). Undoubtedly, attempts have already been made by a number of workers in delineation of ravine land as one of the main wastelands' categories based on conventional and RS data. The scientific community of earth scientists at large is aware that there has been considerable improvement in assessing status of ravine land in particular and degraded lands in general with the availability of better resolution RS data and available digital image processing techniques and softwares to derive more objective spatial data from digital images in conjugation with the field checks at some required scale of mapping. The present study aims mainly to assess sub-watersheds intensification of ravine land in terms of spatial extent with respect to sub-watersheds' areas. Of course, as put forth 'the study of gullied and ravinated areas involves the consideration of identification and mapping of ravine lands measurement and recording of gully morphology, classification of rills and gullies on different considerations' (Singh and Dubey, 2002, p.xi). So far as ravinous land is concerned, there is now a broad agreement that 'the word *ravine* is usually associated not with an isolated gully but a network of gullies formed in deep alluvium and entering nearby river, flowing much lower than the surrounding table lands. The ravines then are extensive system of gullies developed along river courses' (Government of India, 2000, p.4).

Study Area and Data Used

Morel river basin is a sub-catchment of Banas river catchment, which is one of the 500 sub-catchments in India (Watershed Atlas of India, 1990). The study area witnesses predominance of semi-arid climatic conditions 'exhibiting a fluviatile landscape super-imposed by an aeolian landscape' (Doi, 1991, p.41). As also indicated in one of the studies (Sharma, 1980, p.23) it is highlighted that 'among various streams of the Chambal basin, the Morel river is also characterized by ravines and gullies. Generally, the banks of the Morel and its tributaries have been dissected into ravines.' In general, three compact ravine patches are observed in the Survey of India topomaps in the Morel river basin are:

(a) Along the Dhund river between a height of 300-360 metres.
(b) Along the Morel and its tributaries between the Bassi-Jatwara hills and Bonli Lalsot hills at an elevation between 300-360 metres.
(c) Along the lower reaches of the Morel river below 180-190 metres.

The Morel river basin is situated between latitudes 26° 6′ 28″ N - 27° 9′ 28″ N and longitudes 75° 44′ 19″ E - 76° 58′ 49″ E comprising an area of 5892.78 sq. km (Fig. 7.1). The whole study area is covered in 19 SOI topomaps on the scale 1:50,000. Parts of 5 districts of Jaipur, Dausa, Sawai Madhopur, Karauli, Tonk and 17 Tehsils are covered under the study area. The Morel basin is being divided into 5 watersheds. The mean annual rainfall is 675 mm for the study area.

Geocoded FCC on CD-ROM data have been used on the scale 1:50,000. The satellite data products used for the image processing of the study area are as given in Table 7.1.

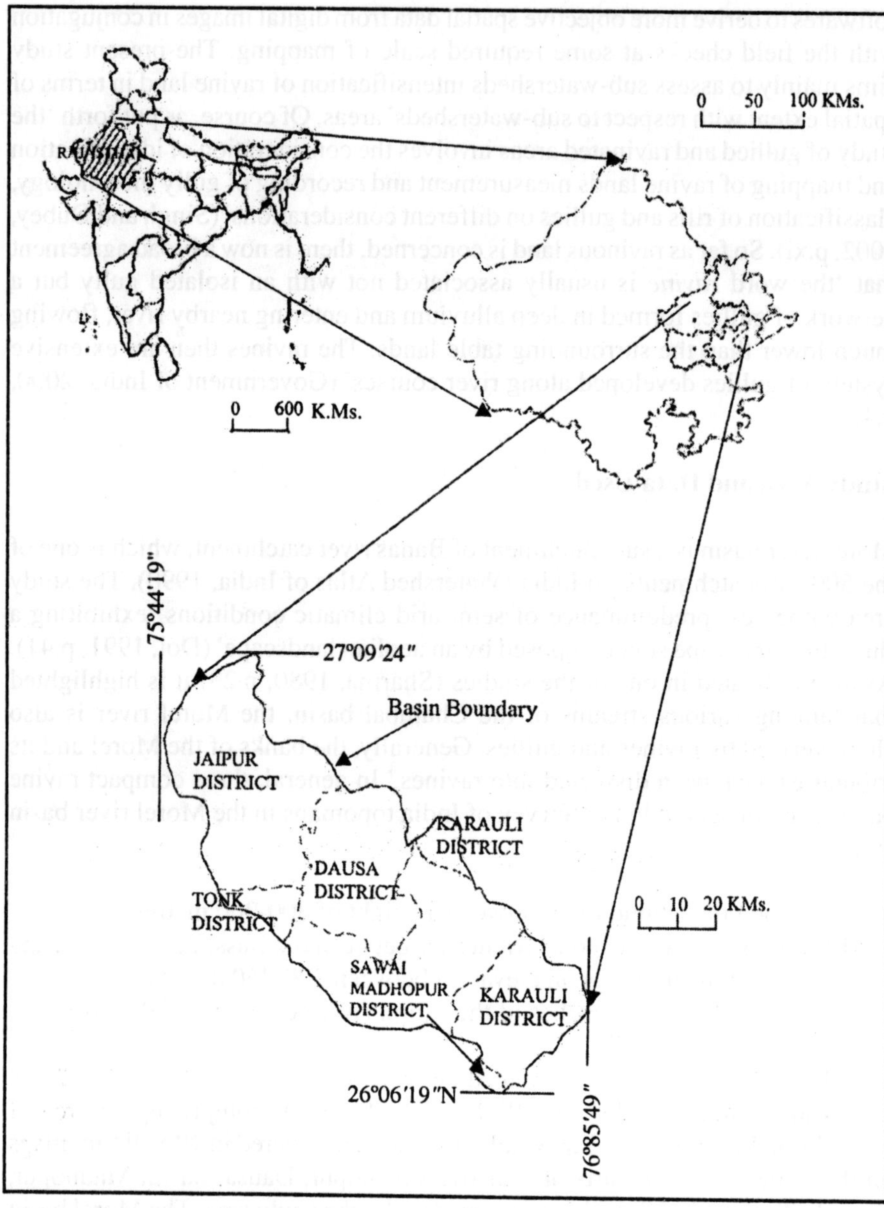

Fig. 7.1 : Location map of the Study Area

SOI topomaps on the scale 1:50,000 have been used to delineate sub-watersheds in the study area. A total of 49 sub-watersheds are delineated based on drainage pattern (Fig.7.2). For example, in coding system 2D2A1a, the first

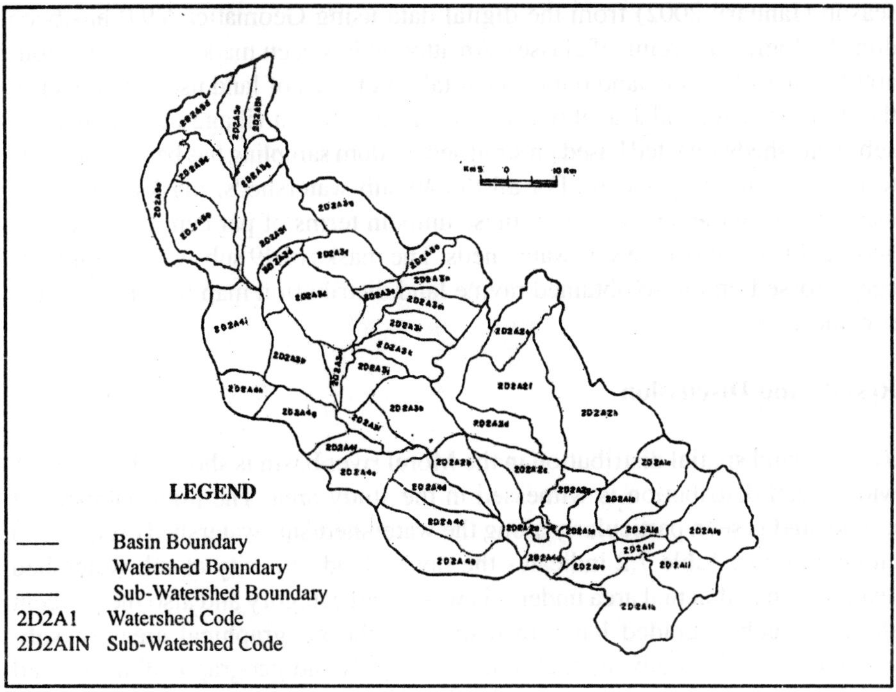

Fig. 7.2 : Sub-watersheds of the Morel river basin

Tabel 7.1: RS Data products of the study area

Satellite	*Sensor*	*Path*	*Row*	*Resolution*	*Date of Pass*
IRS-1D	LISS-III	95	52	23.5m	16.10.2001
IRS-1D	LISS-III	96	52	23.5m	13.10.2001
IRS-1D	LISS-III	95	52	23.5m	24.01.2002
IRS-1D	LISS-III	96	52	23.5m	21.01.2002

Arabic numeral 2 refers to region, the second place letter D represents to the basin. Again the Arabic numeral 2 at the third place refers to the catchment and the fourth capital letter A indicates the sub-catchment. The fifth place Arabic numeral is indicative of watershed and the last small alphabet denotes the sub-watershed. The largest sub-watershed (2D2A2h) and the smallest sub-watershed (2D2A1f) comprise area of 380.64 sq. km and 33.63 sq. km respectively. The watershed 2D2A3 has the maximum of 15 sub-watersheds (i.e. 2D2A3a to 2D2A3o).

Methodology

Aggregate classification of *Kharif* crop season (October, 2001) and *Rabi* crop season (January 2002) from the digital data using Geomatica V9.0 has been done by logical merging of classes. An attempt has been made to separate non-arable class of ravine land out of the total 14 classes of land use/land cover in the study area. A good deal of field checks from different village sites in different sub-watersheds selected based on stratified random sampling has been conducted to improve the output map. For all the 49 sub-watersheds, masks have been used to get ravine land extent in these units in terms of per cent to respective geographical area of the sub-watersheds. The masks of 49 sub-watersheds have been crossed on the so obtained ravine land distribution map to compute area statistics.

Results and Discussion

Ravines land spatial distribution in the Morel river basin is shown (Fig. 7.3). A widespread distribution is witnessed in the study area. The high intensity of ravine land is seen particularly along the watershed/sub-watershed divides and along streams. Table 7.2 indicates the ravine land intensity at sub-watershed level in terms of actual area under this wasteland category and also the per cent share of such degraded lands in respect to the geographical areas of sub-watersheds in the study area. Besides, the codes and geographical area of all the sub-watersheds are also given in Table 7.2. This wasteland category and also the per cent share of such degraded lands in respect to the geographical areas of sub-watersheds in the study area. Besides, the codes and geographical area of all the sub-watersheds are also given in Table 7.2.

It is clear from Table 7.2 that the per cent ravine land to the sub-watersheds' area varies between 1.23 per cent and 41.82 per cent. Both these ranges of per cent are found in the same watershed (2D2A4) during the year 2001-2002. The smallest and the largest per cent share are associated with the sub-watersheds 2D2A4b and 2D2A4h respectively. The average per cent of ravine land is 15.30 for the study area. It is evident from Table 7.2 that the largest sub-watershed (2D2A2h) shares only 3.68 per cent of its geographical area under ravine land category. A total of 6 ravine land classes have been worked out as given in Table 7.3.

Conclusion

An action plan for ravine land development can be formulated after assessing the damage extent to land resource in the study area. A prioritization of sub-watersheds affected in terms of spatial extent as well as by taking in an account the ravine depth criteria as gathered from the field measurements can be determined. Undoubtedly, wide coverage of field checks is necessary to improve

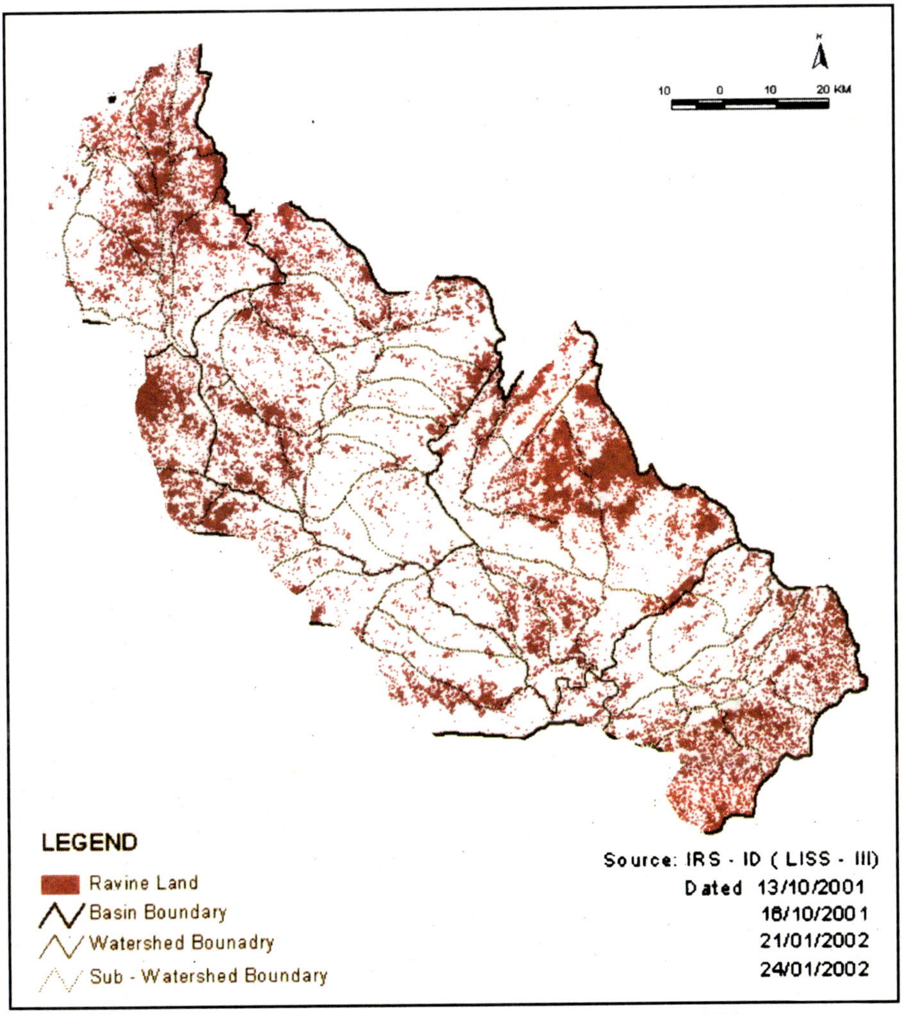

Fig. 7.3 : Spatial distribution of ravinous land in the Morel river basin (2001-02)

the acceptability of data output and to overcome misinterpretability by the machine due to spectral resemblance of signatures. Of course, this kind of data base can be used by agencies concerned with reclamation of such degraded lands on a priority basis because such sub-watershed level input of data is objective to a considerable extent.

Table 7.2 : Sub-watershedwise ravinous land in the Morel river basin (2001-02)

Sl. No.	*Sub-watershed Code*	*Geographical area of sub-watershed (km²)*	*Ravinous land (km²)*	*% of ravinous land to the total sub-watershed's geographical area*
1.	2D2A1a	78.5169	6.40	8.15
2.	2D2A1b	93.5074	9.50	10.16
3.	2D2A1c	99.0609	9.40	9.49
4.	2D2A1d	99.1778	11.00	11.00
5.	2D2A1e	65.7428	13.20	20.00
6.	2D2A1f	33.6319	4.30	12.78
7.	2D2A1g	195.5459	50.30	25.72
8.	2D2A1h	209.6204	68.10	32.48
9.	2D2A1i	113.6952	33.80	29.73
10.	2D2A2a	101.2620	4.80	4.74
11.	2D2A2b	94.0353	3.30	3.50
12.	2D2A2c	122.1062	3.20	2.60
13.	2D2A2d	193.6299	10.50	5.42
14.	2D2A2e	133.7558	31.60	23.60
15.	2D2A2f	245.1434	13.30	5.42
16.	2D2A2g	77.7885	8.30	10.67
17.	2D2A2h	380.6446	14.00	3.68
18.	2D2A3a	36.1893	8.50	23.48
19.	2D2A3b	154.8681	36.60	23.60
20.	2D2A3c	196.6762	31.70	16.12
21.	2D2A3d	57.6489	6.20	10.75
22.	2D2A3e	57.3979	5.90	10.28
23.	2D2A3f	125.5755	21.00	16.72
24.	2D2A3g	170.7404	38.30	22.40
25.	2D2A3h	162.1560	4.90	3.00
26.	2D2A3i	64.0467	2.50	3.90
27.	2D2A3j	90.0150	5.40	6.00
28.	2D2A3k	94.2361	8.90	9.44
29.	2D2A3l	62.3839	7.00	11.22
30.	2D2A3m	91.3139	20.50	22.45
31.	2D2A3n	123.4409	20.40	16.53
32.	2D2A3o	46.5306	11.00	23.64
33.	2D2A4a	53.1330	7.20	13.55
34.	2D2A4b	181.5992	2.30	1.23
35.	2D2A4c	134.7039	2.20	1.63
36.	2D2A4d	101.2383	3.80	3.75
37.	2D2A4e	109.4623	17.80	16.26
38.	2D2A4f	54.4875	7.40	13.58
39.	2D2A4g	118.7668	26.50	22.30
40.	2D2A4h	74.6055	31.20	41.82
41.	2D2A4i	199.0064	71.40	35.88
42.	2D2A5a	156.8286	12.00	7.65
43.	2D2A5b	159.0853	20.20	12.70
44.	2D2A5c	114.4565	19.50	17.04
45.	2D2A5d	110.2639	31.60	28.66
46.	2D2A5e	81.7835	26.10	31.90
47.	2D2A5f	195.9743	42.40	21.63
48.	2D2A5g	103.8537	32.70	31.48
49.	2D2A5h	73.4509	23.30	31.72
		5892.7844	901.40	15.30

Table 7.3 : Ravine land intensity in sub-watersheds of the Morel river basin (2001-2002)

Sl. No.	*% ravine land to the total sub-watershed geographical area*	*Class description*	*No. of sub-watersheds*	*Area covered (Km²)*	*% to the total ravine* land
1.	< 5	Slightly affected	9	41.00	4.50
2.	5-10	Slight to moderately affected	7	65.90	7.30
3.	10-20	Moderate to highly affected	15	197.40	21.90
4.	20-30	Very highly affected	12	344.30	38.20
5.	30-40	Severely affected	5	221.60	24.60
6.	> 40	Very severely affected	1	31.20	3.50
		Total	49	901.40	100.00

Acknowledgements

I am thankful to the UGC, New Delhi for providing funds to purchase RS data. Remote Sensing Division of B.M. Birla Science and Technology Centre, Statue Circle, Jaipur deserves special appreciation for allowing me to avail RS and GIS facilities at reasonable charges.

REFERENCES

Aruchamy, S., Haridas, V.R. and Jagnathan, R. (2001). Characteristics of southwest monsoon wind and its influences on the environment of the Thevarn Valley, Tamil Nadu. *Ind. Jr. of Geom.* Vol 6, No.1, pp.7-18..

Doi, R.D. (1991). *Semi-Arid Land Systems: Use and Capability.* Pointer Publishers, Jaipur, p.41.

Government of India (1990). *Watershed Atlas of India.* All India Soil and Land Use Survey (AISLUS), Dept. of Agriculture and Cooperation, Ministry of Agriculture, New Delhi, plate 10.

Government of India (2000): *Wastelands Atlas of India.* Ministry of Rural Development, Department of Land Resources, New Delhi and NRSA, Department of Space, Hyderabad, p. 4.

Pandey, D., Saxena, R. and Gaur, S.C. (1993). Study of Envrionmental Changes due to Ravine Development along Chambal River near Bhind and Its environs. Conf. by IIRS and ARRAC held at Guwahati, Assam from Nov. 25-27, 1993, pp.9-30.

Pani, P. and Mohapatra, S.N. (2001) Delineation and monitoring of gullied and ravinous lands in a part of lower Chambal valley, India, using remote sensing and GIS. Paper presented 22nd Asian Conf. on Remote Sensing, held at Singapore, from Nov. 5-9, 2001, www:google.com.

Richards, K. (1982). *Rivers – Form and Process in Alluvial Channels.* Methenu, London.

Sharma, H.S. (1980). *Ravine Erosion in India.* Concept Publishing Company, New Delhi, p.23.

Singh, S. and Dubey, A. (2002). Gully Erosion and Management Methods and Application (A field manual), New Academic Publishers, Delhi, pp.xi and 110.

Tolba, M.K. (1982) Rolling back the wasteland, Lecture delivered to the Society for Promotion of Wasteland Development, New Delhi.

SECTION-II

APPLIED GEOMORPHOLOGY

8

Geomorphology in the Appraisal and Integration of Natural Resources for Sustainable Land Use Planning in an Arid Environment

SURENDRA SINGH

Introduction

It is an established fact that the appraisal, assessment, integration and management of the natural resources and associated environmental hazards are of paramount importance for sustainable land use planning, particularly in the arid region. In view of this, the integrated natural resources surveys, involving landforms, soils, surface water, groundwater, vegetation and present land use and also land degradation/desertification processes were conducted in the arid regions of India at the village, block, district and catchment/watershed levels using geomorphic, remote sensing and ground survey techniques. The reliable, comprehensive and systematic information on natural resources and associated hazard has been collected and mapped in more than 0.24 million sq. km area of the total 0.32 million sq.km area of the hot arid regions of India. Landforms being the most conspicuous features are distinctly visible on aerial photographs and satellite imagery and hence geomorphology has significantly contributed in the evaluation and assessment of natural resources. In order to integrate the above natural resources at one place, the concept of Major Land Resource Unit (MLRUs) was developed by Abichandani *et at.* (1975). The information collected of different natural resources has been integrated in the form of resource units (Abichandani *et at.*, 1975; Anonymous, 1982; Shankarnarayan and Singh, 1979; Shankarnarayan and Kar, 1983; Singh, 1982, 1985; Singh and Ghose, 1982; Singh and Shankarnarayan, 1982, 1985; Dhir and Singh, 1986; Chatterji and Kar, 1992; Singh *et at.* 1995, 1996).

Realising the significance of the geomorphology in the evaluation and assessment of physical potentials and limitations of the natural resources and their integration in the form of MLRUs and suitable alternate land use systems for sustainable land use planning and to upgrade the productivity of each resource unit of western Rajasthan have been highlighted in this paper.

Appraisal of Natural Resources

Salient morphological, biophysical and morphohydrological characteristics of the natural resources are discussed below.

Landforms

Fluvial and aeolian landforms of different types have been identified (Singh *et al.,* 1997). Fluvial landforms include varied pediments, flat older alluvial and interdune plains, saline flat older alluvial plains, younger alluvial plains and graded river beds (Fig. 8.1). These landforms are characterised by almost level slope. 00 to 1°, light to medium and medium heavy textured sediments, parallel, dendritic. radial and linear drainage patterns, levees, meander scrolls, points and sand bars.

Aeolian landforms constitute sand dunes, sandy undulating older alluvial and interdune plains and sandy undulating buried pediments. They are characterised by level and irregular slopes, undulations in the form of hummocks, sand dunes, transverse and longitudinal sand ridges, loose and compact sand erosional and depositional features by fluvial and aeolian patterns.

Digital analysis of the Landsat data has enabled to compute quanti-tatively the extent of the fluvial and aeoliar use their biophysical charateristics. Digital data has been proved useful to detect morphological forms and to plan suitable measures to control their further degradation by fluvial and aeolial process.

Soils

The hills and rocky/gravelly pediments are almost devoid of soils except along joints and faults which constitute coarse textured gravelly sand and loamy sand soils (Typic Udipsamments). Sandy to loamy sand and sandy loam to loam soil associated with flat buried pediments, light to medium textured flat older alluvial and interdune plain medium to fine textured flat older alluvial plains constitute loam to clay loam and also silty clay loam soils (Typic Calciorthids). The saline flat older alluvial plains and saline depressions constitute to fine textured loam to clay loam and silty clay loam saline soils (Fig. 8.1).

Aeolian landforms such as sand dunes, sandy undulating older alluvial and interdune plains and sandy undulating buried pediments have deep to very deep well drained fine sands to loamy fine sands (Typic Torripsamments). These soils belong to VIII, VII, III, V, IV land use capability classes, respectively. EC pH and WHC or these soils varied from 0.15 to 35.7 dSm^{-1}, 8.3 to 9.1 and 22 to 40 per cent, respectively (Fig. 8.2). The organic carbon, P_2O_5 and K_2O ranges between 0.04 to 0.5 per cent, 8 to 24 kg/ha and 60 to 30 kg/ha, respectively (Kolarkar *et al.* 1997).

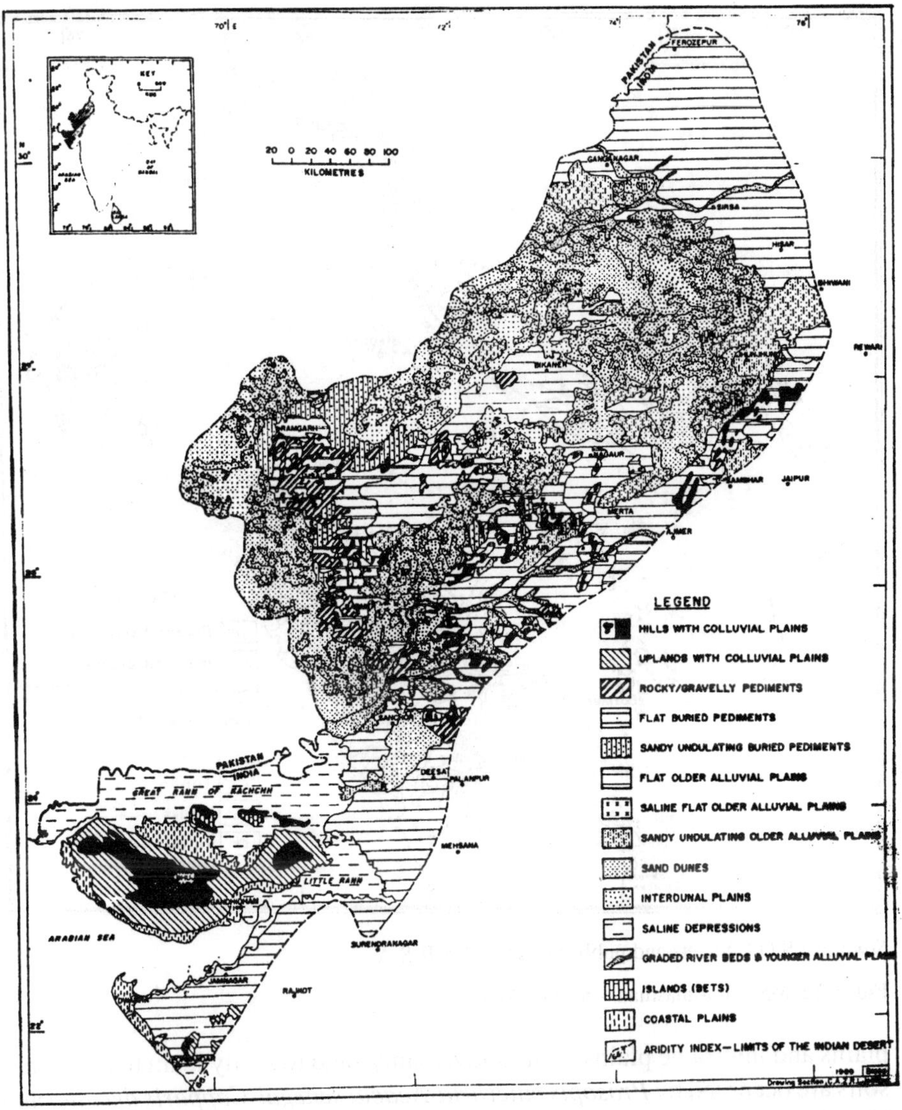

Fig. 8.1 : Indian Desert Landforms

Vegetation

In association with landforms, the broad vegetation types were delineated and mapped, Hills, rocky/gravelly pediments with well developed joints and thin cover of soils support *Euphoribia caducifolia - Acacia senegal - Capparis decidua* and *Maytenus emarginatus-Ziziphus nummularia* (15-600) plants/ha) communities (Fig. 8.3). Flat buried pediments, flat younger and older alluvial

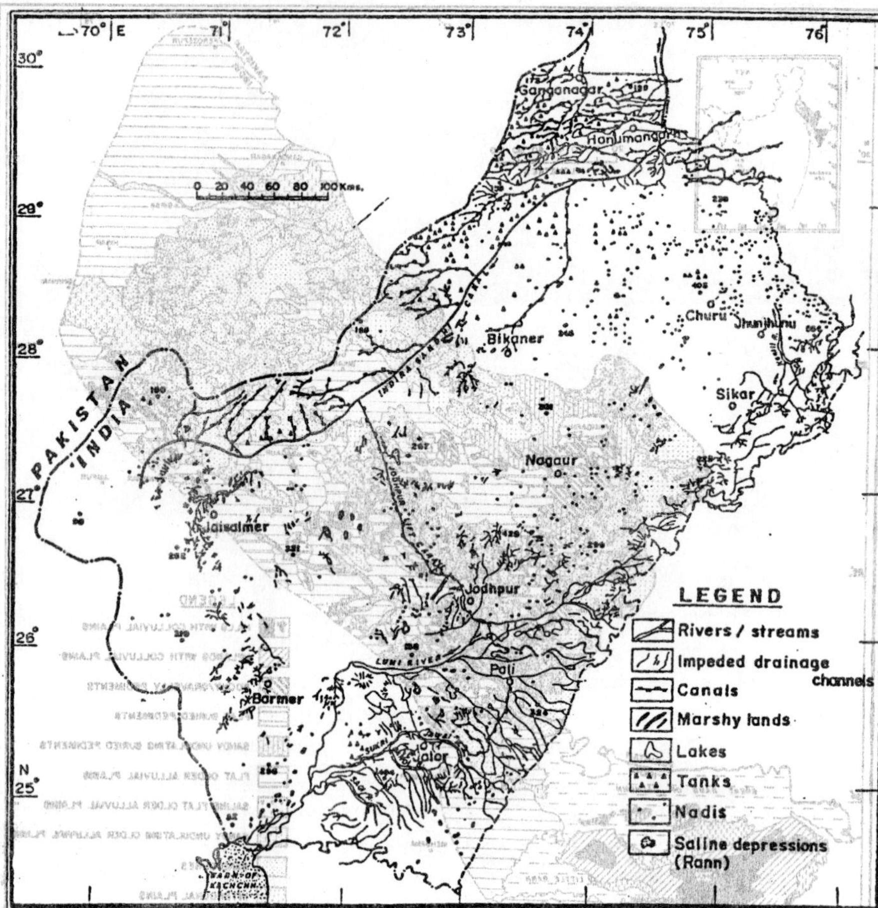

Sources : S.O.I. Sheets and published map of N.R.S.A.

Fig. 8.2 : Western Ranasthan—Surface Water

plains and interdune plains with sandy, loamy sand to sandy loam to clay loam soils are occupied by *Prosopis cineraria-Acacia senegal-Capparis decidua* and *Ziziphus nummularia-Prosopis cineraria-Acacia nilotica indica-Salvadora oleoides* (5-300 plants/ha) communties. Saline flat older plains and saline depressions with highly saline soils are occupied by *Haloxylon salicornicum - Sporobolus marginatus* and *Suaeda fruticosa-Salsola baryosma* (8-50 plants/ ha) communities.

Aeolian landforms, viz., sand dunes, sandy undulating older alluvial and interdune plains and sandy undulating buried pediments with deep to very deep sandy soils support *Acacia senegal-Calligonum polygonoides, Prosopis cineraria-Tecomella undulata* and *Ziziphus nummularia-Calligonum polygonoides*

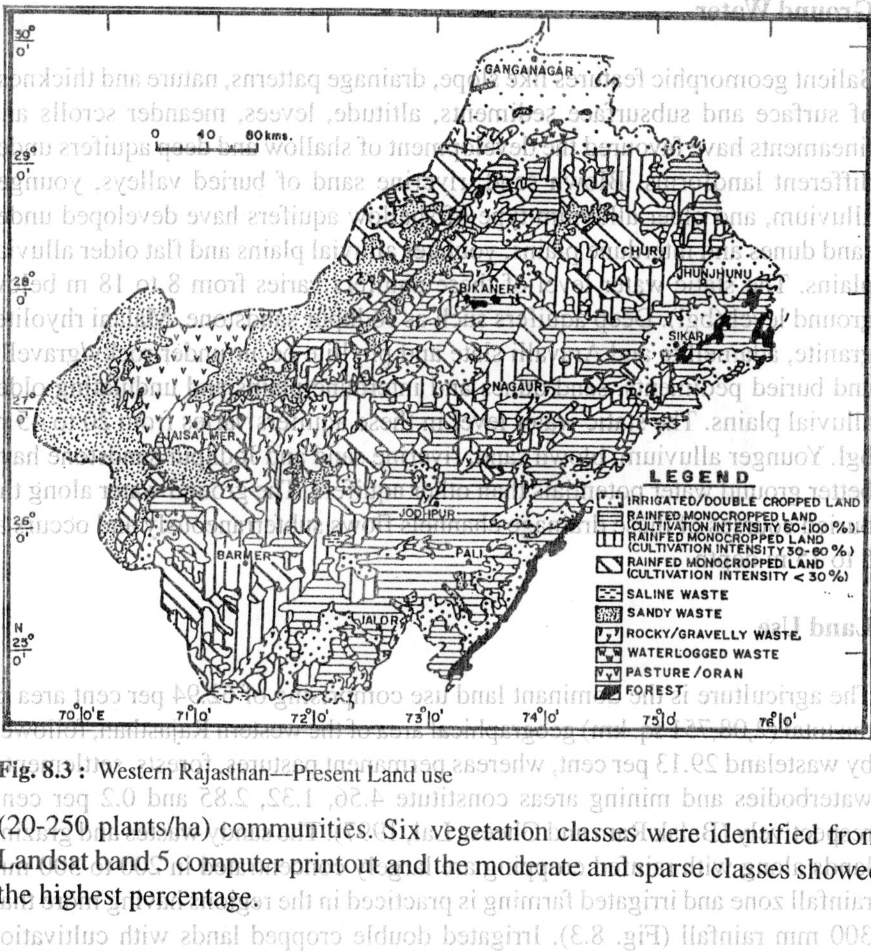

Fig. 8.3 : Western Rajasthan—Present Land use

(20-250 plants/ha) communities. Six vegetation classes were identified from Landsat band 5 computer printout and the moderate and sparse classes showed the highest percentage.

Surface Water

The nadis (village tanks) and reservoirs (Fig. 8.2) located under different landforms are the major source of surface water in western Rajasthan. Due to favourable geomorphic factors, small, medium and large sized reservoirs of 0.10 to 52.84 mcm water storage capacity occur in rocky/ gravely and flat buried pediments, younger alluvial plains and river beds and are generally used for irrigation (Fig. 8.4). The small, medium and large sized nadis (village tanks) are the major source of drinking water for the villagers. The water storage capacity of these nadis varies from 0.19 to 3.12 mcm. The small, medium and large sized nadis with 4.84, 5.08 and 4.40 mcm storage capacity, respectively, have been constructed under different landforms. Due to favourable geomorphic factors, the largest number of nadis occur in the flat older alluvial plains, followed by rocky/gravelly and flat buried pediments.

Ground Water

Salient geomorphic features like slope, drainage patterns, nature and thickness of surface and subsurface sediments, altitude, levees, meander scrolls and lineaments have favoured the development of shallow and deep aquifers under different landforms. Blown sand, riverine sand of buried valleys, younger alluvium, and older alluvium type of shallow aquifers have developed under sand dunes and interdune plains, younger alluvial plains and flat older alluvial plains. The static water level of these aquifers varies from 8 to 18 m below ground level (bgl). Deep aquifers such as Jodhpur sandstone, Malani rhyolite, granite, adamellite and Aravalli slate and phyllite occur under rocky/gravelly and buried pediments, sand dunes and interdune plains and undulating older alluvial plains. The static water level in these aquifers varies from 20 to 45 m bgl. Younger alluvium, blown sand, riverine sand and Jodhpur sandstone have better ground water potentials than other aquifers. The ground water along the buried courses of these drainage channels flows subterraneously and occurs at 5 to 15 m depth.

Land Use

The agriculture is the dominant land use comprising of 62.94 per cent area of the total (2,08,751 sq. km) geographical area of the western Rajasthan, followed by wasteland 29.13 per cent, whereas permanent pastures, forests, settlements, waterbodies and mining areas constitute 4.56, 1.32, 2.85 and 0.2 per cent, respectively (Balak Ram and Gheesa Lal, 1997). The sandy wastes and grazing lands along with rainfed cropping are largely concentrated in 200 to 300 mm rainfall zone and irrigated farming is practiced in the regions having more than 300 mm rainfall (Fig. 8.3). Irrigated double cropped lands with cultivation intensity of 100 per cent and above occur, by and large, in younger alluvial plains, buried pediments, interdune plains and intermontane basins. The major crops grown by irrigation are wheat, mustard, cotton, rice, groundnut, sugarcane, gram, chillies, cumin and isabgol. In some areas due to faulty method of irrigation, salinity problem has developed and only salt tolerant crop's like kharchia wheat is being grown (Fig. 8.5).

Rainfed monocropped lands with cultivation intensity of 80 to 100, 60 to 80 and 30 to 60 per cent are associated with flat older alluvial plains, flat sandy older alluvial plains and sandy undulating older and interdune plains and constitute 18.12, 13.84 and 14.20 per cent of the total area, respectively (Balak Ram and Gheesa Lal, 1997). Moong, moth, guar and bajra are the major crops grown under rainfed conditions (Fig. 8.3).

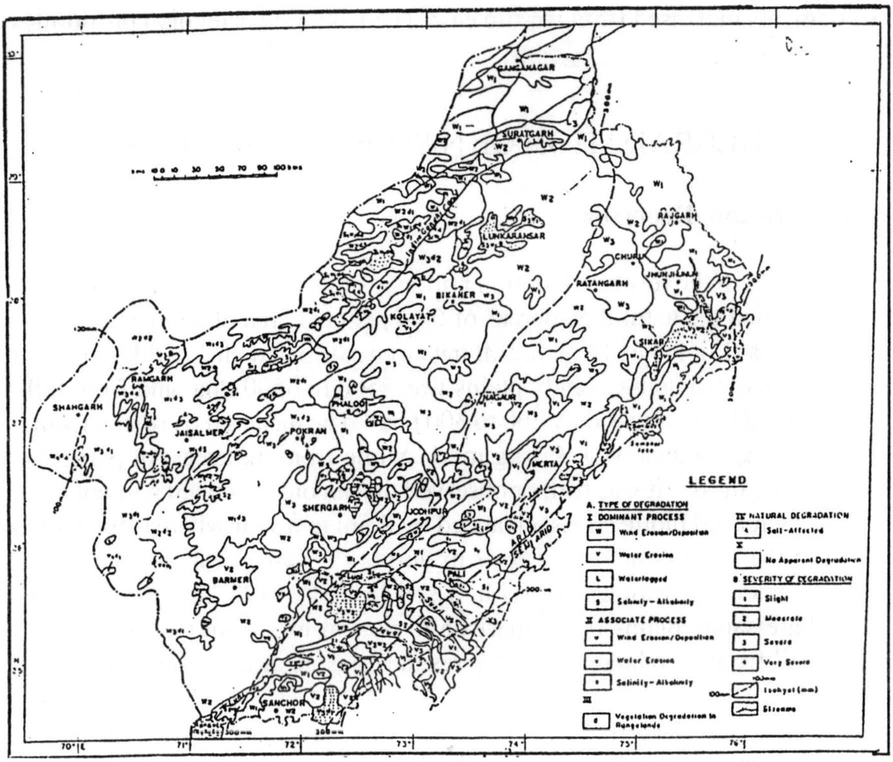

Fig. 8.4 : Western Rajasthan—Desertification Status

Sandy, saline, rocky/stony/gravelly, gullied and mining lands are the major type of wastelands identified in western Rajasthan. Out of the total area under wastelands, sandy waste constitute 62.4 per cent area, followed by rocky/stony/gravelly waste 32.6 per cent, saline waste 4.2 per cent and gullied lands 0.8 per cent.

Forests, pastures/grazing lands, settlements and waterbodies fall under the miscellaneous category of land use. Forests are highly degraded lands constitute 1.32 per cent of the total area. In last three decades, about 22 per cent area has increased under forests because degraded hills, sand dunes and salt affected lands have been acquired and fenced by Forest Department and plantation has been done on them.

Permanent pastures/grazing lands locally known as *gochar*, oran and agar are highly degraded and constitute 4.2 per cent area. Due to increasing livestock population and grazing stress, the extent of these lands shrunk by 8 to 30 per cent.

Settlements and waterbodies comprising of villages, farm steads (*dhanis*), tanks and reservoirs fall under this category of land use. The size and density of settlements increased from west to east. The land used category consutitutes 3 per cent area.

DEGRADATION OF NATURAL RESOURCES

Wind Erosion /Deposition

This hazard in the form of deflation hollows, sand dunes and sandy hummocks has degraded the natural resources of 68.4 per cent area of aeolian and fluvial ecosystems (Fig.8.4 and Fig. 8.2).It has been estimated that soil loss from the deeply tractor ploughed sandy plains was 2500 to 3500 t/ha and from fallow and untilled lands it was only 200 to 300 t ha^{-1} The mean sand erosion from the crest of barchan dune was 351 kg $m^{-2}d^{-1}$ while from the creast of parabolic it was 325 kg $m^{-2}d^{-1}$. It was further observed that an old vegetated parabolic dune due to increased biotic stress advanced by 0.5m in a year whereas the barchan dune in the vicinity advanced by 7 m during the same year. The accelerated wind erosion/deposition resulted in a soil loss of 1449 t ha^{-1} (April to June) from bare sandy plains in comparison to no soil loss from pasture lands and 225 t ha^{-1} from crop lands with crop residue on soil surface (Gupta and Raina, 1994).

Water Erosion, Flood and Industrial Effluent

This hazard degraded 11.2 per cent area of the fluvial and aeolian econsystems in the form of rills, gullies, dissections, siltation and surface crust. In the Luni basin, the accelerated water erosion removed 0.2 to 53.5 g^{-1} sediments in the hilly terrain of Aravallis and 20.5 to 453.6 g^{-1} in the lower reaches of the basin (Sharma *et al.*, 1984). The rills and gullies of 0.5 to 10 m depth and 1.2 to 15.3m width created by fluvial action degraded the hills, rocky/gravelly and buried pediments and flat older alluvial plains. If has also been estimated that over a period of 28 years (1958-86) gullied area increased from 199 ha to 242 has (Singh *et al.*, 1990). It has been estimated that during the flash floods of 1990, large quantity of sediments were the form of sand sheets of 20 to 60 cm thickness and sand mounds and sand ridges of 0.7 to 2.5 m letion of biological productivity of original sediments (Fig. 8.6).

Industrial effluent along the Jojri, the Bandi and the Luni rivers in Jodhpur, Pali and Barmer districts has affected 34,500 ha area of river bed, younger and older alluvial plains. Due to hazard, the ground water has been very severely, moderately and slightly polluted (EC 2.5-45.6 dSm^{-1}). The irrigation by this polluted water has changed the morphological and physico-chemical properties of the soils, resulting in the reduction in the reduction in their infiltration rate and fertility status. The biomass, density and cover of the plants in the affected areas also declined by 3 to 5 times. The cropping intensity, crop yield and crop residue declined by 60 to 70, 40 to 50 and 30 to 30 per cent, respectively.

Wind Erosion/Deposition and Water Erosion

The combined impact of this hazard has affected the natural resources of 7.3 per cent area of sandy undulating buried pediments and obstacle dunes. Very severe wind erosion/deposition and severe water created the sand dunes and hummocks of 5 to 30 m height and gullies of 6 to 15 m depth and 8 to 25m width, resulting in the degradation of large acreage of the above ecosystems. Moderate to severe wind erosion/deposition and moderate water erosion hazards were responsible for creating sand dunes of 3 to 12 m height, sand sheets of 250 to 450 cm thickness and gullies of 2.5 to 5.6 m depth and 5.5 to 10.5 m width. Slight to moderate wind erosion/deposition and water hazards created sand dunes and hummocks of 1.5 to 5 m height and gullies of 0.5 to 3.5 m depth and 2.5 to 4.5 m width (Singh *et al.*, 1994).

Salinity/Alkalinity and Water Logging

This problem developed in the Indira Gandhi canal Command area and the Luni-Jawai drainage system due to faulty methods of irrigation, presence of subsurface barrier at shallow depth and construction of tanks and reservoirs along the courses of palaeo drainage channels has affected the natural resource of 3.1 per cent area of different fluvial and aeolian ecosystems (Singh *et al.*, 1992). Recent field investigations revealed that in Suratgarh tehsil of Ganganager district and Rawatsar tehsil of Hanumangarh district, 4500 ha and 5,000 ha agricultural lands, respectively have been desertified by salinity/alkalinity and waterlogging problem. Due to this problem, natural cover of *Haloxylan salicornicum* and *Lasiurus sindicus* is being depleted at a faster rate. In the catchment area of the Sardar Samand and Jaswant Sagar reservoirs, 9600 and 6170 ha area, respectively have been degraded due to seepage and suspended salts and productivity of large average of agricultural lands has been depleted.

Natural desertification hazard occurring in the west of 100 mm rainfall zone affected the natural resources of 2.1 per cent area of aeolian landforms. Remaining 8.1 per cent area is free from such hazards (Singh *et al.* 1992).

Integration, Assessment and Management of Natural Resources for Sustainable Landuse Planning

Integration, assessment and management of natural resources in the form of

Major Land Resource Units (MLRUs) are of paramount importance (Fig. 8.5) for sustainable land use planning. Suitable land use management practices for different fluvial and aeolian land resources units falling under two rainfall zones are suggested below.

Less than 300 mm Rainfall Zone

In this rainfall zone, fluvial and aeolian land resource units like sand dunes and hummocks, sandy undulating older alluvial and interdune plains and sandy undulating burried pediments and characterised by undulating topography, irregular slope, inherent vulnerability to wind erosion/deposition, loose, structureless sand to loamy sand soils with poor fertility and low water holding capacity. Frequent droughts, low and erratic rainfall and cultivation on marginal lands aggrevate the sand drifting activities and removal of nutrients from the soils. These land resource units, therefore, should not be used for arable farming except Indira Gandhi Canal Command area but they may be developed into silvipastures, hortipastures, pastures and animal husbandry by using suitable management techniques. Silvipastoral management need fencing and mulching of these resource units to check the wind erosion/deposition activities. Reseeding with *Cenchrus ciliaris* (CAZRI 358) and *Cenchrus setigerus* (CAZRI 175) and *Lasiurus sindicus* at the rate of 5 to 7 kg/ha will be necessary. Concurrently top feed trees such *as Prosopis cineraria, Acacia tortilis, Ziziphus nummularia* and *Acacia jacquemontii* may be planted at 5×5 m distance in such a way as to provide 40 to 60 trees/ha. After establishment of the silvipastures, deferred-rotational grazing should be followed. The forage production from the present level of 175 to 400 kg/ha will increase to 650 to 1100 kg/ha by using above management techniques. In the flat areas of these land resource units, hortipastures using fruit trees like *Ziziphus mauritiana, Cordia myxa* and grasses like *Lasiurus sindicus* and *Cenchrus ciliaris* could be developed.

Flat older alluvial and interdune plains and flat buried pediments of this rainfall zone may be developed into hortipastures by growing Cenchrus ciliaris in association with *Ziziphus rotundifolia* and *Ziziphus nummularia* and later budded with *Ziziphus mauritiana* variety by using proper soil conservation measures like wind breaks, shelterbelts, contour, field bundings etc. These management techniques will enable to increases the present forage yield from 135 to 450 kg/ha 600 to 1200 kg/ha. The present productivity of bajra and wheat under khadin system of cultivation could be increased from 500 to 600 kg/ha to 1000 to 1200 kg/ha by using suitable soil and water conservation measures (Vashishtha, 1997). Agroforestry and agri-hortipastoral systems using crops, trees, fruit plants and grasses such as moong, moth, guar, *Acacia senegal, Tecomella undulata, Ziziphus mauritiana, Cenchrus ciliaris* and *Cenchrus*

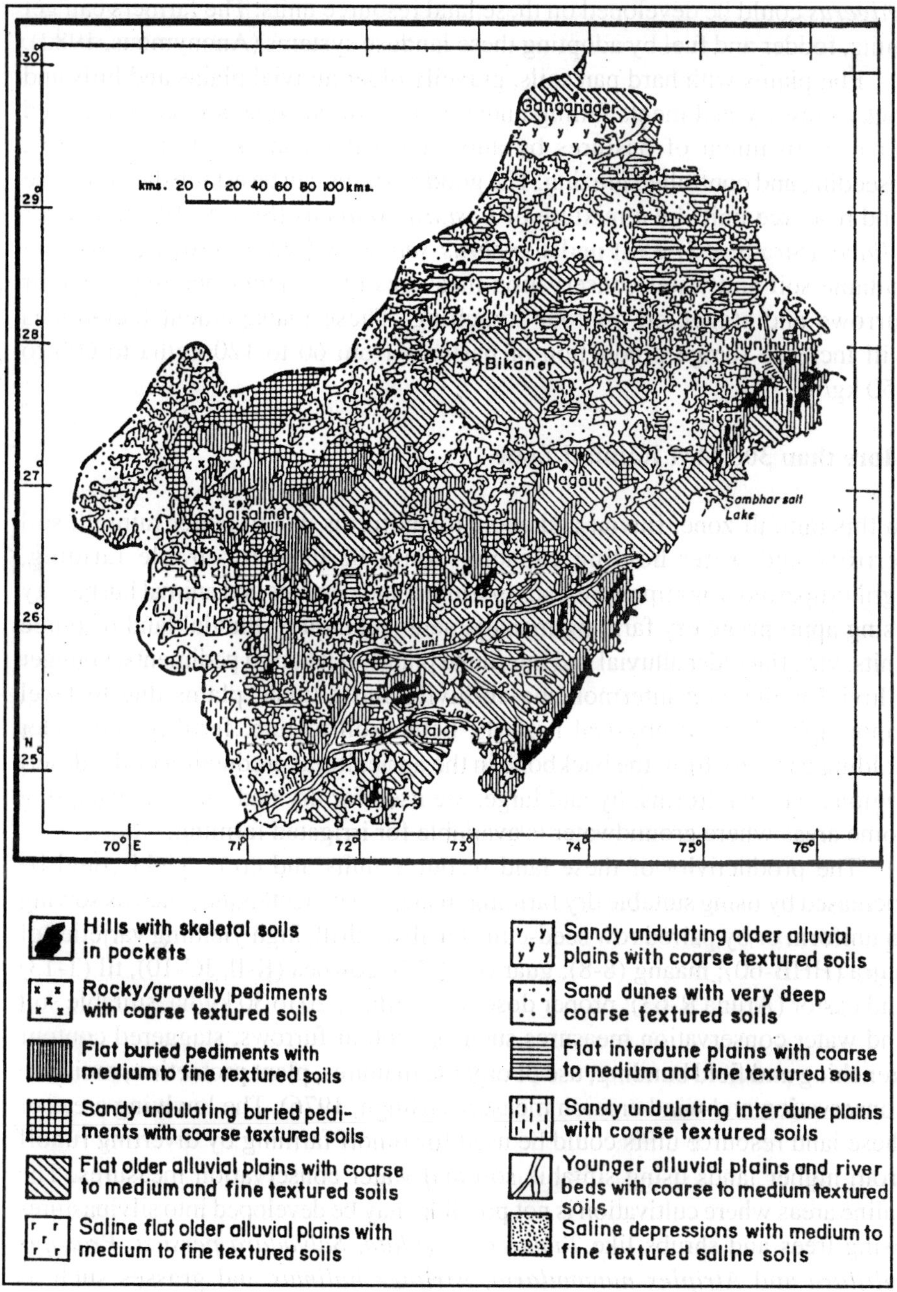

Fig. 8.5 : Western Rajasthan Major Land Resources Units (MLRUs)

setigerus could be developed on these land resource units. The farmers can get fruits, fodder and fuel by adopting these landuse systems (Anonymous, 1982).

Flat plains with hard pan soils, gravelly older alluvial plains and hills and rocky/gravelly pediments falling under above rainfall zone are recommended for establishment of pastures by adopting water conser-vation measures, reseeding and controlled grazing. The good pastures can be established on these land resource units by a mixture of *Lasiurus sindicus* (Strain 319); *Cenchrus ciliaris* (Strain 357, 358) and *Cenchrus setigerus* (Strain 176) by adopting suitable soil and water conservation measures like contour bunding, contour furrows and staggered contour trenching. These management techniques will increase forage production of pastures from 60 to 120 kg/ha to 600 to 850 kg/ha.

More than 300 MM Rainfall Zone

In this rainfall zone, fluvial and aeolian landforms due to better rainfall, soil fertility and water holding capacity could be used for arable farming, agrihortipastures, hortipastures, silvipastures, and pastures and animal husbandry using appropriate dry farming management techniques. Pluvial land resource units, viz., flat older alluvial and interdune plains, flat buried pediments, younger alluvial plains and intermontane colluvial and alluvial plains due to level topography, limited physical hazards, better soil texture, fertility and water holding capacity form the backbone in the agricultural production of the desert regions. The landforms, by and large, are used for dryland farming excepting some areas where groundwater is available for irrigated farming.

The productivity of these land resource units and crops yield could be increased by using suitable dry farming management techniques such as sowing in uniform or by paired row seed-cum-fertilizer drill, high yielding varieties of hajra (HHB-60); maang (8-8); guar (F8-277); cowpea (K-ll, JC-10), til (T-13) and castor (Aruna R-63); proper doses of fertilizer 20 to 30 kg/ha, suitable soil and water conservation measures such as contour furrows, staggered contour trenching and field bunding, use of organic manures, plant protection, moisture conservation and wind erosion measures (Singh, 1976). The lowlying areas of these land resource units could be used for runoff farming by diverting runoff from higher lands using suitable soil and water conservation measures. The saline areas where cultivation is not possible may be developed into silvipastures using trees and shrubs like *Tamarix articulata, Salvadara persica, Prasapis julijlara* and *Atriplex nummularia, Atriplex halimus* and grasses such as *Dichanthium annulatum, Sparahalus marginatus* and *Sparahalus helvalus* (Shankarnarayan and Kar, 1983).

Bajra is the staple food of arid zone but it involves the risks of failure. In order to avoid total failure of hajra crop, it is suggested that hajra may be grown on above land resource units in intercropping system with green gram, mothbean and clusterbean. This system of intercropping not only increases the productivity but also provides protein rich food to human beings and the fodder to the animals (Daulay, 1987).

Flat areas of the sandy undulating older alluvial and interdune plains and sandy undulating buried pediments with sandy soils may be used for arable farming by using proper soil and water conservation measures, viz., contour bunding, gully plugging, wind breaks, shelterbelts and stubble mulching. In order to upgrade the productivity of these land resource units, wind strip cropping system by growing crops in 30 m strips and grasses in 5 m strips at the right angle to the prevailing south-west winds is suggested. The fruit crops like pomegranate, gala and mundia varieties of her may be grown under these resource units.

The uplands of the above land resource units, sand dunes and hummocks falling under this rainfall zone should also not be used for cultivation but they could be developed into silvipastures by using the similar dryland agriculture techniques as suggested for such land resource units occurring in less than 300 mm rainfall zone. Due to better rainfall, the uplands of these land resource units will provide higher forage yield varying from 750 to 1500 kg/ha.

Hills and rocky/gravelly pediments due to physical limitations are not fit for cultivation but the areas of these land resource units with shallow soil cover may be developed into good pastures and woodlands. The present forage productivity of the degraded pastures could be upgraded from 150 to 250 kg/ha to 650 to 950 kg/ha by contour furrowing and staggered trenching, followed by reseeding with *Dichanthium annulatum, Cenchrus ciliaris* and *Cenchrus setigerus* and proper protection and controlled grazing. These land resource units due to protection and controlled grazing can support 65 to 110 trees/ha of *Acacia senegal* and *Anogeissus pendula* and one tree could provide 120 gm of gum in addition to 350 to 550 kg/ha forage production.

Conclusion

Fluvial and aeolian landforms identified in the western Rajasthan, have different physical potentials and limitations. Remote sensing techniques in conjunction with ground truth have proved very useful in the appraisal and management of natural resources and associated environmental problems in the form of Major Land Resource Units (MLRUs) for sustainable land use planning using alternate landuse systems like pasture, silvipasture, agroforestry, agrihorticulture, hortipasture and hortisilvipasture.

REFERENCES

Abichandani, C.T., Singh, S., Saxena, S.K., Kolarkar, A.S. and Sen, A.K. 1975. Assessment and management of the biophysical resources in district Bikaner, western Rajasthan. *Annals of Arid Zone* 14 (4) : 292-301.

Anonymous, 1982. *Basic and Human Resources of Jodhpur District (Rajasthan).* CAZRI, Jodhpur, 142 p.

Balak Ram and Gheesa Lal, 1997. *Landuse and agriculture in the Indian arid ecosystem.* In *Desertification Control in the Arid Ecosystem of India for Sustainable Development* (Eds. Surendra Singh and Amal Kar), pp. 80-92, Agro Botanical Publishers, Bikaner.

Chatterji, P.C. and Kar, A. (Eds.) 1992. *Integrated Natural and Human Resources Appraisal of Jaisalmer District.* CAZRI Publ. 30, CAZRI, Jodhpur, 156 p.

Dauley, H.S. 1987. Cropping systems for pearl millet in arid agroclimate. Production Technology of Bajara and Minor Millets (Mimeog.), pp. 79-84, CAZRI. Jodhpur.

Dhir, R.P. and Singh, S. 1986. Integrated natural resource survey using remote sensing. In Proceedings of International Seminar on Photogrammetry and Remote Sensing in Developing Countries, New Delhi. T.5, pp. 14.1-14.13.

Gupta, J.P. and Raina, P. 1994. *Wind erosion and its control in hot arid areas of Rajasthan, India.* In Proceedings of International Symposium on Wind Erosion in West Africa: The Problem and its Control (Eds. B. Bunkerf; F.E., Allisin and M. Yon Oppen), pp. 209-218. University of Hohenhein, Germany.

Kolarkar, A.S., Joshi, D.C. and Singh, N. 1997. Soils of Indian arid ecosystem. In *Desertification Control in the Arid Ecosystem of India or Sustainable Development* (Eds. Surendra Singh and Amal Kar), pp. 46-57. Agro Botanical Publishers (India), Bikaner.

Sharma, K.D., Yangani, N.S. and Chaudhary, J.S. 1984. Sediment transport characteristics of the desert streams in India. *Journal of Hydrology*, 67: 261-272.

Shankarnarayan, K.A. and Singh, S. 1997. *Application of Landsat Data for Natural Resources Inventory and Monitoring of Desertification.* SDSU-RSI- 79-17, Remote Sensing Institute, Brookings, S.D. USA, 136 p.

Shankarnarayan, K.A. and Kar, A. (Eds.) 1983. *Upper Luni Basin—An Integrated Analysis and Human Resources for Development Planning.* CAZRI Jodhpur, 577 p.

Singh, R.P. (Ed.) 1976. Improved Dryland Agriculture for Western Rajasthan, *CAZRI Technical Bulletin* No.1.

Singh, S. 1982. Role of photogeomorphological techniques in the integrated natural resource survey of the central Luni basin, western Rajasthan (India). In *Perspectives in Geomorphology* (Ed. H.S. Shanna), Vol. 3, pp. 279-296. Concept Publ. Co., New Delhi.

Singh, S. 1985. The role of geomorphology in management of land and water resources in an arid environment—A case study of Jodhpur district. In *Dry Land Agriculture in India* (Eds. M. Shafi and Mahdi Raza), pp. 173-190. Rawat Publications, Jaipur.

Singh, S. and Ghose, B. 1982. Role of photogeomorphology in reclamation and development of Rajasthan desert. In *Proceedings of Workshop on Prob of Desert in India*, pp. 1-10. GSI Miscellaneous Publication No.49, Jaipur.

Singh, S. and Shankarnarayan, K.A. 1982. Role of remote sensing in appraisal of natural resources for the development planning of Indian Transitions of Indian Society of Desert Technology. 7 (2) : 89-90.

Singh, S. and Shankarnarayan, K.A. 1985. Remote sensing in monitoring natural resouces and environmental hazards in Indian desert. In *Proceedings Asian Conference on Remote Sensing*, pp. 34-41. NRSA, Hyderabad.

Singh, S., Shanna, K.D. and Bohra, D.N. 1990. Monitoring change detection in uplands and gullied areas through satellite remote sensing. *Annals of Arid Zone* 29 (3) : 171-177.

Singh, S., Kar, A., Joshi, D.C., Balak Ram, Suresh Kumar, Vats, P.C., Singh, N., Raina, P., Kolarkar, A.S. and Dhir, R.P. 1992. Desertification main western Rajasthan. *Annals of Arid Zone* 31 (4) : 237-246.

Singh, S., Kar, A., Joshi, D.C., Suresh Kumar and Shanna, K.D. 1994. Desertification problem in western Rajasthan. *Annals of Arid Zone* 33 (4) : 191.

Singh, S., Shanna, K.D. and Joshi, D.C. (Eds.) 1995. *Integrated Natural and Human Resources Appraisal for the Development of Jalor District, Jodhpur,* 108 p.

Singh, S., Vangani, N.S. and Singh, N. (Eds.) 1996. *Integrated Natural and Human Resources Appraisal for the Development of Sikar District, Jodhpur,* 172 p.

Singh, S., Singh, H.P. and Kar, A. 1997. Integration and assessment of resources for the management of arid ecosystem. In *Desertification Control in A, Ecosystem of India for Sustainable Development* (Eds. Surendra Singh and Arnal Kar), pp. 211-218. Agro Botanical Publishers (India), Bikaner.

Vashishtha, B.B. 1997. Role of arid horticulture in combating desertification. In *Desertification Control in Arid Ecosystem of India for sustainable Development* (Eds. Surendra Singh and Amal Kar), pp. 211-218. Agro Botanical Publishers (India), Bikaner.

9

Applied Geomorphological Study on Eastern Coastal Zone with Particular Reference to the Sundarbans, India

S.C. MUKHOPADHYAY

Introduction

Present paper is concerned with the study on applied geographical problems of Eastern Coastal Zone with particular reference to the Sundarbans Mangrove Forest, India (Figs. 9.1-9.4). The coastal zones of West Bengal, i.e. the lower 'Bengal Basin' comprises the district 24-Parganas (South and North) and the Medinipore, which includes about 73 per cent areas of the Sundarbans and rest 27 per cent of the coastal area of the Medinipore district. These coastal areas within West Bangal lie between longitude 86°E and 89° 10′E and latitude 21° 30′ N and 22° 30′ N. The entire Mangrove habitat area is demarcated as the CRZ-1 (Coastal Regulatory Zone-1), while the Kulpi-Kakdip area of the South 24-Parganas is demarcated as CRZ-II. Total length of this West Bengal Coastline from west to east is about 220 km, extending from the eastern part of the river Subarnarekha or New Digha coast to the river Harinbhanga or Herobhanga of the India-Bangladesh border. This entire coastal stretch is intersected by the rivers Haldi and Rasulpur in the Medinipore and the 24-Parganas south. On the other hand, six major tidal rivers, viz. Baratala (Murigana), Saptamukhi, Thakura/Jamira, Matia, Goasaba, Herobhanga or Harinbhanga flow from north to south and meet the bay in their southern extremities. Most of these rivers in the Indian Sundarbans have silted up as the upstream connections with the river Ganga have choked since the three to four centuries.

Environments of coastal Zones (West Bengal)

Mixing of the upstream freshwater into the bay or sea, the ecotone or the buffer zone between freshwater of the upstream and saltwater of the sea is called estuary. The estuarine, mouths of the mighty river Ganga meet the bay of Bengal and in this transitory zone at the confluence of the river and the bay from several large to small islands having with dense growth of several important mangroves, mangroves associates and back mangal species. Besides these unique floral successions in the inter-tidal lands, good number of algae, fungus, aquatic-terrestrial amphibian- Avi fauna are found as the most common dwellers in the estuarine rivers, canals, creeks and wetland of this complex mangrove ecosystem. The glamour and fame of Sunderbans mangal is worldwide, not only for its world renowned 'Royal Bengal Tiger' but for its unique floral succession and large groups of avian, aquatic, mud dwelling and semi-aquatic or amphibian fauna. The present day estuarine boundary of the river Ganga in India is well known as the Hugli-Matla estuary, which also include the east tract of inter-tidal mangrove swamps and large number of medium to vast India and popularly known as Sunderbans.

Besides these, the large-scale human interaction and deforestation activities, *vis-à-vis*, premature-reclamation and renovation of the mangrove forestland of Sunderbans resulted in silt deposition on these estuarine riverbeds. This also accelerated land accretion in these coastal areas. Furthermore, good number of broad to narrow rivers, rivulets, tidal canals and creeks, viz., the river Bidya, Raimangal, Kalindi, Kartal, Hogol, Jamuna, Ichhamati, Gomor, Ajmalmari are being blocked and number of tidal canals and creeks intersect all these major estuarine rivers and ultimately form several large and small islands are changed ecologically. All these were formed a close network system with anastomosing tidal rivers, canals and creeks. During high tidy and sea surges most of these coastal zones, islands and delta lands coastal phases along with especially and the Sunderbans Mangals (mangroves) get inundated or flooded with saline sea water. In this saline zones, the mangrove trees, shrubs and the halophyte herbs, shrubs and climber mostly grow naturally and step by step and naturalised, during the early period. This author has also made an attempt to consider the major elements, e.g. coastal systems, resources, and environments, coastal issues institutional arrangements, planning and management techniques. These are found to be involved managing effective by Sundarban coastal resources and environments (Mukhopadhyay, 2003).

Interegrated Coastal Zone Management

It relates to the fulfilment of the recent concept of Integrated Coastal Zone Management (ICZM) of Sunderbans under the prevailing fluvio-marine environments of the Purba Medinipur, North and South 24-Paraganas districts as very coastal zone of West Bengal in general and Sundrbans in particular. Development approach may solve the problems one has to address. Even if it remains restricted to the relatively narrow coastal strip, a successful application of integrated coastal zone management (ICM) requires above all the establishing

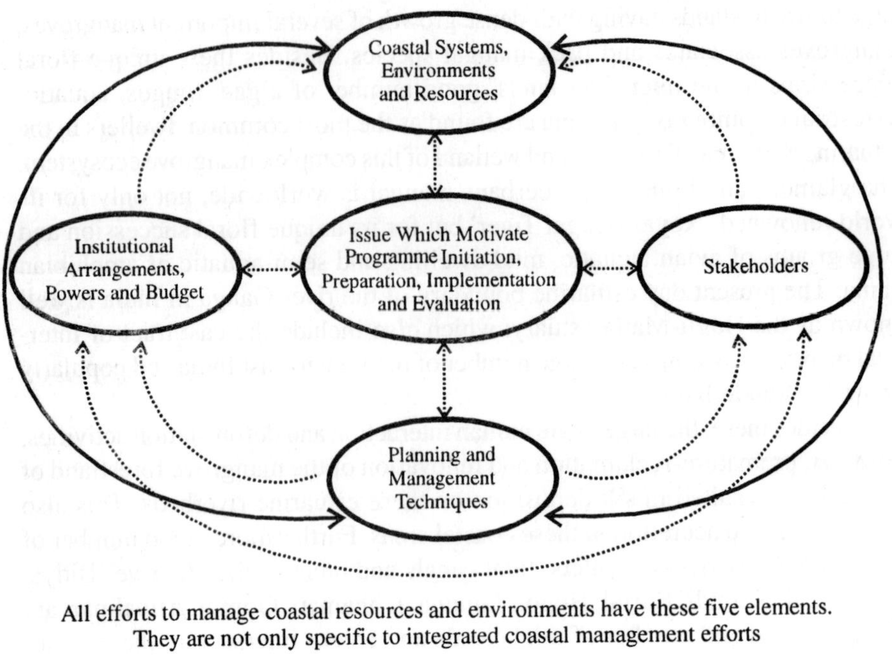

Fig. 9.1 : Elements involved in managing coastal resources and environments (Sorensen, 1997)

of a common work platform for all its actors. This illustrates the nature of co-operation of various actors in integrated coastal zone management (ICZM Fig.9.1). The fisherman will hardly be interested in the soil erosion he causes by felling trees in the hinterland for boat building. Indeed, he may even think: how would he be affected if the fields erode ? or the topsoil of the woodlands floats away ? He can scarcely imagine that the reefs in the sea, which constitute the nurseries for young fish are covered by sedimentation deposits as with a shroud. The farmer, on the other hand, cutting timber in the seemingly useless mangrove swamps or admitting the large-scale construction of ponds for shrimp farming on the common, may just be able to realize that silting-up destroys the breeding grounds for fish, though he himself is not directly affected, but that might cause a change in the micro-climate, which in turn has a negative effect on the amount of rainfall in the region, and thus on this crop, is no longer within his comprehension (Fig. 9.2).

In a theoretical sense co-operation between various factors (Fig. 9.2) in the integrated coastal zone management can be presented as a Jigsaw, which make sense only when it is fitted together correctly (Lohmeyer, 1998). The figure shows that it makes sense first to examine the dependencies of the particular parts—the 'actors', ecosystems, economic dependencies and living habits of people and then to proceed to join the individual parts so as to form a stable

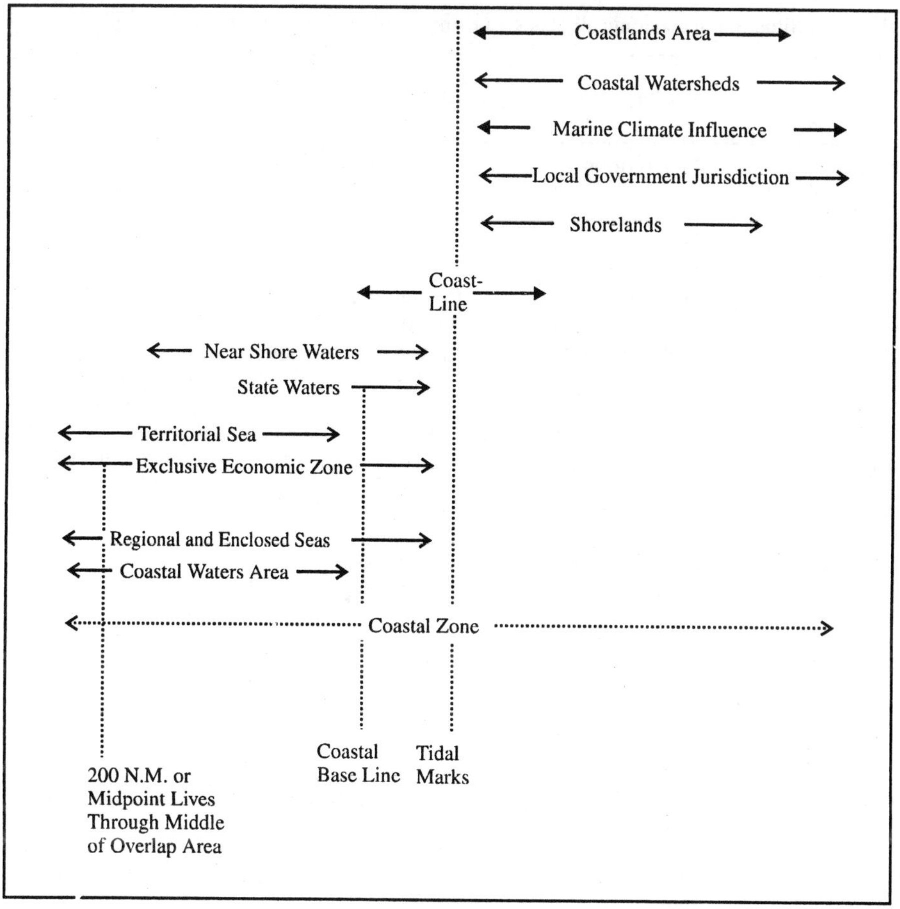

Fig. 9.2 : Relationship between types of Coastal areas and a coastal zone (Sorensen, 1997)

mosaic. The diagram also illustrates three essential characteristics of coastal zone management, which are to secure an optimal utilization of the resources (Table 9.1).

(a) General dependencies on other actors vary, but they are of equal importance for the stability of the biosphere (mosaic),
(b) There is not only one relationship with different actors nor are they limited to the core zone of the coastal region, and
(c) Only the co-operation of all the actors results in the desired stable ecological, economic and social cohesion.

Actually, the core zone is the living area of the coastal inhabitants. In rural areas, they are fishermen, small traders, farmers, hunters and gatherers. They

Table 9.1 : Situation in economically backward-interior (rural) areas in coastal zones

ICM Sector Resource destruction	*Health*	*Effects / Production*	*Aesthetics*
Overfishing	Malnutrition	Production deficits	Loss of biodiversity
Timber felling	Lack of basic nutrition from hunting	Loss of production from forest and wildlife	Loss of natural beauty
Intensive water use	Health-risks from poor quality water	Water shortage through salinization/ pollution	—
Sand extraction	Risks from storm and flood damage	Loss of land through erosion	—
Loss of habitats	—	—	—
Destruction of coral and sea weed	Loss of nutritional base	Decline in fish catches and tourism	Loss of bio-diversity and beauty of natural environment
Transformation of mangrove areas (shrimp ponds, settlement for area)	Loss of nutritional base for hunters and gatherers as well as danger of storm flood	Loss of production from these sectors, lack of coastal production	—
Environmental pollution	—	—	
Water pollution	Poisoned fish	Loss of production area due to salinization	—
Sedimentation	Infected drinking water and danger of epidemics	Loss of production area due to salinization	—
Reduced availability of fresh water	Lack of recreational value/swimming	Decline of aquaculture recession in tourist industry	Dirt and unpleasant smells
Rubbish dump materials	Risk of toxic substances	—	—

(Based on Lohmeyer 1998, Scura *et al.* 1992)

benefits from ICM in many different ways, as may be explained in the usual manner involve the major points like: sustainable growth on the basis of the utilization of natural resources, protection of the natural habitats and the variety of species, controlling the entering of harmful substances and changes to coastal land and beaches, control in the area of water inflow (deposits of sediments and harmful substances), controlling the removal of corals, chances to lagoons and the sea-bed, rehabilitation of destroyed areas and provision of technologies and relational allocation of resources etc. Now two essential prevailing conditions seem to be indispensable for the optimal utilization by an ICM:

(a) Governments should provide legal basic which allows harmful side-effects caused by firms to be regulated by appropriate environmental conditions in accordance with principle that the person who causes damage would wear the coast, so that development efforts of adjacent sectors do not impede one another;

(b) The rights of use must be classified in order to find investors for natural resource utilization; and

(c) The active participation of the beneficiaries and of the population concerned and are regarded as the most important factor for an effective management for sustainable development of the coastal region. Further, the basis of good management must be provided by the traditional habits of the coastal areas inhabitants, their sense of what is right and wrong and traditions of how, when, and against which 'legitimate' background they make use of the resource for themselves. The binding regulation of rights, however, is just as important as the duties which integrated management entails. It would be illusory to rely solely on the ability of states to enforce the necessary regulations. There are too many possibilities for circumventing government regulations, when the realization of one's own benefits does not motivate the observances of the law. Development schemes increasingly realize that responsibility for sustainable utilization of resources can only effectively take place through co-operation (co-management) of all the groups who partake in the benefits (Lohmeyer, 1998).

It seems appropriate to refer Herbert (1997) some important conditions should be taken into consideration if coastal zone management is to be successful. These prevailing conditions include :

(i) Safeguarding the coastal economy,

(ii) Policy working towards an integrated marine and coastal management, national and international co-operation care,

(iii) Protection and rehabilitation of coastal ecology information management,

(iv) Applied research and technological development for resource development education and training in all areas, and

(v) Involvement of the public, of organisations for environmental production and representatives of interests of user groups (co-management), and the like.

In a practical sense, this amounts to a challenge for technical co-operation to contribute by way of collaboration in order to preserve the biosphere, which is threatened by increasing degradation of the environment, biodiversity and

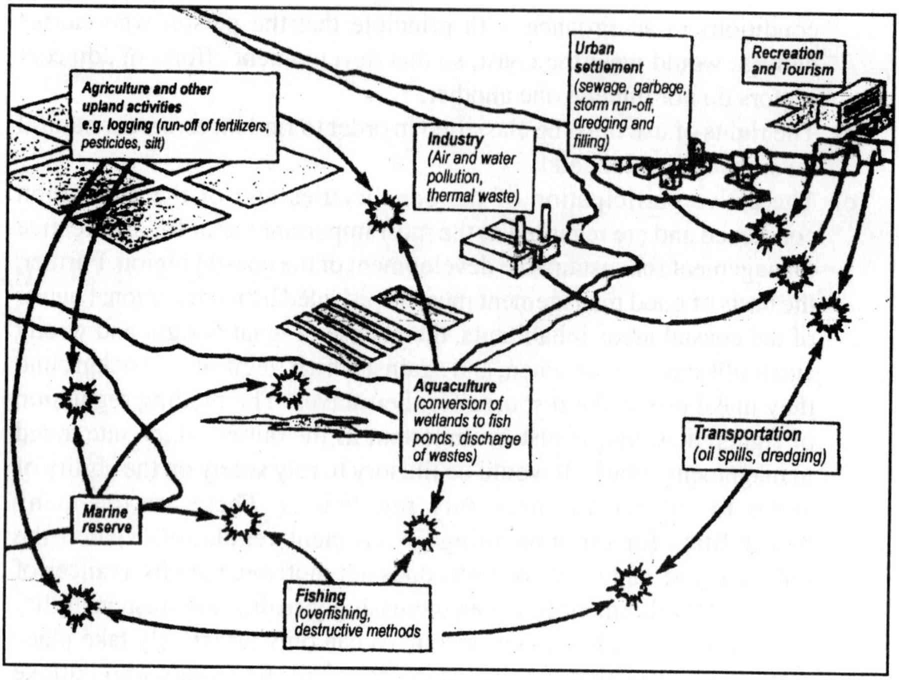

(*Source* : Jain, 2003).

Fig. 9.3 : Schematic diagram depicting Environmental Effects of Development on Coastal Resources.

natural resources, and thus to safeguard an optimal and sustainable utilization for the benefit of often extremely poor inhabitants. It appears that possibilities of employment and livelihood for immigrants lacking any means occur above all in the potential of aquaculture and tourism. The environmental effects of development on coastal resources are illustrated in Fig. 9.3.

It may be said that the use of the coastal zone is as diverse as the interactions, which arise through its utilization. The meanwhile recognised value of the coastal zone for securing nutrition, the preservation of its inherent biodiversity, and the global dimension of the concepts necessary for sustainable utilization, raise hopes that a critical awareness will be generated in the worldwide public as has been the case with the tropical forest, and that those who take the decision in politics, economics, science and technology and population policies will meet in a concerted course of action, which is necessary and indispensable if a successful, integrated coastal zone management is to be established. With these brief discussion on integrated coastal zone management (ICZM) let the salient aspects of Sunderbans, West Bangal be presented in the following pages.

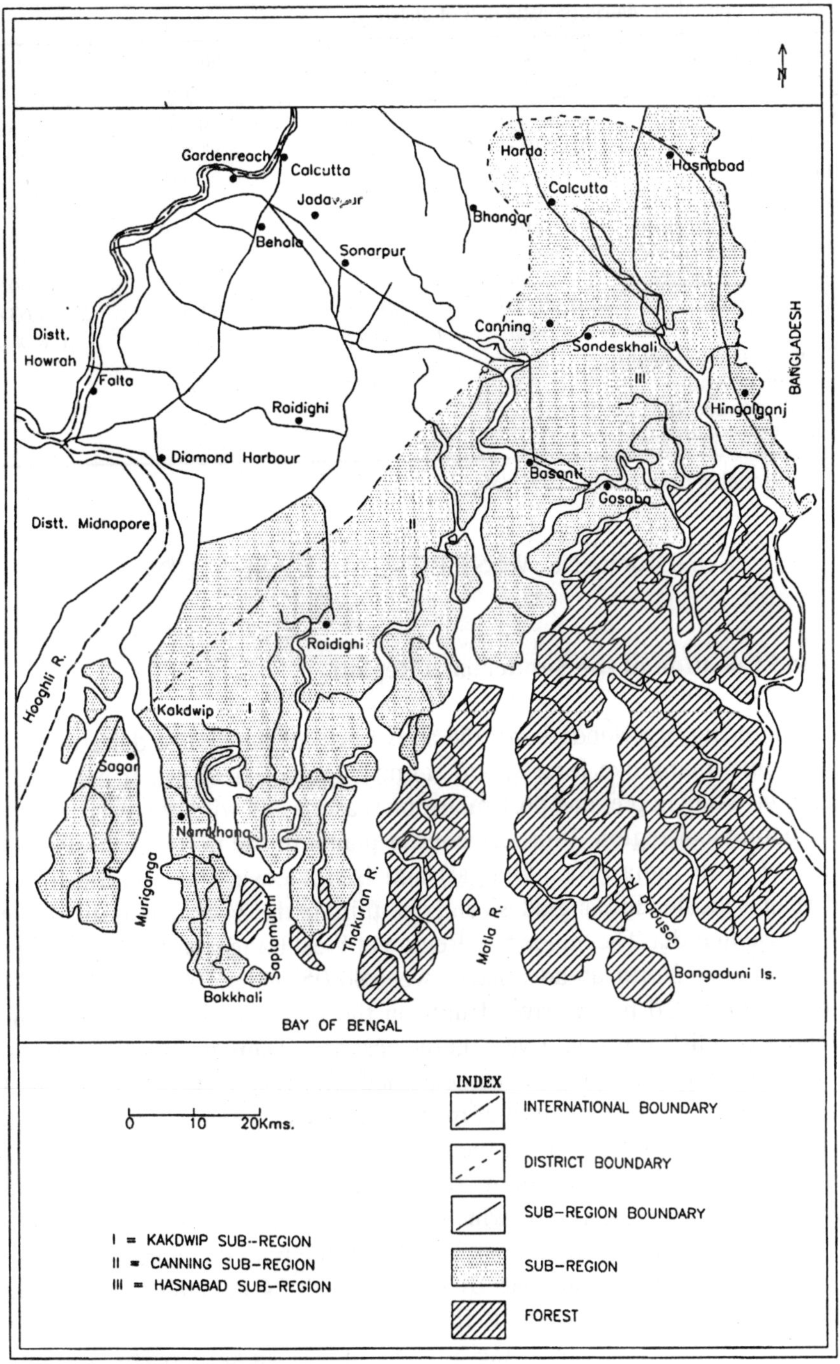

Fig. 9.4 : Map of Sundarbans showing the Mangrove Forest Region and adjacent subregions, West Bengal, India

Table 9.2 : Population Matrix (coastal part of districts)

Sl. No.	*Description*	*South 24-Parganas*	*North 24-Parganas*	*Purba Medinipur*	*Total*
1.	Population	19,46,411	5,26,670	17,09,350	41,82,431
2.	(a) Male	10,00,117	2,71,854	8,81,832	21,53,803
	(b) Female	9,46,294	2,54,816	8,27,518	20,28,628
	(c) Sex Ratio	962	937	838	942
3.	(a) Scheduled Catstes	7,68,982	2,53,239	3,10,120	13,32,341
	(b) Scheduled Tribes	59,245	92,352	4,283	1,55,880
4.	Density of population	590	627	867	694
5.	No. of literable	7,90,765 (39%)	1,90,765 (36%)	10,53,329 (62%)	20,09,637 (48%)
6.	Growth rate per annum based on decennial growth between 1981 and 1991 census (in%)	3.02	3.17	2.36	

Source : Compiled from census, 1991 data.

SUNDERBANS, WEST BENGAL : CASE STUDY

Area and Location of the Sunderbans Region

As shown in Fig. 9.5 Sunderbans, the active face of the Ganga delta with its tidal systems, mangrove forests, soil salinity, lack of fresh water resources, cyclones invading in pre- and post-monsoon seasons and wild life reserves- is a unique ecological region. It covers coastal plains in the north of the Bay of Bengal, of both India (West Bengal) and Bangladesh. Early two-third of Sunderban is located in Bangladesh and one-third in the 24-Parganas district of West Bangal in India. The Sunderban Region of West Bangal is located between 21°12′ and 22°44′ North latitudes and between 80°05′ and 89°00′ East longitude. The region bounded by the river Hugli on the west. Bay of Bengal on the Ichamati-Kalindi-Raimangal rivers on the east and 'Dampier-Hodge Line' on the north. The Dampier and Hodge line named after two Surveyors-General of India delimited the southern limit of mangrove forests and swamps of Mangal delta as existed till 1830-33. Since then to colonisation the forest boundary has retreated substantially to the south. The geographical area of the Sunderban Region is about 9630 sq.km. of which nearly 4185 sq. km. is under reserve forest. Administratively, Sunderbans is a part of the both the North and South 24-Paraganas districts. Sunderban Blocks of North 24-Paraganas district are Minakhan, Haroa, Sandeshkhali IR-II, Hasnabad and Hingalganj (6 Blocks). In South 24 Paraganas District, the Sunderban Blocks are Gosaba, Basanti, Canning-I, Canning-II, Joynagar-I, Joynagar-II, Kultuli, Patharpratima,

Table 9.3 : Recent indicative changes in land use pattern of the districts of South and North 24-Parganas Purba Medinipur

Sl. No.	*Major areas of changes*	*Deltaic plains and coastal block*
1.	Agriculture Land to Pisiciculture	1. Canning I and II , Gosaba, Mathurapur I and II Sagar, Namkhana, Kakdwip and Pathapratima (district South 24-Paraganas) 2. Minakhana, Hasnabad, Hingalgung and Sandeshkhali–I (District Purba Medinipur) Ramnagar I and II, Contai-III, Egra-II, Bagawanpur-II, Potashpur-II (District Purba Medinipur)
2.	Salt Pans to pisciculture	1. Ramnagar-II, Contai-I (District Purba Medinipur)
3.	Mangrove to pisciculture	1. Basanti, Sagar, Patharatima (District South 24-Parganas)
4.	Mud flat to pisciculture	1. Basanti, Kultali, Mathurapur I and II, Patharatima (District South 24-Parganas) 2. Hasnababad, Sandeshkhali I and II, Hingalgung (North 24-Parganas)

Source : Pilot study on Integrated Costal Zone management Plan for west Bengal (July 2000) Central Pollution Control Board, Zonal Office (East), Kolkata.

Namkhana, Sagar, Kakdwip and Mathurapur-II (13 blocks). However, in comparison to the north, its extent is much larger in South 24-Paraganas districts (Lahiri, 1998). The remaining forests of the region are wholely located in South 24-Paraganas. The region had a total population of about 3.1 million in 1991 (Tables 9.2 and 9.3).

Some Proposals for Development

Present author likes to convey the following proposals for development especially in terms of an integrated management after National Biodiversity Strategy and Action Plan (NBSAP), West Bangal (Chowdhury, 2002). These are:

(i) Preparation of data base and regular upgradation particularly in the gap area, and policy formation, and (ii) Floating/encouragement project proposal related to the gap area, and proper monitoring of the role of introduced species due to ballast discharge in the coastal waters. (iii) Establishment of a monitoring committee in which researchers, academicians along with bureaucrats would be members. The main task of this committee would be to see the progress of the projects and identify the zone of overlapping and intersecting between two or more scientific experiments/technology. (iv) Critical identification and analysis of environmental and social issues, and selection of key issues for institutional focus. (v) Adoption of training programme and workshop (particularly in local language) to inject the essence of conservation in the grassroots level people. Arrangement of awareness campaign and enforcement of the ban on the plastic carry bags within the CRZ in particular. (v) Survey of economically important untapped biotic resources (like oyster, seaweed, horseshoe crab for lection and hemolymph etc.) from a professional angle so that their culture can be linked to economic viability and promotion and popularization of high yielding and salt tolerant rice varieties. (vi) Low cost technology to culture oyster and seaweed tough tiles string and rope culture method in an eco-friendly manner needs to be developed. For a horseshoe crab conservation rearing and breeding in enclosures in natural habitats needs to be promoted. (vii) Encouragement to form self-help group particularly in the field of seaweed culture, oyster culture, crab culture and the concept of co-management may be introduced where primary stock holders can interact on the responsibilities rights and roles based on a set of consensus data to be followed by shared implementation and regular evaluation by the community. The entitlements and benefits to be shared should be clearly defined and the procedures for managing the resources should be brought under an institution's framework (viii) Conservation of micro-organisms can be effectively done through cell line culture or establishment genebank. The micro-organisms like cynobacteria can be cultured in mass scale (owing to its nitrogen fixing property) as a source biofertiliser in the coastal zone where chemical fertilisers often become inactive due to high salinity. (ix) (a) Port authority to prohibit discharge of waste water from port based industries (b) Status report on pollunation related activities to be submitted by the lease holder industries in the port area, to the SPCB, (c) Port authority to demand safety and preventive measures from the user industries, (d) Demand for PUC from the vehicle entering the port. DG (Shipping) to ensure pollution control measures by shipping industries using port water during navigation and cargo handing, (e) Rational utilization of coastal land by the port authorities and other port-based industries, and (f) Ballast water treatment plant and green bell channel to be made strictly operational under MARPOL (73/78) norms.

Broadly speaking, harmony between development and conservation is crucial specially for Sundrenbans where human hand has spoiled many of the properties of its unique bio-diversity. Human hand is to repair it and discharge the responsibility adequately. In this regard the guideline including step was adopted at the Rio Earth Summit in 1992, when the following demand to the community of the states was made (Agenda 21, Chapter 17), protection of the ocean, all kinds of seas including enclosed and semi-enclosed seas, coastal area and the protection, rational use the development of their living aquatic resources. This author like to stress much on the integration of the above mentioned activities particularly of all the government departments/Ministry and N.G.O. for total healthy management of the coastal zone environment particularly the Sunderbans region under consideration.

A summary of land and water use changes in the different coastal blocks of South and North 24-Parganas and Purba Medinipur districts is given in Table 9.3.

Conclusion

Since the industrial revolution with increasing population and their ever increasing demand on the natural resources, the human-being are exerting tremendous pressure on this Fragile Ecosystem creating irreversible damage to the system. Realising the importance of this eco-system various scientific policies have been and are being adopted at the national and international level for the development and integrated coastal zone management (ICZM) so that further damage through over exploitation and unplanned development are arrested forthwith. As regard the Sunderban region—it forms one of the largest deltaic eco-system in the proposal climatic zone with luxuriant growth of interracial halophytic vegetation. The preordering delta Environment having numerous channels and carrying voluminous silts and sediments with very good amount of nutrients systems the diverse and rich biological domains here making it globally very important.

Acknowledgements

The author is indebted to professor S.P. Dasgupta, (Retd.) Director, NAMTO, Government of India, Calcutta, Professor Aamlesh Choudhary (Retd.) Head of the Department of Marine Science, Calcutta University for their helpful discussion and advice. He is grateful to Shri D.R.Nandy, (Retd.) Director, Geological Survey of India, and member of the Department of Science and Technology (DST), Government of India, New Delhi his constant encouragement and valuable suggestions during the preparation of this interdisciplinary paper.

REFERENCES

Choudhuri, A.B. and Choudhary, A. (1994) *Mangrove of the Sunderbans*, IUCN the World Conservation Union,Vol. I (India), Dhaka, pp. 4-67, 201-206.

Chodhury, A. (2002) *Coastal Ecosystem, Biodiversity, Strategy Action Plan, West Bangal,* Ramkrishan Mission, Narendrapur, and Department of Environment, Kolkata, pp. 57-67, 80-85.

Jain, A. (2003) *Ecology and Natural Resource Management*, M-Publication, Deharadun. pp. 103-105.

Lohmeyer Uwe, (1998) *Integrated Coastal Management - a new term of a concept of the sustainable utilization of coastal regions.* Animal Research and Development Inst. Scientific Co-operation, Germany, Vol. 48, Tubingen - pp 58-60.

Lahiri, K. (1998), The Sunderbans Region of West Bengal : Environmental problems. *Indian Jr. of Landscape Systems and Ecological Studies*, ILEE, Vol. 21 (2), Kolkata, 54-58.

Naskar, K. P. and Guha Bakshi, N. (1987) *Mangrove Swamp of the Sunderbans - An Ecological Perspectives*, Naya Prakash, Kolkata, pp. 1-264.

Mukhopadhyay, S.C (1996) Hydrogeomorphology of the Brahmaputra- Barak-Manipur Basins with special reference to Neotectonics *Ind. Jr. of Landscape Systems and Ecological Studies, ILEE*, Vol. 19 (2) Kolkata, pp. 136-157.

Sorensen, Jens (1997) National and Internastion Efforts at Integrated Coastal Management: Definition, Achievement, and Lessons. 25th anniversary invited paper, *Journal of Coastal Management,* Jaylore and Francis, Vol. 25, pp. 3-41.

10

Hydro Geomorphological and Arable Land Use Characteristics with the Aid of Satellite Data for Sustainable Development

A Case Study of Jhunjhunu District, (Rajasthan)

Ripunjaya Singh

Introduction

Land use data represent fairly a good deal of varied development activities that take place in a region. The repetitive availability of satellite data for all crop seasons as well as for the desired stages of crop growth, the study of land use pattern, especially its spatial distribution has become easy and could be accomplished in a short time with a meaningful accuracy.

Study of the satellite data of two or more seasons of an agricultural year given an accurate spatial distribution of the arable land use categories such as single crops / double crops (kharif and rabi), fallow lands, culturable waste lands, forest lands, water bodies, habitation etc.

The landforms in the district are alluvial plains, flood plains, eolian plains, sandy plains, buried pediments, pediments, intermontane valley, valley fills, obstacle dunes, structural/denudational hills and linear ridges. Most of the pediments and obstacle dunes are culturable wastelands. Large part of the alluvial plains/flood plains/eolion plains/sandy plains, buried pediments are marginally double cropped, whereas in other landforms, either single cropping is dominant or it remains culturable waste land.

The Study Region

Jhunjhunu district located in between 27°38′ and 28°31′ North latitude and 75°02′ and 76°06′ East longitude, covering a geographical area of 5928 sq. kms. It is bounded on the north-west by Churu district, on the north-east by Hissar and Mahendragarh districts of Haryana state and on the west, south and south-east by Sikar district.

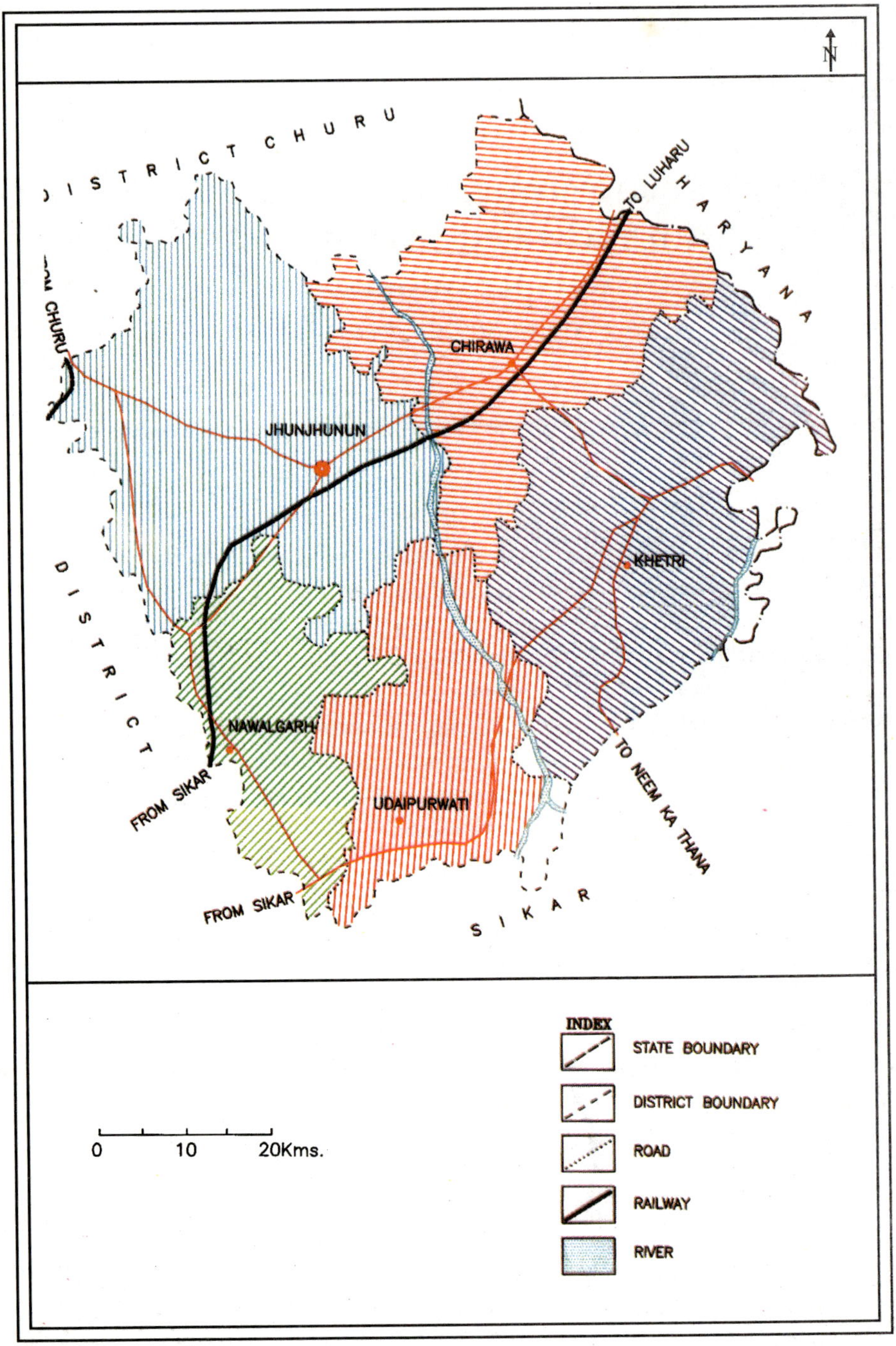

Fig. 10.1 : Administrative Setup—Jhunjhunu District

Administratively Jhunjhunu district is a part of Jaipur division. The district is divided into 5 tehsils, namely (1) Jhunjhunu, (2) Chirawa, (3) Khetri, (4) Navalgarh, (5) Udaipurwati.

The district has a dry climate with hot summers. The average annual rainfall at the district headquarter is 444.5 mm. The summer sand storms are characteristic features of the district. The minimum and maximum temperature recorded are 1°C and 45°C respectively, while mean temperature is 23°C.

The major part of the district is covered by recent to sub-recent wind blown sand, however, the small area of eastern part is occupied exclusively by the rocks of Delhi Super Group and later intrusive. The Delhi Super Group is represented by quartzite, marble, schist and phyllite of Alwar Group and quartzite, mica schist, gneiss assemblages of the succeeding Ajabgarh Group. The rocks have been intruded by amphibolite, granite and rhyolite which are considered to have provided the material for the base metal mineralization in the area.

The district is well known as copper district in the country. Copper deposits are located at various places in the district, specially in Khetri tehsil, silver, gold, nickel and cobalt are accessory minerals.

North-eastern part of the district consists of mass rolling of hills and its remaining part near to the south-eastern border contains some offshoots of Aravalli range of hills, running in the south-eastern direction. General elevation above mean sea level is between 300 to 450 metres.

The drainage is inland and is related to the Kantli river system which is a non-perennial river. Beside this, there are four major streams, namely—Dohan, Chandravati, Lohagarh-ki-nadi and Sukh nadi.

Jhunjhunu district is sandy with frequent rolling topography soils. They are brown to reddish brown loamy fine and generally non-calcareous and deep to very deep. Physio-chemical properties of these soils are generally moderate in nutrient status and poor to moderate in water holding capacity. These soils are classified as Torrifluvents great group of Entisoils order. In eastern part of the district, soils are brown, fine sand to loamy sand, deep, non-calcareous and belong to Eutrisol order. The forest area of the district is only 34,332 hectares, i.e. 5-80 per cent of the total area of the district.

Method and Data Base

The primary source of data used in the present study is Survey of India toposheets on the scale of 1:50,000. These were used for preparation of base maps on 1:50,000 scale which were further reduced. The aid of Geo-coded Satellite data 1:50,000 scale of 2 main crop seasons, i.e. Kharif October 1998 and Rabi February, 1999 of IRS-ID LISS-II FCC imagery were used visually, recognition

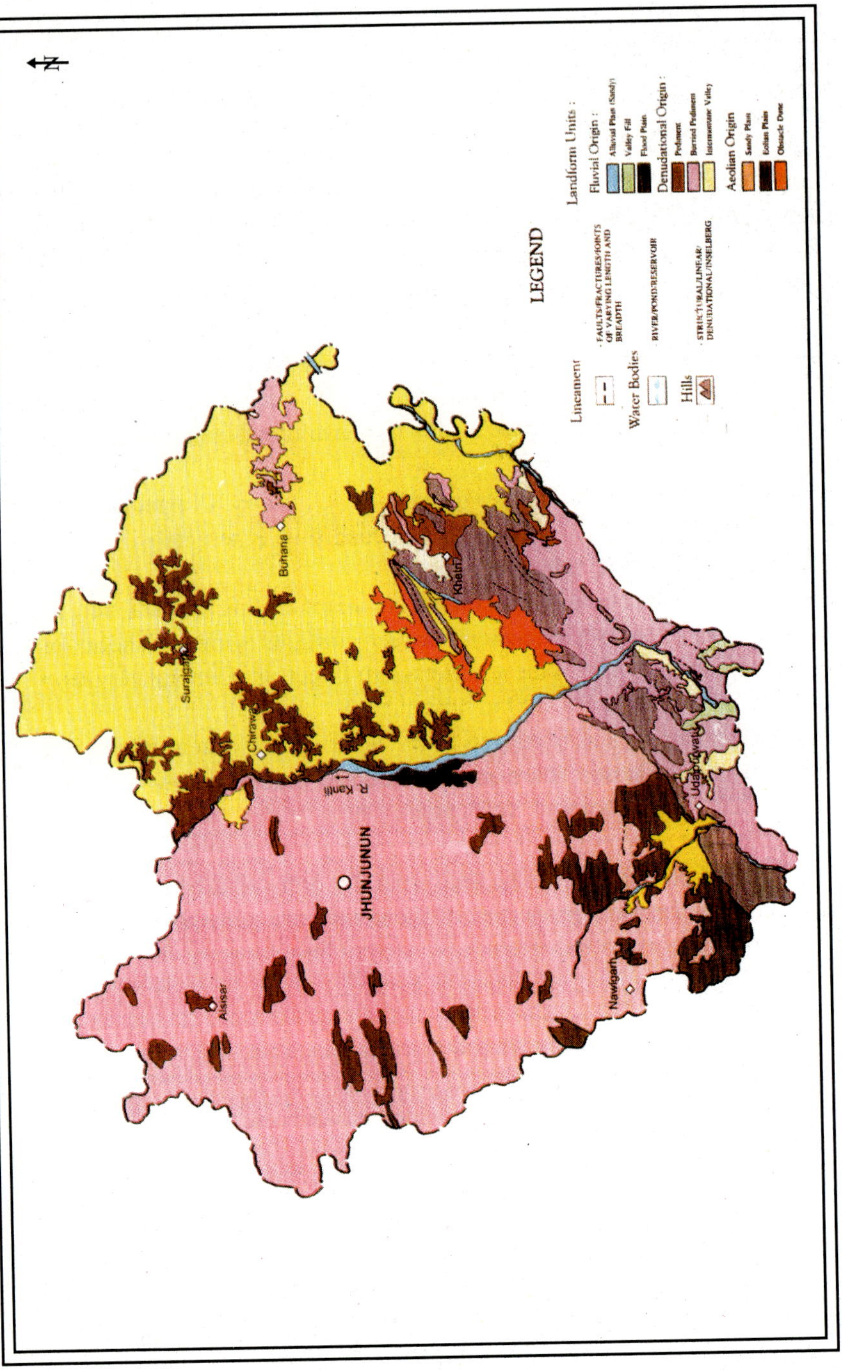

Fig. 10.2 : Jhunjhunu District—Geomorphology

elements such as tone, texture, pattern, shape, size, location etc. and Geo-technical elements such as landforms, drainage vegetation etc. Different hydro geomorphological units were identified and mapped to prepare hydro geomorphological map. Similarly spatial distribution of different identified land use categories were also identified.

Secondary data were collected from agriculture, groundwater and irrigation department for additional support.

The hydro geomophological characteristics and arable land use prepared with the interpretation of satellite data is superimposed to suggest scheme for optimum utilization of land units.

Inerpretation

Relationship between geomorphological units and arable land is given in Table 10.1.

Table 10.1 : Relationship between geomorphological units and arable land

Landform Unit (geomorphology)	*Occurrence, Lithology and Material*	*Land use/Land cover*
FLUVIAL ORIGIN		
Alluvial Plain (Sandy)	East of river Kantli, around Pilani, Bamanwas, Jetpur habitations. Flat to gentle undulating plain formed due to fluvial activity, mainly consists of gravels, sand, silt and clay with unconsolidated material of varying lithology, predominantly sand.	Double crop, single crop (Kharif/Rabi), open scrub, fallow.
Valley Fill	Near Khetari, Mandawara, Chirana towns. Formed by fluvial activity, usually at lower topographic locations, comprising of boulders, cobbles, pebbles, gravels, sand, silt and clay. The unit has consolidated sediment deposits.	Double crop area.
Flood Plain	West bank of river Kantli near Hansalsar, Rharunda village. The surface or strip of relatively smooth land, adjacent to river channel, formed by river and covered with water when river overflows its bank. Normally subject to periodic flooding.	Double crop, single crop (Kharif/Rabi), fallow.
DENUDATIONAL ORIGIN		
Pediment	Mainly in south-east part near Chironi, Manor villages. Broad gently sloping rock floor, erosional surface of low relief between hill and plain, comprised of varied lithology, criss-crossed by fractures and faults.	Marginal kharif crop, open scrub, fallow.

(Contd...)

Table 10.1 *(Contd.)*

Burried pediment	South and south-east around Udaipurwati and Ponkh village. Pediment covered essentialy with relatively thicker alluvial, colluvial or weathered materials.	Double crop, single crop (Kharif/Rabi), fallow land with or without scrub.
Intermontane Valley	Scattered in south and south-east hilly region of district. Depression between mountains, generally broad and linear, filled with colluvial deposit.	Double crop, single crop (Kharif/Rabi), fallow land with or without scrub.
AEOLIAN ORIGIN		
Sandy Plain	West of river Kantli, covers almost half of district. Formed by aeolian activity, wind blown sand with gentle sloping on undulating plain, comprising of coarse sand, fine sand, silt and clay. Generally cultivated.	Single crop (Kharif), fallow, open scrub.
Eolian Plain	Scattered uniformly in entire district. Formed by aeolien activity, with sand dunes of varying heights, size, slope, Long stretches of sand sheet. Gentle sloping flat to undulating plain, comprised of fine to medium grained sand and silt. Also scattered xerophytic vegetations.	Single crop (Kharif), fallow land with or without scrub.
Obstacle Dune	West of Aravali ranges. Formed on windward/leeward sides of obstacle like isolated hills or continuous chain of hills, due to obstruction in path of sand ladden winds. Badly dissected well cemented and vegetated.	Fallow land with or without scrub.
HILLS		
Structural Hills	In south and south-east part (Aravalli ranges). Consists of varying lithology with folding and faulting, linear to accurate hills, showing definite trend lines.	Forest, mining, open scrub.
Inselburg	Near Jhunjhunu town. Residual isolated hill, usually smooth and round rising abruptly above the surrounding pediplain, generally barren rocky, partly buried by debris.	Barren with negligible scrub.
Denudational Hills	South-east part (Aravalli ranges). Steep sides, relict hills undergone denudation, comprising of varying lithology with joints, fractures and lineaments.	Forest, mining, open scrub.

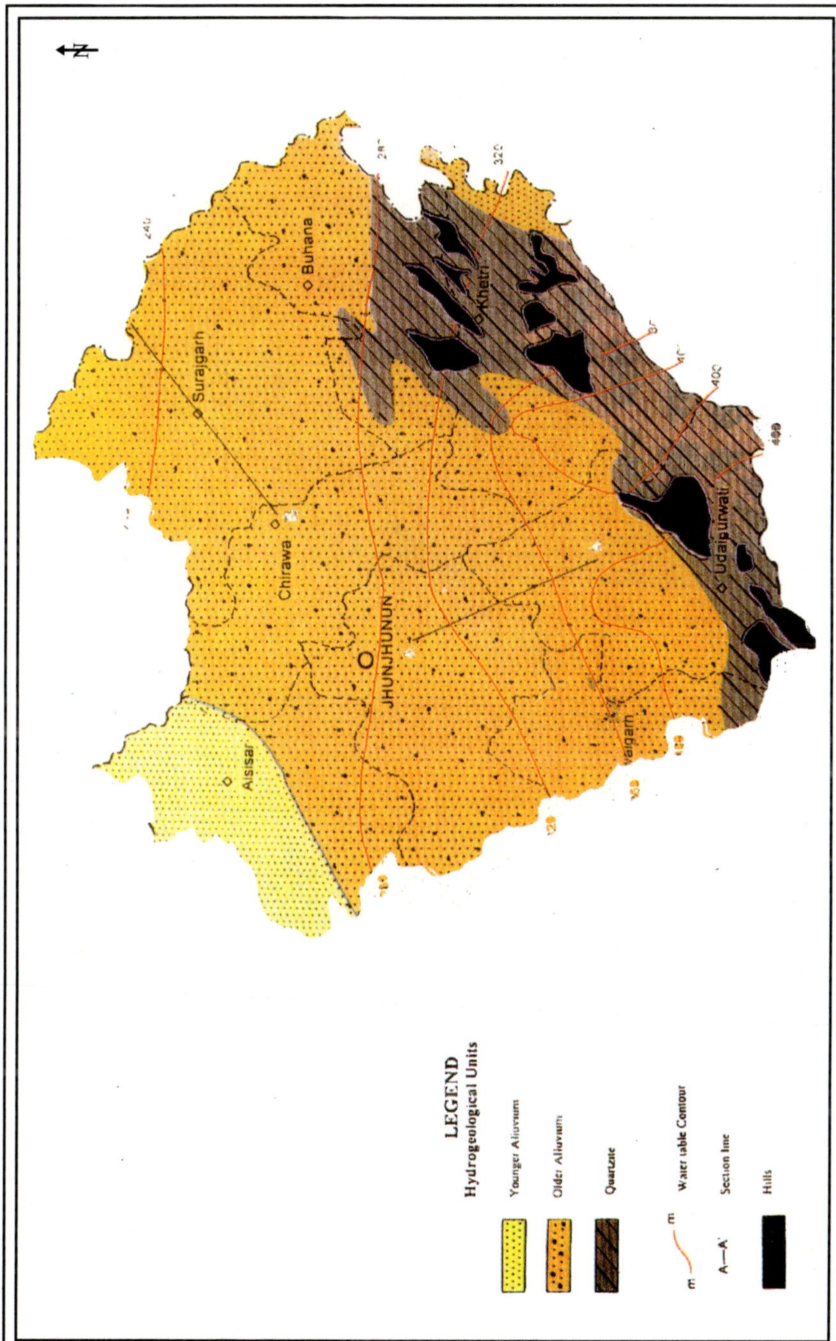

Fig. 10.3 : Jhunjhunu District—Hydrology

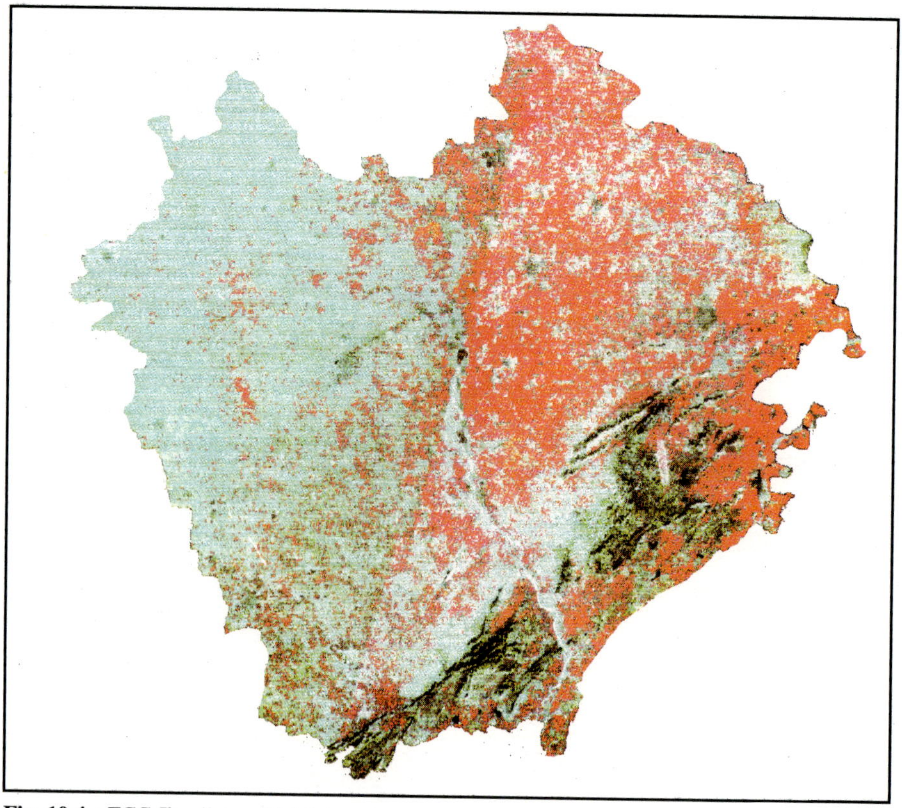

Fig. 10.4 : FCC Jhunjhunu District

Hydrogeology

Hydrogeological unit	*Description of the unit/Geological section*	*Occurrence*
Younger Alluvium (Quaternary)	It comprises unconsolidated fine grained sand and stream laid deposits; clay, silt, gravel, pebbles, boulders etc. Thickness of the litho unit varies upto 50 m.	The litho unit occupies flood plains and drainage courses of river Kantli and Dhon rivers and their tributaries. Udaipurwati, Chirwa and Khetri blocks. Younger alluvium cover nearly 6 per cent potential area.
Older, Alluvium (Quaternary)	It is heterogeneous assemblage of unconsolidated to loosely consolidated sand, silt, clay and *kankar* which at places contains lenses of coarse sand and gravel. Cross section in the north western part reveals maximum thickness of about 100 m. Other sections show alluvium thickness between 80-100 m.	The litho unit cover major part of the district. South-eastern hilly region comprising Khetri block and small pockets in Buhana, Nawalgarh and Udaipurwati blocks are exception, where hard rock occupies the area. Older alluvium encompasses nearly 80 per cent potential area.

(Contd....)

(Contd....)

Quartzite, Schist, Phyllite, Limestone and Granite (Post Delhi Intrusives and Delhi Super Group)	Post Delhi intrusives mainly comprise amphibolite, granite, pegmatites and quartz veins. Granite is of dark grey coloured, coarse grained very hard and compact and at places jointed. Schist and phyllite are well foliated, cleaved metamorphic rocks of green to dark grey coloured. Quartzite is pink to brown coloured metamorphic rock.	These are mainly confined to south-eastern region in part of Khetri, Udaipurwati, Buhana and Nawalgarh blocks. Quartzite cover nearly 14 per cent potential area.

Ground Water Flow

Ground water flow follows surface drainage pattern. In major part of the area, ground water flow has been inferred south to north. Hydraulic gradient is more in south and south-eastern part and gradually decreases northward. In southern part Kantli river is effluent while in northward, it becomes influent.

Water Level Trends

Depth of Water Level		*Change in Wather Level (1984-1997)*	
Range in Metres	*Area*	*Range in Metres*	*Area*
< 20	Part of Khetri and Udaipurwati blocks situated in south-eastern part of the district has shallow water level less than 20m.	0 to 3	A pocket northwest of Alsisar show rise in water level within the range.
20 to 40	Major part of the area west of Kantli river and some pockets in eastern part has depth to water level between the range.	0 to 3	Area between Alsisar-Mandawa Jhunjhunu located in western part of the district exhibit marginal depletion water level between the range.
> 40	Area around Mandawa and northern part around Chirasa extending upto northern boundary has deep water level more than 40 m.	-3 to -7	Major part of the district, leaving aside south-eastern peripheral area extending upto Chirawa and area west of Jhunjhunu, show depletion in water level between the range.
		-7 to -15	South-western peripheral area between Nawalgarh and Buhana extending upto Chirawa in central part exhibit steep depletion in water level between the range.

Scheme for Theme Integration

S. No.	*Land Form*	*Ground-Water Prospect*	*Present Land use*	*Suggested Utilisation*
1.	Alluvial plains	Good	Double cropped area and marginally double cropped	Efficient utilisation of groundwater recommended
2.	Alluvial plains	Good	Single cropped area	Agro-horticulture/Intensive agriculture with efficient use of groundwater
3.	Alluvial plains	Good	Fallow	Agro-horticulture
4.	Alluvial plains	Good	Culturable wastelands	Fodder and fuel wood plantation
5.	Flood plains	Good	Double cropped area and marginally double cropped	Efficient utilisation of groundwater recommended
6.	Flood plains	Good	Single cropped area	Agro-forestry
7.	Flood plains	Good	Fallow	Agro-forestry
8.	Flood plains	Good	Culturable wastelands	Fodder and fuel wood plantation
9.	Eolian plains/ sandy plains	Moderate	Double cropped area and marginally double cropped	Efficient utilisation of groundwater recommended
10.	Eolian plains/ sandy plains	Moderate	Single cropped area	Agro-forestry
11.	Eolian plains/ sandy plains	Moderate	Fallow	Agro-forestry
12.	Eolian plains/ sandy plains	Moderate	Culturable wastelands	Fodder and fuelwood plantation/silvipasture
13.	Obstacle dunes	Poor	Single cropped area	Agro-forestry
14.	Obstacle dunes	Poor	Fallow	Agro-forestry
15.	Obstacle dunes	Poor	Culturable wasteland	Fodder and fuelwood plantation
16.	Buried pediment	Moderate	Double cropped area and marginally double cropped	Efficient utilisation of groundwater recommended/ Agro-horticulture
17.	Buried pediment	Moderate	Single cropped area	Agro-forestry
18.	Buried pediment	Moderate	Fallow	Agro-forestry
19.	Buried pediment	Moderate	Culturable wasteland	Fodder and fuelwood plantation
20.	Pediment	Poor	Single cropped area	Agro-forestry
21.	Pediment	Poor	Culturable wasteland	Fodder and fuelwood plantation
22.	Intermontane valley/valley fill	Moderate to good	Double cropped area and marginally double cropped	Efficient use of ground water recommended
23.	Intermontane valley/valley fill	Moderate to good	Single cropped area	Intensive agriculture with efficient utilisation of groundwater/agrohorticulture
24.	Intermontane valley/valley fill	Moderate to good	Culturable wasteland	Fodder and fuelwood plantation
25.	Structural hill/ denudational hill/linear ridges	Poor	Culturable wasteland	Fodder and fuelwood plantation

REFERENCES

Agriculture Statistics of Rajasthan. Deptt. of Agriculture. GOR 1998.

AGI (1972) Margaret Gary, Robert Mc Afee Jr. and Carol L. Wolf (editors) *Glossary of Geology*, American Geological Institute, Washington, D.C.

Anon. (1988) Preparation of Hydrogeomorphological Maps of India on 1:250,000 scale using satellite imagery; Project Report, Deptt. of Space, Banglore.

Bhu-jal (1991) Special issue on Ground Water Statistics, *Bhu-jal News*, Vol. 5, No. 4. October-December, 1990.

CGWB (1994) Hydrogeological Atlas of Rajasthan, Central Ground Water Board, Ministry of Water, *Resource*, Govt. of India.

Distt. Census Handbook. Jhunjhunu (24) Cencus of India 2001.

Ground Water Atlas of Rajasthan SRSAC. Deptt. of Science & Technology GOR Jodhpur, 1999.

Ground Water Resource Atlas of Rajasthan Ground Water Department GOR Jodhpur, 1997.

Parker, Sybil, P. (1988) (Editor in Chief) *McGraw Hill Encylopaedia of the Geological Sciences*, 2 Edition, McGraw Hill Book Company, New York.

Rajasthan (1991) *Some Facts about Rajasthan 1991*. Directorate of Economic and Statistics, Rajasthan, Jaipur.

SAC (1986) Report on Ground Water Potential Maps of Rajsthan (based on visual interpretation of Landsat Thematic Mapper Data). SAC Ahmedabad, July 1986.

11

Aestheticism and Landscape—A Few Issues Concerning the SCENIC Evaluation of Landforms for Recreation

A Study of Khammam District

S.PADMAJA AND G. SOMIAH

From immemorial times, the landscapes have not only provided a virtual bliss to humankind in terms of material values but also have provided suitable and amicable natural environment for seeking solace and recreating oneself. The impact of natural surrounding on human mind is phenomenal and exemplary. No wonder that environmental geographers have become aware of this potentiality of landforms as the main recreational sources or avenues and have accordingly named them as either recreational parks or natural museums and so on and so forth. Once the relevance and importance is known of a landscape, it becomes important to know the methodology of evaluating the landscapes.

Landscapes—an agglomeration of various kinds of landforms have their own relevance and connotation in various environments. They vary from the most rugged mountainous terrain to the most lowlying coastal lands, both are equally fascinating and infatuating. However, the landforms existing in-between are also equally eye catching and enormous. They range in their variety from a few gently sloping hillocks, escarpments, plateaus, cuestas to vales, rolling lands etc. The variety of landforms is as exhaustive as landforms themselves. No wonder that each land unit exhibits its own unique ID. Correspondingly, the effect and impact, which they can exhibit, is also equally as unique as the variety itself, but the most common impact is in terms of scenic appeal. Therefore, the pertinent question is how to evaluate, assess and quantify the characters that exhibit scenic value. For this, a number of evaluation techniques have been worked out comparatively.

The studies pertaining to landscape evaluation for recreation purpose are scarce. In India, the scenic evaluations of landforms has not been done for consumer level, but have only been assessed in terms of their land capability, in North America and Europe a few works do exist.

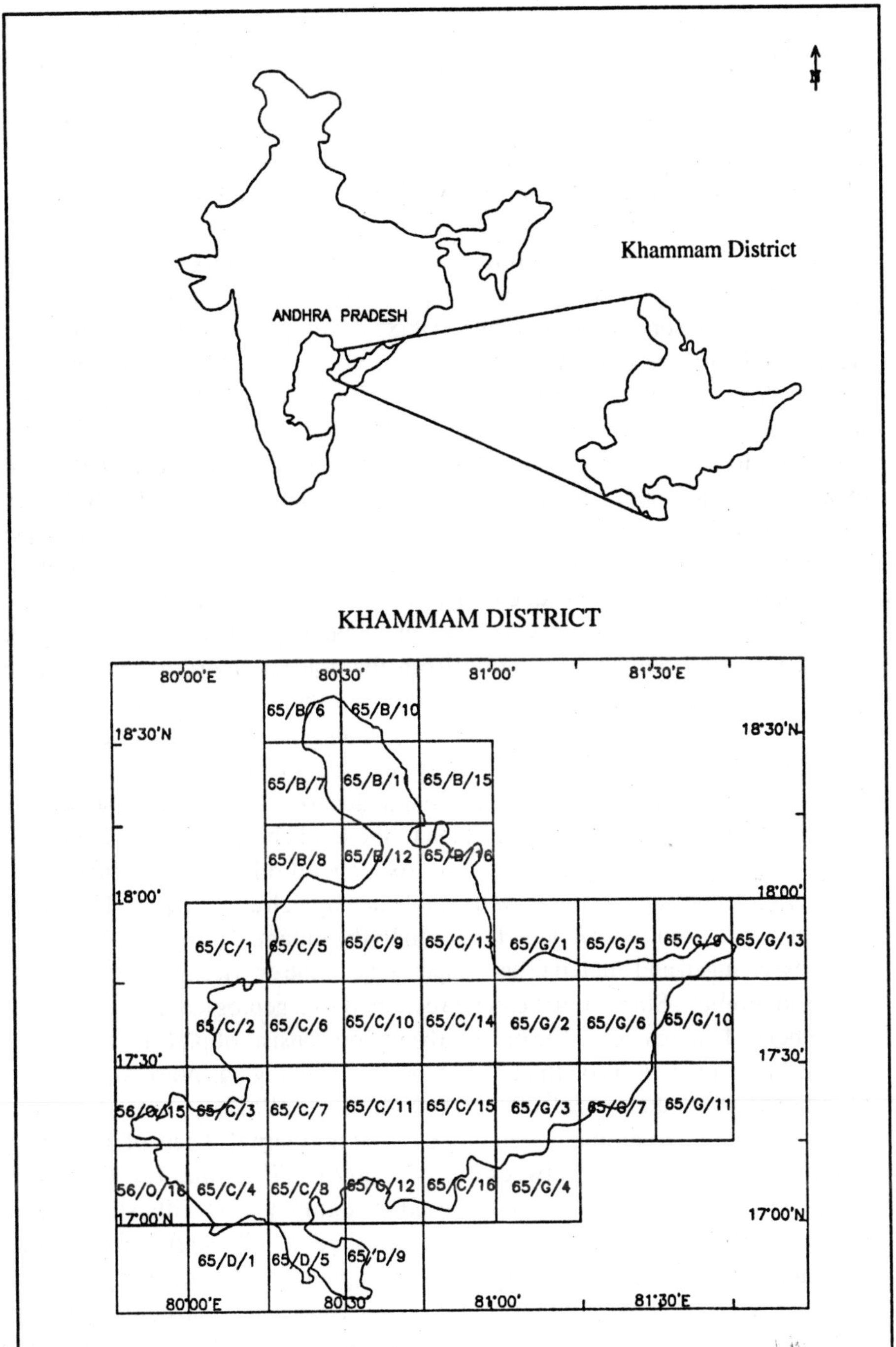

Fig. 11.1 : Location Map of Khammam District

Linton (1968) developed a technique for the assessment of scenic resources in Scotland that provided an aerial classification of landscape. The two categories of sceneries are Landform Landscapes and Land use Landscapes. Leopold (1969) attempted to quantify aesthetic features of the proposed site, together with those of eleven other possible dam sites in Idaho and sites of renowned beauty elsewhere in US. Leopold's approach was founded on the assumption that there is some benefit to society from the existence of 'unchanged' landscapes. A unique landscape is more useful to society than a common one, and unique qualities enhancing landscape value are those having some aesthetic, scenic or human-interest connotations. Fines (1968) developed a survey technique in which subjective responses to view are classified on a single, comprehensive and predetermined scale and after converting these to land surface values, landscape evaluation map is produced. The scale was produced using the responses to a set of photographs. Linton (1968, 1973) developed a method for evaluating landscape. Shafter (1969, 1970), Isachenko (1973), Chubb and Bauman (1977), Smith and Small (1980) developed a set of evaluative techniques to derive a scenic map (Padmaja, 1998) attempted scenic evaluation through a combination of select technique to derive urban recreation centres.

India is a unique country with a set of land systems, representative of contrasting environments—ranging from the coldest environment to that of hottest environment and at the same time from the wettest patch to the driest patch. All these variations in the climate will automatically lead to changes in the forms of the land and hence a variety of them to pick and choose from. In fact every region in India has a uniqueness added to its landscape, amalgamated by cultural ethos and variety. This adds to the total character of the landscape.

Andhra Pradesh is no exception to this general rule and the state exhibits rich variety of physical features well established by rich cultural past. The variety of landscapes is quite well depicted through the untamed terrain of Adiabad to the most changed and altered urban scapes. One of the districts, namely Khamma (Khammammett, Stambagiri) can be reckoned to a geological museum. Here, a number of features exist. The bulk of the region consists of plateaus, developed on the Archean gneissis on the north. However faulting has preserved a belt of Gondwana along the lower Godavari trough (Spate, 1967).

Study Area : Khammam, being one of the northern districts of Telangana, has a geographical extent between 79° 47′ and 81° 47′ (E) and 16° 45′ and 18° 35′ N. (Fig.11.1). This region has varied topography with hills stretching in central and eastern parts of the district. A range of hills stretch from Parkal (Warangal district) and Singareni to Ashwaraopet in the SE bounding the lower Godavari Valley. The most important hills of the district are Marrigutta, Rajagutta and Yerragutta. The plain land includes mandals of Bhadrachalam, Borgam, Kothagudem, Pudimalla, Yellandu and Ashwaraopet with patches of deciduous

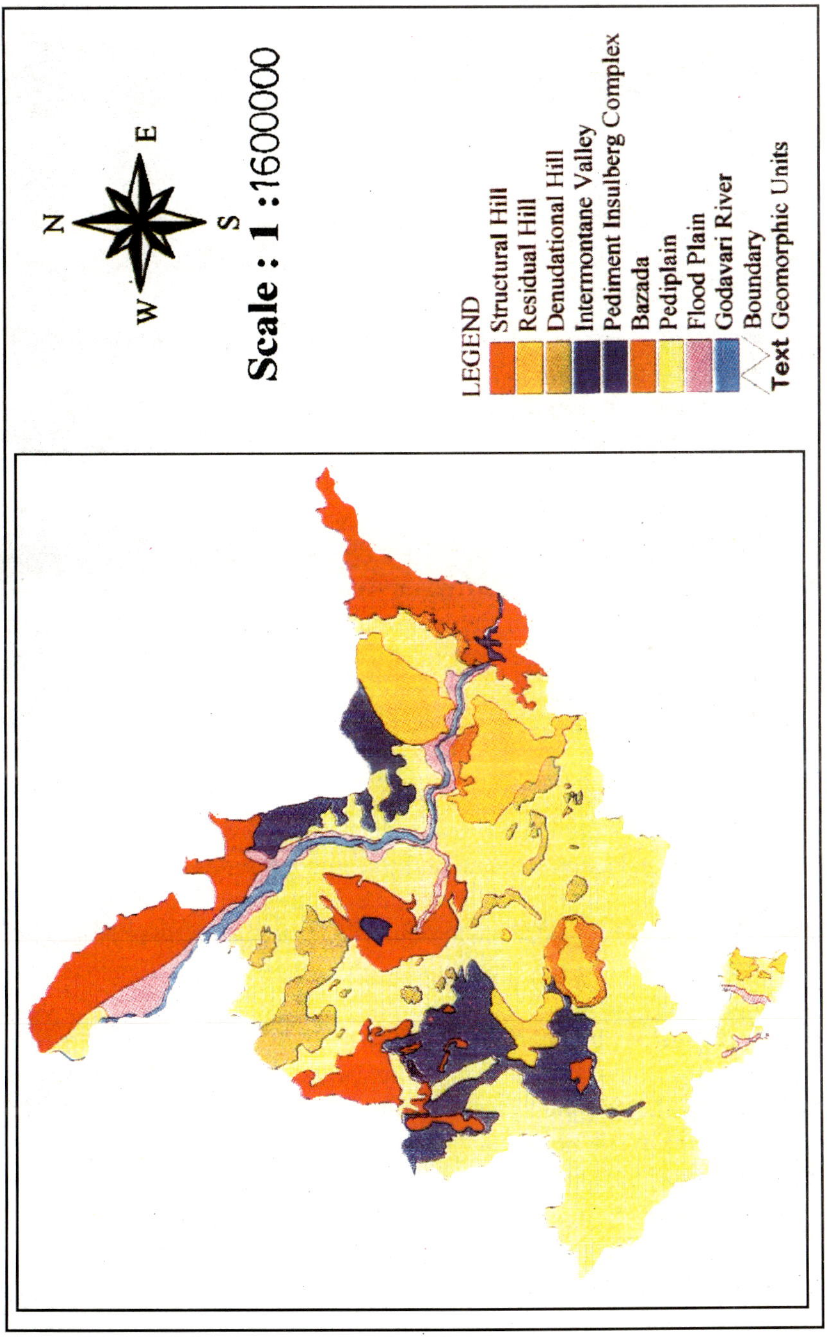

Fig. 11.2 : Khammam District—Land System Iso-Region Map

forest. The study area accordingly is demarcated into following categories of landforms arranged in a system (Fig. 11.2), namely

I. HILL SYSTEM
 (a) Structural (S)
 (b) Residual (E)
 (c) Denudational (E)
 (d) Valley (S)
II. PEDIMENT SYSTEM
 (e) Pediment Inselberg Complex (E)
 (f) Bajada (D)
III. PLAIN SYSTEM
 (A) Pediplain (E)
  (i) Shallow Weathered
  (ii) Moderately Weathered
 (B) Flood Plain (D)
  E = Erosional, D = Depositional, S = Structural

The main rivers which flow through this district are the Godavari, Sabari Kinnerasani, Munnar, Paler, Akal and Wyra. Approximately 64 per cent of the geographical area is covered with the forests within the altitude ranges from 250′ to 2200′ above sea level. The geoglobal formations range from archean to subrecent to recent comprising of quartzites, schists, gneisses, phyllites, slates, marbles, limestone, sandstone, boulder beds, shales and alluvium.

The objectives of the study are well founded which are as follows:

Firstly to identify the landforms of Khammam district.
Secondly to analyse and quantify the scenic attributes of landscape.
Thirdly to apply Leopold's, Fine's site evaluation technique to evaluate consumers' value of landscape through enquiries, maps and photographs and lastly to prepare landscape evaluation map of regional recreation.

These objectives are built on the hypotheses that are to be tested in the region which are to quote—Firstly, Unused landscapes are of more recreational value than the used landscapes. Secondly, the landscape attributes and water qualities have more positive impact on recreational landuse. Thirdly, the arrangement of density of flora with respect, to slope aspect has a positive imprint on recreation.

For all this to be achieved, the data is derived mainly from

(a) SOl toposheets of (560, 658, C, D and G series on l:50000 Scale).

(b) Imageries IRS lC/Landsat -TM
(c) Landuse Maps
(d) Reports and Memoirs
(e) Records
(f) Field observation and reports

Methodology

The methodology adopted is exhaustive as well as interesting. Evaluative techniques of landscape falling between two extremes of measurable components of landscape deemed to be representative of scenic quality are evaluated quantitative or semi-quantitive terms. At the other extreme, the emphasis is on consumers' perception of scenic quality.

The scenic beauty of landscape evaluation of the site is measured on the basis of landforms. Morphometric, Non-morphometric, landuse/cover and landforms supplement the evaluation of recreation sites. 25 km^2 grids are chosen representative of all land systems and landforms in the region which are identified for their uniqueness value. This is obtained by giving a weightage to a morphometric attribute in a grid by giving the values as 1, 2 & 3 which are Low, Medium and High respectively. The individual weightages of the grid are calculated by dividing the weightage value by its total number of occurrences which means that if a particular value of an attribute is occurring in only one grid, then the grid is unique, whereas if many grids share the same value, then the uniqueness of the attribute is lost.

A few morphometric attributes having scenic relevance selected for the study area (a) Average Slope, (b) Relative relief, (c) Intensify of relief, (d) Dissection Index, (e) Ruggedity, (f) Contour crenulation, (g) Isolation of hills from neighbouring area, (h) Width of valley, (i) Stream Order, (l) River Pattern and (m) Vegetative Index.

The non-morphometric attributes are basically those which do show relevance but need not essentially be numerical in their assessment. These are waterfalls, lakes/reservoirs, accessibility, nearby settlements, mines, sanctuaries, religious sites and historical sites. Unit weightage value is given for each attribute. If the attribute exists in more than one selected grid, the assigned unit value for each grid are added up or if a particular grid has an occurrence of more than one attribute which is more or less an usual case, the unit values assigned for each grid are added up. Ultimately, the higher the grid value, the more is the recreation potential. Thirdly, the number of landforms existing in the region as depicted through profiles are counted. Subsequently, if a particular grid possesses more number of landforms, more weightage is given, because of

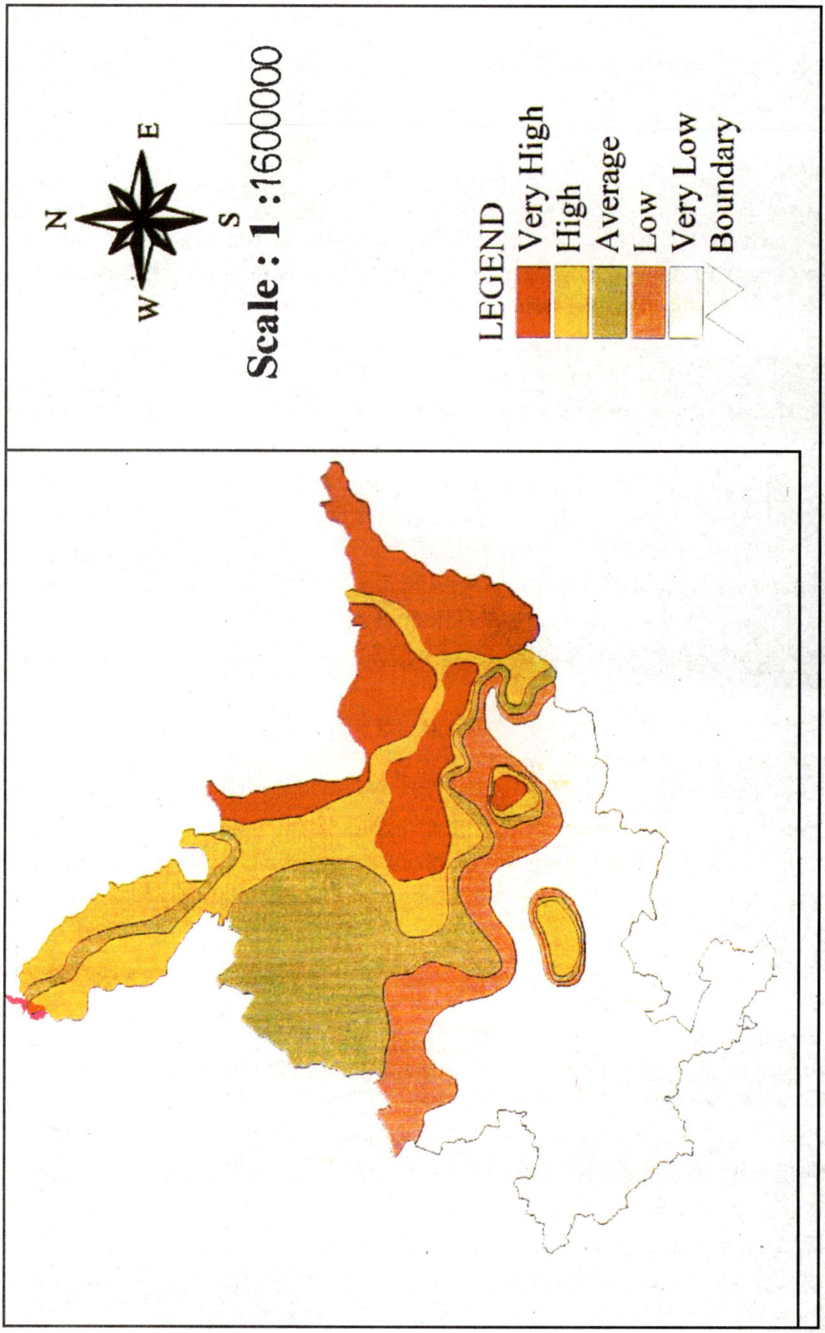

Fig. 11.3 : Khammam District—Morphometric ISO-Region Map

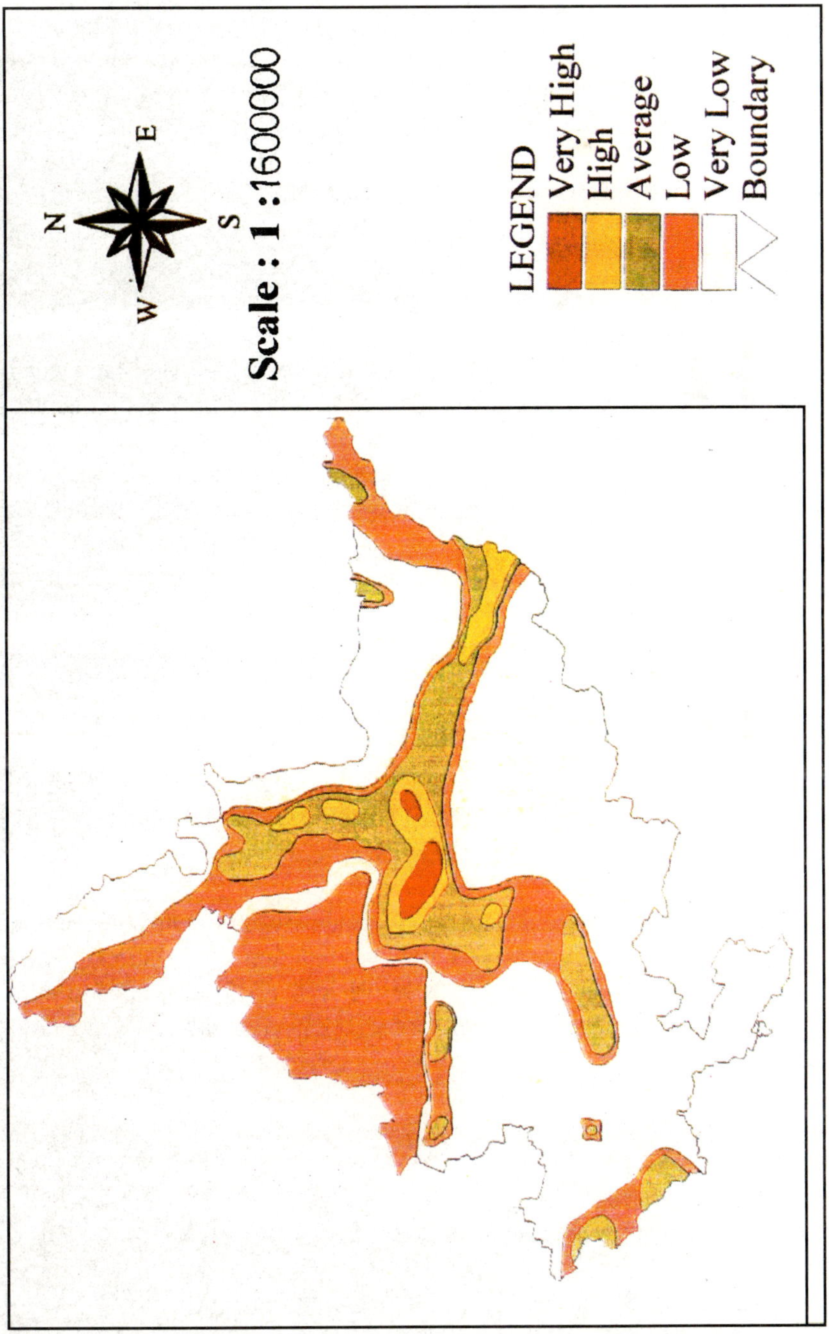

Fig. 11.4 : Khammam District—Non-Morphometric ISO-Region Map

the fact that each landform is assigned a unit value. Fourthly, landuse/landcover obtained from IRS IC data is classified into three major categories. Namely forestry and water bodies as category one, wasteland and agriculture land as two and three respectively. The weightages as per the tree cover are given as follows. Forest-3, Wasteland-2 and Agriculture-1. The percentage of landuse/ land cover is calculated with the area covered in each grid with respect to its weightages. For example, in a grid, if area covered under forest is 40 per cent, wasteland is 20 per cent, agriculture is 40 per cent, then it is calculated as follows :

Lastly, photographs of the selected site were circulated to consumers to get their response towards the site as a possible recreation spot. The photographs were rated according to a scale ranging from 1-5, which corresponds to poor to excellent.

This methodology is carried on for 44 grids for which the data is generated representative of the various land systems pertaining to the region. Subsequently, all the weightages are added up, and a cumulative weightage is taken into consideration, from which Iso-regions are demarcated on a range basis with designations as very good, good, average, low and very low. With the help of this classification scheme scenic evaluation map is prepared.

Discussion and Analysis

Aesthetics and landforms seem to go hand in hand. Therefore, scenic value and recreation can match each other as landscapes can be counted as potential sites for recreational purpose. Based on the methodology described earlier, scenic attributes are selected as morphometric (13), non-morphometric (8), landuse (3), landform (8) and lastly photographic perception (41) (Tables 11.1-11.6). The selected morphometric variables are listed as below :

> (a) Average Slope (b) Relative Relief (c) Intensity of Relief (d) Dissection Index (e) Ruggedity Index (f) Contour Crenulation (g) Isolation of Hills from neighbouring area (h) Valley Width (i) Stream Frequency (j) Drainage Intensity (k) Stream Order (l) River Pattern (m) Vegetation Intensity.

Morphometric Scenic Evauation

The 13 variables have been given uniqueness values as discussed in methodology. The total morphometric weightages composited or added have given rise to a minimum of 0.9 to a maximum of 4.47 (Table 11.1). The Iso-morpho map (Map 11.3) depicting the same values is distributed accordingly. The high

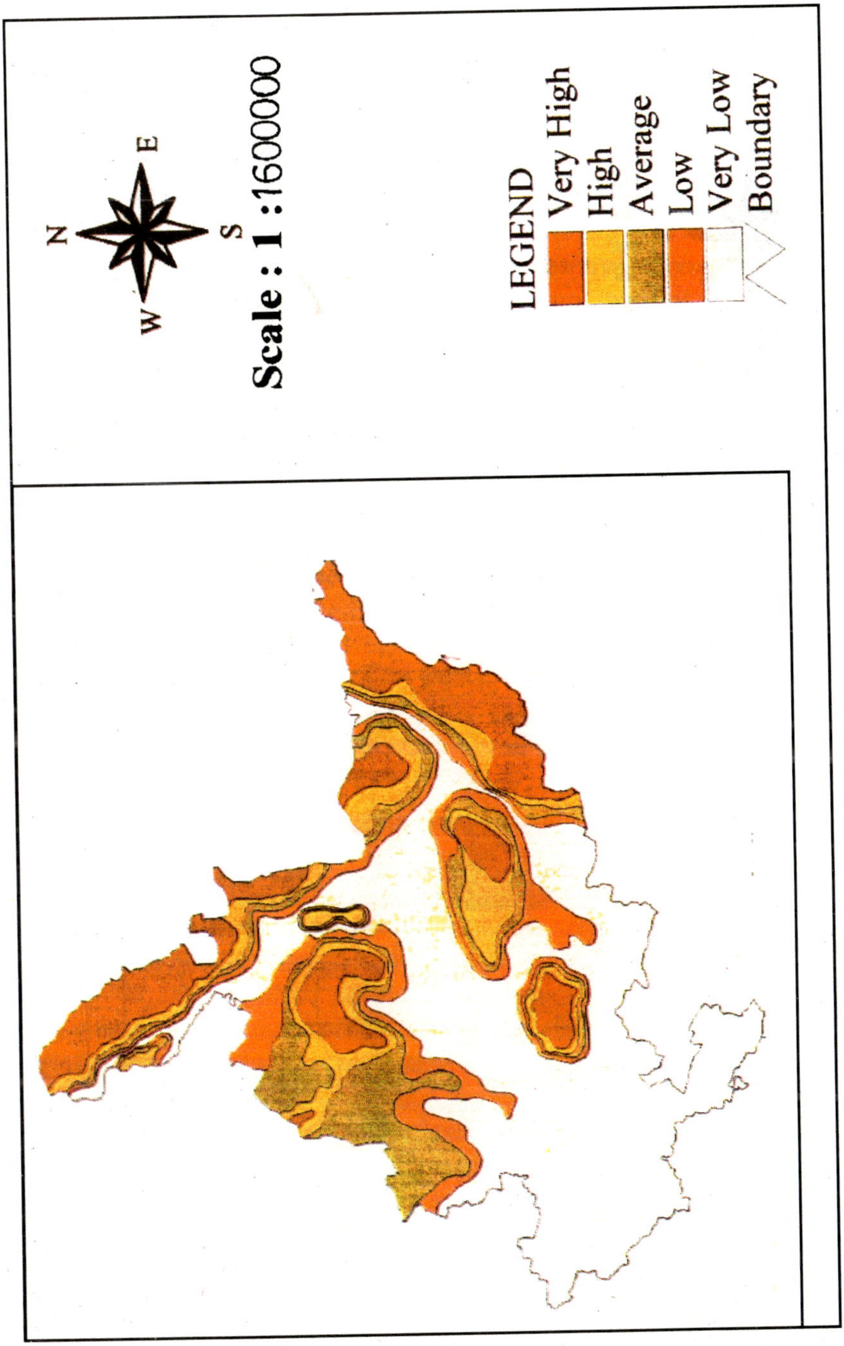

Fig. 11.5 : Khammam District—Land use and Land cover ISO-Region Map

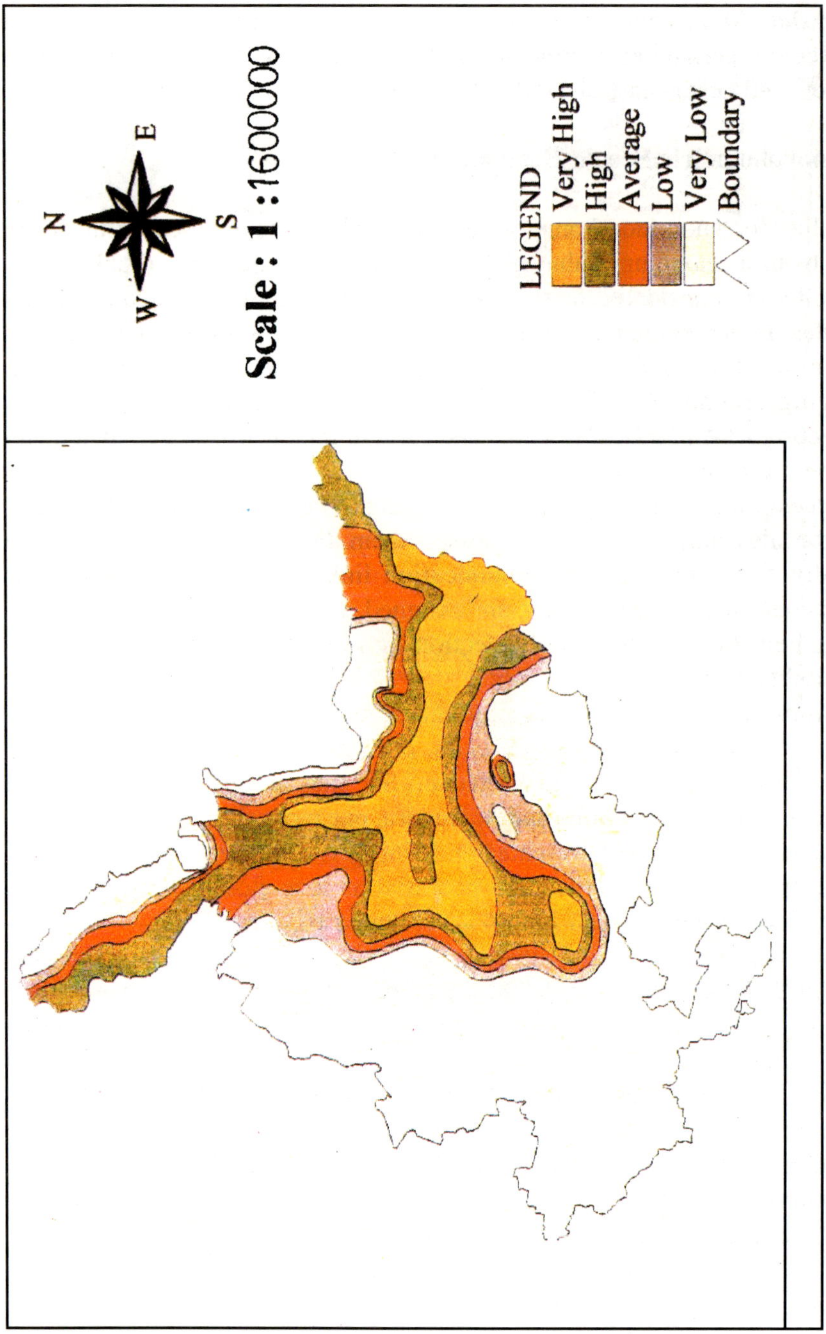

Fig. 11.7 : Khammam District—Composite Index ISO-Region Map

Table 11.1 : Khamman District—Morphometric Weightages

Selected	*Average Slope*			*Relative Relief*			*Intensity of Relief*			*Dissection Index*			*Ruggedity*		
Toposheet Grid Reference	*Low*	*Medium* 1	*High*	*Low*	*Medium* 2	*High*	*Low*	*Medium* 3	*High*	*Low*	*Medium* 4	*High*	*Low*	*Medium* 5	*High*
65 B/11 1/1	–	0.1	–	0.043	–	–	0.043	–	–	–	0.133	–	0.053	–	–
65 B/11 2/3	–	0.1	–	0.043	–	–	–	0.182	–	–	0.133	–	0.053	–	–
65 B/11 ¾	–	0.1	–	–	–	0.3	–	–	0.3	–	–	0.2	–	0.154	–
65 B/11 3/5	0.071	–	–	–	0.182	–	0.043	–	–	–	0.133	–	–	–	0.25
65 C/9 1/5	–	–	0.3	0.043	–	–	0.043	–	–	–	0.133	–	–	0.154	–
65 C/9 2/2	–	–	0.3	0.043	–	–	0.043	–	–	0.071	–	–	–	0.154	–
65 C/9 3/5	0.071	–	–	0.043	–	–	0.043	–	–	–	0.133	–	–	0.154	–
65 C/9 5/3	–	0.1	–	0.043	–	–	0.043	–	–	0.071	–	–	0.053	–	–
65 C/9 5/5	–	0.1	–	0.043	–	–	0.043	–	–	0.071	–	–	0.053	–	–
65 C/10 1/1	–	0.1	–	–	0.182	–	0.043	–	–	0.071	–	–	0.053	–	–
65 C/10 1/3	0.071	–	–	0.043	–	–	0.043	–	–	0.071	–	–	0.053	–	–
65 C/10 4/2	0.071	–	–	–	–	–	0.043	–	–	–	–	–	0.053	–	–
65 C/10 5/1	0.071	–	–	–	0.182	–	–	0.182	–	–	0.133	–	–	0.154	–
65 C/10 5/4	0.071	–	–	0.043	–	–	–	0.182	–	–	0.133	–	0.053	–	–
65 C/10 5/5	–	0.1	–	–	0.182	–	–	0.182	–	–	–	0.2	–	0.154	–
65 C/11 1/1	0.071	–	–	–	0.182	–	–	0.182	–	–	0.133	–	–	0.154	–
65 C/11 1/5	–	–	0.3	–	–	0.3	–	–	0.3	–	0.133	–	–	–	0.25
65 C/11 3/5	–	–	0.3	–	0.182	–	–	0.182	–	–	0.133	–	–	–	0.25
65 C/11 4/1	0.071	–	–	0.043	–	–	0.043	–	–	0.071	–	–	0.053	–	–
65 C/11 5/2	0.071	–	–	0.043	–	–	0.043	–	–	0.071	–	–	0.053	–	–

(Contd...)

Table 11.1 *(Contd.)*

65 C/14 ¼	–	0.1	–	–	0.182	–	–	0.182	–	–	–	0.2	0.053	–	–
65 C/14 4/2	0.071	–	–	0.043	–	–	0.043	–	–	0.071	–	–	0.053	–	–
65 C/14 5/5	0.071	–	–	0.043	–	–	0.043	–	–	–	–	0.2	0.053	–	–
65 C/15 2/2	0.071	–	–	0.043	–	–	0.043	–	–	0.071	–	–	0.053	–	–
65 C/15 4/3	–	0.1	–	0.43	–	–	0.043	–	–	0.071	–	–	–	–	0.25
65 C/15 5/1	–	0.1	–	–	–	0.3	0.043	–	–	0.071	–	–	–	0.154	–
65 C/15 5/2	–	–	0.3	–	0.182	–	–	0.182	–	–	0.133	–	–	–	0.25
65 G/2 1/5	–	–	0.3	–	0.182	–	–	0.182	–	–	–	0.2	0.053	–	–
65 G/2 2/5	–	0.1	–	–	0.182	–	–	0.182	–	–	–	0.2	0.053	–	–
65 G/2 4/1	–	0.1	–	–	0.182	–	–	0.182	–	–	–	0.2	0.053	–	–
65 G/2 4/2	–	0.1	–	–	–	0.3	–	–	0.3	–	–	0.2	–	0.154	–
65 G/2 5/3	–	0.1	–	–	–	0.3	–	–	0.3	–	–	0.2	–	–	0.25
65 G/3 1/1	–	–	0.3	0.043	–	–	0.043	–	–		0.133	–	–	0.154	–
65 G/3 1/3	–	0.1	–	0.043	–	–	0.043	–	–	0.071	–	–	0.053	–	–
65 G/3 3/1	0.071	–	–	0.043	–	–	0.043	–	–	0.071	–	–	0.053	–	–
65 G/3 5/2	–	0.1	–	0.043	–	–	0.043	–	–	–	0.133	–	–	0.154	–
65 G/6 2/5	0.071	–	–	–	–	0.3	–	–	0.3	–	–	0.2	–	–	0.25
65 G/6 4/4	–	–	0.3	–	–	0.3	–	–	0.3	–	0.133	–	–	–	0.25
65 G/6 4/5	--	–	0.3	–	–	0.3	–	–	0.3	–	–	0.2	–	–	0.25
65 G/6 5/2	–	0.1	–	–	–	0.3	–	–	0.3	–	–	0.2	–	–	0.25
65 G/7 1/1	–	0.1	–	0.043	–	–	0.043	–	–	–	–	0.2	–	0.154	–
65 G/7 2/2	–	–	0.3	–	0.182	–	–	0.182	–	–	0.133	–	–	0.154	–
65 G/7 3/1	–	0.1	–	–	–	0.3	–	–	0.3	–	–	–	–	–	0.25
65 G/7 5/1	–	0.1	–	–	–	0.3	–	–	0.3	–	–	0.2	–	–	0.25

(Contd.)

Table 11.1 *(Contd.)*

Selected	*Contourcrenulation*			*Isolation of Hill Neibour Area*			*Width of Valley*			*Stream Frequency*		
Toposheet Grid Reference	*Low*	*Medium* 6	*High*	*Low*	*Medium* 7	*High*	*Low*	*Medium* 8	*High*	*Low*	*Medium* 9	*High*
65 B/11 1/1	0.06	–	–	–	–	–	0.063	–	–	–	0.1	–
65 B/11 2/3	0.06	–	–	–	–	–	–	0.14	–	–	0.1	–
65 B/11 3/4	0.06	–	–	–	–	–	0.063	–	–	0.0667	–	–
65 B11 3/5	–	0.13	–	0.06	–	–	–	0.14	–	0.0667	–	–
65 C/9 1/5	–	–	0.27	0.06	–	–	–	0.14	–	–	0.1	–
65 C/9 2/2	–	–	0.27	0.06	–	–	–	–	0.22	–	0.1	–
65 C/9 3/5	–	0.13	–	–	–	–	–	0.14	–	–	0.1	–
65 C/9 5/3	–	0.13	–	–	–	–	–	0.14	–	–	0.1	–
65 C/9 5/5	–	–	0.27	–	–	–	–	0.14	–	–	–	0.333
65 C/10 1/1	0.06	–	–	–	–	–	–	0.14	–	–	0.1	–
65 C/10 1/3	–	0.13	–	–	0.25	–	–	–	0.22	0.0667	–	–
65 C/10 4/2	–	0.13	–	–	0.25	–	–	–	0.22	0.0667	–	–
65 C/10 5/1	–	0.13	–	0.06	–	–	–	–	0.22	–	0.1	–
65 C/10 5/4	0.06	–	–	–	–	0.5	–	–	0.22	–	0.1	–
65 C/10 5/5	0.06	–	–	–	–	–	–	–	0.22	0.0667	–	–
65 C/11 1/1	–	0.13	–	–	–	–	–	0.14	–	0.0667	–	–
65 C/11 1/5	–	0.13	–	–	–	–	0.063	–	–	–	0.1	–
65 C/11 3/5	–	0.13	–	0.06	–	–	–	0.14	–	–	0.1	–
65 C/11 4/1	0.06	–	–	–	–	0.5	–	0.14	–	0.0667	–	–
65 C/11 5/2	0.06	–	–	–	–	–	–	–	0.22	0.0667	–	–
65 C/14 1/4	0.06	–	–	–	–	0.5	–	–	0.22	0.0667	–	–
65 C/14 4/2	0.06	–	–	–	–	0.5	–	–	0.22	0.0667	–	–
65 C/14 5/5	–	0.13	–	–	0.25	–	–	–	0.22	–	0.1	–

(Contd.)

Table 11.1 *(Contd.)*

		6			7			8			9	
65 C/15 2/2	–	–	0.27	–	0.25	–	–	0.14	–	–	0.1	–
65 C/15 4/3	–	–	0.27	0.06	–	–	0.063	–	–	–	–	0.333
65 C/15 5/1	–	0.13	–	–	–	–	0.063	–	–	–	–	0.333
65 C/15 5/2	–	–	0.27	0.06	–	–	0.063	–	–	–	–	0.333
65 G/2 1/5	–	–	0.27	–	–	0.5	–	0.14	–	0.0667	–	–
65 G/2 2/5	–	0.13	–	–	0.25	–	–	–	0.22	0.0667	–	–
65 G/2 4/1	0.06	–	–	0.06	–	–	–	–	0.22	0.0667	–	–
65 G/2 4/2	0.06	–	–	–	0.25	–	–	–	0.22	0.0667	–	–
65 G/2 5/3	0.06	–	–	–	0.25	–	–	–	0.22	–	0.1	–
65 G/3 1/1	0.06	–	–	–	0.25	–	0.063	–	–	–	0.1	–
65 G/3 1/3	–	0.13	–	0.06	–	–	0.063	–	–	–	0.1	–
65 G/3 3/1	0.06	–	–	0.06	–	–	–	0.14	–	0.0667	–	–
65 G/3 5/2	–	0.13	–	0.06	–	–	0.063	–	–	–	0.1	–
65 G/6 2/5	–	–	0.27	0.06	–	–	0.063	–	–	–	0.1	–
65 G/6 4/4	–	–	0.27	0.06	–	–	0.063	–	–	–	–	0.333
65 G/6 4/5	–	–	0.27	–	–	–	0.063	–	–	–	–	0.333
65 G/6 5/2	–	0.13	–	–	–	0.5	–	0.14	–	–	0.1	–
65 G/17 1/1	0.06	–	–	0.06	–	–	0.063	–	–	–	0.1	–
65 G/7 2/2	–	0.13	–	0.06	–	–	0.063	–	–	–	–	0.333
65 G/7 3/1	0.06	–	–	0.06	–	–	0.063	–	–	–	–	0.333
65 G/7 5/1	–	–	0.27	–	–	–	0.063	–	–	–	–	0.333

(Contd.)

Table 11.1 *(Contd.)*

Selected Toposheet Grid Reference	*Drainage Density* Low	Medium 10	High	*Stream order* Low	Medium 11	High	*River Pattern* Low	Medium 12	High	*Vegetation* Low	Medium 13	High	*Total Grids Morphometrs C Weight Ages*
65 B/11 1/1	0.053	–	–	–	0.074	–	–	0.29	–	–	–	0.176	1.19
65 B/11 2/3	0.053	–	–	0.067	–	–	0.03	–	–	–	0.118	–	0.95
65 B/11 3/4	0.053	–	–	–	0.074	–	–	0.29	–	–	–	0.176	1.84
65 B/11 3/5	–	0.188	–	–	0.074	–	0.03	–	–	–	–	0.176	1.53
65 C/9 1/5	–	0.118	–	–	0.074	–	0.03	–	–	–	–	0.176	1.86
65 C/9 2/2	–	0.118	–	–	0.074	–	0.03	–	–	–	–	0.176	0.61
65 C/9 3/5	–	–	0.375	–	0.074	–	0.03	–	–	0.1	–	–	1.39
65 C/9 5/3	0.053	–	–	0.067	–	–	0.03	–	–	–	–	0.176	0.91
65 C/9 5/5	–	0.118	–	0.067	–	–	0.03	–	–	–	0.118	–	1.15
65 C/10 1/1	–	0.118	–	–	0.074	–	0.03	–	–	–	–	0.176	1.01
65 C/10 1/3	–	0.118	–	–	0.074	–	–	0.29	–	0.1	–	–	1.53
65 C/10 4/2	0.053	–	–	–	0.074	–	–	–	0.6	0.1	–	–	1.77
65 C/10 5/1	–	0.118	–	–	0.074	–	0.03	–	–	–	–	0.176	1.55
65 C/10 5/4	0.053	–	–	–	0.074	–	–	0.29	–	0.1	–	–	1.74
65 C/10 5/5	0.053	–	–	–	0.074	–	–	0.29	–	0.1	–	–	1.63
65 C/11 1/1	0.053	–	–	–	0.067	–	–	0.29	–	0.1	–	–	1.52
65 C/11 1/5	–	0.118	–	0.067	–	–	0.03	–	–	–	–	0.176	1.97
65 C/11 3/5	–	0.118	–	–	0.074	–	0.03	–	–	–	–	0.176	1.80
65 C/11 4/1	0.053	–	–	0.067	–	–	0.03	–	–	0.1	–	–	1.30
65 C/11 5/2	0.053	–	–	–	0.074	–	0.03	–	–	–	0.118	–	0.90
65 C/14 1/4	0.053	–	–	–	0.074	–	–	–	0.6	–	0.118	–	2.22
65 C/14 4/2	–	0.118	–	–	0.074	–	–	–	0.6	–	0.118	–	2.04
65 C/14 5/5	0.053	–	–	–	0.074	–	0.03	–	–	–	–	0.176	1.44

(Contd.)

Table 11.1 *(Contd.)*

	10			*11*			*12*				*13*		
65 B/11 1/1	0.053	–	–	–	0.074	–	–	0.29	–	–	–	0.176	1.19
65 B/11 2/3	0.053	–	–	0.067	–	–	0.03	–	–	–	0.118	–	0.95
65 B/11 3/4	0.053	–	–	–	0.074	–	–	0.29	–	–	–	0.176	1.84
65 B/11 3/5	–	0.118	–	–	0.074	–	0.03	–	–	–	–	0.176	1.53
65 C/9 1/5	–	0.118	–	–	0.074	–	0.03	–	–	–	–	0.176	1.86
65 C/9 2/2	–	0.118	–	–	0.074	–	0.03	–	–	–	–	0.176	0.61
65 C/9 3/5	–	–	0.375	–	0.074	–	0.03	–	–	0.1	–	–	1.39
65 C/9 5/3	0.053	–	–	0.067	–	–	0.03	–	–	–	–	0.176	0.91
65 C/9 5/5	–	0.118	–	0.067	–	–	0.03	–	–	–	0.118	–	1.15
65 C/10 1/1	–	0.118	–	–	0.074	–	0.03	–	–	–	–	0.176	1.01
65 C/10 1/3	–	0.118	–	–	0.074	–	–	0.29	–	0.1	–	–	1.53
65 C/10 4/2	0.053	–	–	–	0.074	–	–	–	0.6	0.1	–	–	1.77
65 C/10 5/1	–	0.118	–	–	0.074	–	0.03	–	–	–	–	0.176	1.55
65 C/10 5/4	0.053	–	–	–	0.074	–	–	0.29	–	0.1	–	–	1.74
65 C/10 5/5	0.053	–	–	–	0.074	–	–	0.29	–	0.1	–	–	1.63
65 C/11 1/1	0.053	–	–	–	0.074	–	–	0.29	–	0.1	–	–	1.52
65 C/11 1/5	–	0.118	–	0.067	–	–	0.03	–	–	–	–	0.176	1.97
65 C/11 3/5	–	0.118	–	–	0.074	–	0.03	–	–	–	–	0.176	1.80
65 C/11 4/1	0.053	–	–	0.067	–	–	0.03	–	–	0.1	–	–	1.30
65 C/11 5/2	0.053	–	–	–	0.074	–	0.03	–	–	–	0.118	–	0.90
65 C/14 1/4	0.053	–	–	–	0.074	–	–	–	0.6	–	0.118	–	2.22
65 C/14 4/2	–	0.118	–	–	0.074	–	–	–	0.6	–	0.118	–	2.04
65 C/14 5/5	0.053	–	–	–	0.074	–	0.03	–	–	–	–	0.176	1.44

(Contd.)

Table 11.1 *(Contd.)*

		10			*11*			*12*			*13*		
65 C/15 2/2	–	0.118	–	–	0.074	–	0.03	–	–	0.1	–	–	1.36
65 C/15 4/3	–	–	0.375	–	0.074	–	0.03	–	–	–	0.118	–	1.79
65 C/15 5/1	–	–	0.375	–	0.074	–	0.03	–	–	–	0.118	–	1.53
65 C/15 5/2	–	–	0.375	0.067	–	–	0.03	–	–	–	0.118	–	2.31
65 G/2 1/5	0.053	–	–	–	0.074	–	0.03	–	–	–	0.118	–	2.14
65 G/2 2/5	0.053	–	–	–	0.074	–	0.03	–	–	–	–	0.176	1.68
65 G/2 4/1	0.053	–	–	–	0.074	–	0.03	–	–	–	0.118	–	1.37
65 G/2 4/2	0.053	–	–	0.067	–	–	0.03	–	–	–	0.118	–	1.92
65 G/2 5/3	0.053	–	–	0.067	–	–	0.03	–	–	–	0.118	–	1.70
65 G/3 1/1	–	0.118	–	0.067	–	–	0.03	–	–	–	–	0.176	1.52
65 G/3 1/3	0.053	–	–	–	0.074	–	0.03	–	–	–	0.118	–	1.03
65 G/3 3/1	0.053	–	–	–	0.074	–	–	0.29	–	0.1	–	–	1.07
65 G/3 5/2	–	0.118	–	0.067	–	–	0.03	–	–	–	–	0.176	1.77
65 G/6 2/5	–	0.118	–	0.067	–	–	0.03	–	–	–	0.118	–	1.97
65 G/6 4/4	–	–	0.375	–	0.074	–	0.03	–	–	–	–	0.176	2.57
65 G/6 4/5	–	–	0.375	0.067	–	–	0.03	–	–	–	–	0.176	2.67
65 G/6 5/2	–	0.118	–	–	0.074	–	0.03	–	–	–	0.118	–	2.36
65 G/7 1/1	–	0.118	–	–	0.074	–	0.03	–	–	–	0.118	–	1.17
65 G/7 2/2	–	0.118	–	–	0.074	–	0.03	–	–	–	–	0.176	1.85
65 G/7 3/1	–	–	0.375	–	–	1.5	–	–	0.6	0.1	–	–	4.24
65 G/7 5/1	–	–	0.375	–	–	1.5	–	–	0.6	–	0.118	–	4.47

Table 11. 2 : Khammam District—Non-Morphometric Weightages

Selected Toposheet Grid Reference	*Water Falls 1*	*Lakes/ Reservoir 2*	*Accessi-bility 3*	*Nearby Settlements 4*	*Mines 5*	*Sancturies 6*	*Religious Sites 7*	*Historical Sites 8*	*Total Weightage*
65 B/11 1/1	1	–	1	1	–	–	–	–	3
65 B/11 2/3	1		1	1	–	–	–	–	3
65 B/11 3/4	–	–	–	–	–	–	–	–	–
65 B/11 3/5	–	–	1	1	1	–	–	–	3
65 C/9 1/5	–	–	1	1	–	–	–	–	2
65 C/9 2/2	–	–	1	–	–	–	–	–	1
65 C/9 3/5	–	–	1	1	–	–	–	–	2
65 C/9 5/3	–	1	1	1	–	–	–	–	3
65 C/9 5/5	–	–	1	1	–	–	–	–	2
65 C/10 1/1	–	–	1	–	–	–	–	–	1
65 C/10 1/3	–	–	1	1	–	–	–	–	2
65 C/10 4/2	–	1	1	1	1	1	–	–	5
65 C/10 5/1	–	–	1	1	1	–	1	1	5
65 C/10 5/4	–	–	1	1	1	–	1	1	5
65 C/10 5/5	–	–	1	1	1	–	–	–	3
65 C/11 1/1	1	1	1	1	–	–	–	–	4
65 C/11 1/5	1	1	1	1	–	–	–	–	4
65 C/11 3/5	2	–	1	1	–	–	–	–	4
65 C/11 4/1	1	–	1	1	–	–	–	1	4
65 C/11 5/2	–	1	1	1	–	–	1	–	4

(Contd.)

Table 11.2 *(Contd.)*

	1	*2*	*3*	*4*	*5*	*6*	*7*	*8*	
65 C/14 1/4	–	–	1	1	1	–	1	–	4
65 C/14 4/2	–	1	1	1	1	–	1	1	6
65 C/14 5/5	–	–	1	1	–	–	–	–	2
65 C/15 2/2	–	–	1	1	–	–	–	–	2
65 C/15 4/3	–	–	1	1	–	–	–	–	2
65 C/15 5/1	–	–	1	1	–	–	–	–	2
65 C/15 5/2	–	–	1	1	–	–	–	–	2
65 G/2 1/5	–	–	1	1	–	–	–	–	2
65 G/2 2/5	–	–	1	1	–	–	–	–	2
65 G/2 4/1	–	–	1	1	–	–	–	–	2
65 G/2 4/2	–	–	1	1	–	–	–	–	2
65 G/2 5/3	–	–	1	1	–	–	–	–	2
65 G/3 1/1	–	–	1	1	–	–	–	–	2
65 G/3 1/3	–	–	1	1	–	–	–	–	2
65 G/3 3/1	–	–	1	1	–	–	–	–	2
65 G/3 5/2	–	–	1	–	–	–	–	–	1
65 G/6 2/5	–	1	1	1	–	–	1	–	4
65 G/6 4/4	–	–	–	1	–	–	–	–	1
65 G/6 4/5	–	–	1	1	–	–	1	1	4
65 G/6 5/2	–	–	1	1	–	–	–	–	2
65 G/7 1/1	–	–	1	1	–	–	–	–	2
65 G/7 2/2	–	–	1	–	–	–	–	–	1
65 G/7 3/1	1	1	1	1	–	–	–	–	4
65 G/7 5/1	1	1	1	1	–	–	1	1	6

Kinnerasani dam and as evidenced from the sites, this is the only site with given weightage. Out of total weightages compiled for all the 8 parameters, the values vary from a maximum of 6 to minimum of 0 (Table 11.2) while rest of the values fluctuate around 2. Kinnerasani region is ranked high, whereas one site near Venkatapuram is assigned 0.

The Iso-non-morpho map (Map 11.4) clearly depicts the regional distribution of values. An isolated pocket near Kinnerasani and Papikondalu have higher values above 5, therefore, are considered as higher scenic regions running in a mid-eastern region to mid-central region. It grades down slowly in all sides depicting regions of lesser values. It is mostly the central region, which has higher scenic value.

Land use/Cover Scenic Evaluation

The three land use categories identified in the region have been given the weightages as per their recreational potential. Agricultural land being the most utilized is considered unsuitable from scenic point of view, hence is given a weightage of 1. Wasteland's being neither fully utilized nor underutilized, at the same time assuming a rugged appearance is assigned a value of 2. In the end forests and water bodies as such are important and are assigned higher values of 3 (Table 11.3) depicts the derived values which range from a maximum of 3.00 to minimum of 0.05, wherever the values around 3.00 appear, it pertains to only unused landscape.

Iso-Landuse landcover (Fig. 11.5) depicts a regular and continuous appearance whereby the highest value greater than 2.75 occurs in clusters around eastern, north-eastern and mid-north-western part, south-eastern part and southern part. These are basically over Hill system followed by pediplain and pediment inselberg complex.

Landforms and Scenic Evaluation

Landforms enhance the scenic value. With this assumption; the landforms have been counted in numbers based on relief cross-sections. A maximum of 6 landforms are identified over 3 sites representative of structural hills, flood plains and Pedi plain (Table 11.4).

The Iso-landforms map (Fig. 11.6) depicts very high region around Godavari floodplain, with structural hills abutting it. Another pocket of high spreads over Mallapur, located over intermontane valley and structural hills. A pocket of medium values of 4 and 3 centre round Bhadrachalam, Burgampad, Kothagudem, Manugum and Aswapuram. The least preferred zone with low values is around south-western and western region which mainly comprises of Pedi plain under canal command area and agriculture landuse.

Table 11.3 : Khammam District—land use and land cover weightage

Selected Toposheet Grid Reference	*Low (Agriculture Land) 1*	*Medium (Wasteland Land) 2*	*High (Fresh Water Bodies) 3*	*Total Weightage*
65 B/11 1/1	0.05	–	2.85	2.90
65 B/11 2/3	0.05	0.60	1.95	2.50
65 B/11 3/4	–	–	3.00	3.00
65 B/11 3/5	–	0.40	2.40	2.80
65 C/9 1/5	0.05	–	2.85	2.90
65 C/9 2/2	–	0.10	2.85	2.95
65 C/9 3/5	0.05	–	1.50	2.00
65 C/9 5/3	0.05	–	2.85	2.90
65 C/9 5/5	0.05	–	2.85	2.90
65 C/10 1/1	0.10	–	2.70	2.80
65 C/10 1/3	0.60	0.20	0.90	1.70
65 C/10 4/2	0.40	0.40	1.20	2.00
65 C/10 5/1	0.20	–	2.40	2.60
65 C/10 5/4	0.50	0.40	0.90	1.80
65 C/11 1/1	0.75	–	0.75	1.50
65 C/11 1/5	–	0.20	2.70	2.90
65 C/11 3/5	–	–	3.00	3.00
65 C/11 4/1	0.40	–	1.80	2.20
65 C/11 5/2	0.20	–	2.40	2.60
65 C/14 1/4	0.65	–	1.05	1.70
65 C/14 4/2	0.60	0.20	0.60	1.40
65 C/14 5/5	0.30	–	2.10	2.40
65 C/15 2/2	0.60	–	1.20	1.80
65 C/15 4/3	0.25	–	2.25	2.50
65 C/15 5/1	0.20	–	2.40	2.60
65 C/15 5/2	0.10	–	2.70	2.80
65 G/2 1/5	0.05	–	1.85	1.90
65 G/2 2/5	0.10	–	2.70	2.80
65 G/2 4/1	0.15	0.20	2.25	2.60
65 G/2 4/2	0.20	–	2.40	2.60
65 G/2 5/3	0.35	–	1.95	2.30
65 G/3 1/1	–	–	3.00	3.00
65 G/3 1/3	0.35	0.50	1.20	2.05
65 G/3 3/1	0.35	0.30	1.50	2.15
65 G/3 5/2	–	–	3.00	3.00
65 G/6 2/5	0.20	0.10	2.25	2.55
65 G/6 4/4	–	–	3.00	3.00
65 G/6 4/5	–	–	3.00	3.00
65 G/6 5/2	0.05	0.60	1.95	2.60
65 G/7 1/1	0.20	–	2.40	2.60
65 G/7 2/2	–	–	3.00	3.00
65 G/7 3/1	0.15	–	2.65	2.80
65 G/7 5/1	0.10	–	2.70	2.80

Composite Index and Scenic Evaluation

From the obtained composite index (Table 11.5) values a scenic evaluation map (Fig. 11.7) is prepared. The Isoline of 12 passes through a zone comprising of Papikondalu, Perantapalli, Koyada, Rudramakota, Sri Ramgiri, Kunavaram, Bhadrachalam, Kineerasani, Dummagudem and Parnasala. This is a highly scenic region having the composite index values of above 12, the maximum being 17.3. The values diminish towards east and westward manner depicting the values around 9, the minimum being 7. The places are Dammapeta, Pusukunta, Morkodu, Madira and Kalluru. It is evident from the map (Fig.11.7) that the river scape, along with landscape, supported by the other non-morphometric, landforms characters get higher values. These land systems broadly fall under hill system with accompanying intermediate zone. Otherwise, the Pedi plain system overall is stated to have lower values, categorized as low scenic value. This is an area, which can broadly be categorized as used landscape. In contrast, majority of the unused landscape to the eastern and south-eastern region have higher range of values.

Consumer Response Through Photograph Analysis

The response to the scenery is one of the most desired aspects of the study as it provides a scope and clue to acquire the status of the sites in terms of its customization and appreciation as a recreation spot. This can be achieved through photographing the sites and trying to get or acquire the responses of the consumers for whose purpose the recreation spots have been identified in the first place. 200 respondents have been asked to assess the photographs selected from the accessible sites and are asked to grade the photographs on a 5 point scale, comprising of 'Excellent', 'Very Good', 'Good' , Average' and 'Poor' with the descending order of weightages respectively (Table 11.6). From the table, it is clear that majority of the sites, which are ranked as 'Excellent' are also ranked as only 'Very Good' and 'Good' and do not get a ranking below the rank of 'Good'. This proves the fact that there has been a clear choice by majority of the respondents in an overwhelming manner towards a particular site. There has been 100 per cent response to Papikondalu region, followed by nearly 90 per cent of response to Kennerasani region followed closely by Dummagudem and Parnasala. All these three sites are rated highly. Interestingly all the respondents have rated no site as 'Poor'. Rest of the photographs get an 'Average' to 'Good' rank.

These findings seem to be in Good stead as far as the recreational sites of Khammam district are concerned. It has been amply proved that unused

Table 11. 4 : Khammam District—Land Forms Weightage

Selected Toposheet Grid Reference	*Structural Hills*	*Denudational Hills*	*Residual Hills*	*Inter Montain Valley*	*Pedi Mental Inselberg*	*Bazadas*	*Flood Plain*	*Shallow Pedilain(SP) & SP Canal Command Land*	*Moderate Pediplain (MP) & MP Canal Command Land*	*Land Forms From Profiles*	*Total*
65 B/11 1/1	1	–	–	–	–	–	–	–	1	2	4
65 B/11 2/3	1	–	–	–	–	–	1	–	–	2	4
65 B/11 3/4	1	–	–	–	–	–	–	–	–	2	3
65 B/11 3/5	1	–	–	–	–	–	1	–	–	2	4
65 C/9 1/5	–	–	–	–	–	–	–	–	1	2	3
65 C/9 2/2	–	1	–	–	–	–	–	–	1	2	4
65 C/9 3/5	–	1	–	1	1	–	–	–	–	3	6
65 C/9 5/3	–	1	–	–	–	–	–	–	1	2	4
65 C/9 5/5	–	1	–	–	–	–	–	–	–	2	3
65 C/10 1/1	–	1	–	–	–	–	–	–	1	1	3
65 C/10 1/3	–	1	–	–	–	–	–	–	1	2	4
65 C/10 4/2	1	–	–	–	–	–	1	1	–	1	4
65 C/10 5/1	1	–	–	–	–	–	–	1		2	4
65 C/10 5/4	–	–	1	–	–	–	–	–	1	1	3
65 C/10 5/5	–	1	1	–	–	–	–	–	1	2	5
65 C/11 1/1	–	–	1	–	–	–	–	1	1	2	5
65 C/11 1/5	–	–	1	–	–	–	–	–	–	2	3
65 C/11 3/5	–	–	1	–	–	1	–	–	–	2	4
65 C/11 4/1	–	–	1	–	–	–	–	1	1	1	4
65 C/11 5/2	–	1	–	–	–	–	–	–	1	2	4
65 C/14 1/4	–	–	1	–	–	–	–	–	1	2	4
65 C/14 4/2	–	–	–	–	–	–	1	1	–	1	3
65 C/14 5/5	–	1	1	–	–	–	1	–	–	2	5

(Contd...)

Table 11.4 *(Contd.)*

65 C/15 2/2	–	1	–	–	–	–	–	–	1	1	3
65 C/15 4/3	–	1	–	–	–	–	–	–	1	1	3
65 C/15 5/1	–	1	–	–	–	–	–	–	1	2	4
65 C/15 5/2	–	1	–	–	–	–	–	–	1	2	4
65 G/2 1/5	–	1	–	–	–	1	–	–	–	2	4
65 G/2 2/5	–	1	–	–	–	1	–	–	–	2	4
65 G/2 4/1	–	–	1	–	1	–	–	–	–	3	5
65 G/2 4/2	–	–	1	–	–	–	–	–	–	2	3
65 G/2 5/3	–	–	1	–	–	–	1	–	1	2	5
65 G/3 1/1	–	–	1	–	–	–	–	–	–	2	3
65 G/3 1/3	–	–	1	–	–	–	–	–	–	1	2
65 /G/3 3/1	–	–	1	–	–	–	–	–	–	2	3
65 G/3 5/2	–	–	1	–	–	–	–	1	1	2	5
65 G/6 2/5	1	–	–	–	–	1	–	1	–	3	6
65 G/6 4/4	1	–	–	–	–	1	–	–	–	3	5
65 G/6 4/5	1	–	–	–	–	1	–	–	–	3	5
65 G/6 5/2	1	–	–	–	–	1	–	–	–	2	4
65 G/7 1/1	–	–	1	–	–	–	1	1	1	2	6
65 G/7 2/2	–	–	1	1	–	–	–	–	–	3	5
65 G/7 3/1	1	–	–	1	–	–	–	–	–	2	4
65 G/7 5/1	1	–	–	–	–	–	1	–	–	2	4

Table 11.5 : Composite Index Weightage of Khammam District

Selected Toposheet Grid Reference	*Morphometric Values* 1	*Non-Morphometric Values* 2	*Land Form Values* 3	*Land Use Values* 4	*Composite Value* 5
65 B/11 1/1	1.19	3	4	2.9	11.1
65 B/11 2/3	0.95	3	4	2.5	10.5
65 B/11 3/4	1.84	0	3	3.0	7.8
65 B/11 3/5	1.53	3	4	2.8	11.3
65 C/9 1/5	1.86	2	3	2.9	9.8
65 C/9 2/2	1.66	1	4	3.0	9.6
65 C/9 3/5	1.39	2	6	2.0	11.4
65 C/9 5/3	1.01	3	4	2.9	10.9
65 C/9 5/5	1.15	2	3	2.9	9.1
65 C/10 1/1	1.01	1	3	2.8	7.8
65 C/10 1/3	1.53	2	4	1.7	9.2
65 C/10 4/2	1.77	7	4	2.0	14.8
65 C/10 5/1	1.55	5	4	2.6	13.2
65 C/10 5/4	1.74	5	3	1.8	11.5
65 C/10 5/5	1.63	3	5	2.8	12.4
65 C/11 1/1	1.52	4	5	1.5	12.0
65 C/11 1/5	1.97	4	3	2.9	11.9
65 C/11 3/5	1.80	4	4	3.0	12.8
65 C/11 4/1	1.30	5	4	2.2	12.5
65 C/11 5/2	0.90	4	4	2.6	11.5
65 C/14 1/4	2.22	4	4	1.7	11.9
65 C/14 4/2	2.04	6	3	1.4	12.4
65 C/14 5/5	1.44	2	5	2.4	10.8
65 C/15 2/2	1.36	2	3	1.8	8.2
65 C/15 4/3	1.79	2	3	2.5	9.3
65 C/15 5/1	1.53	2	4	2.4	9.9
65 C/15 5/2	2.31	2	4	2.8	11.1
65 G/2 1/5	2.14	2	4	1.9	10.0
65 G/2 2/5	1.68	2	4	2.8	10.5
65 G/2 4/1	1.37	2	5	2.6	11.0
65 G/2 4/2	1.92	2	3	2.6	9.5
65 G/2 5/3	1.70	2	5	2.3	11.0
65 G/3 1/1	1.52	2	3	3.0	9.5
65 G/3 1/3	1.03	2	2	2.1	7.1
65 G/3 3/1	1.13	2	3	2.2	8.3
65 G/3 5/2	1.77	1	5	3.0	10.8

(Contd.)

Table 11.5 *(Contd.)*

	1	2	*3*	*4*	*5*
65 G/6 2/5	1.97	5	6	2.6	15.5
65 G/6 4/4	2.57	1	5	3.0	11.6
65 G/6 4/5	2.67	4	5	3.0	14.7
65 G/6 5/2	2.36	3	4	2.6	12.0
65 G/7 1/1	1.17	2	6	2.6	11.8
65 G/7 2/2	1.85	2	5	3.0	11.9
65 G/7 3/1	4.24	3	4	2.8	14.0
65 G/7 5/1	4.47	6	4	2.8	17.3

landscapes in terms of natural terrain supported by a few more additional factors of accessibility, Reservoirs, religious or historical factors enhance the scenic value of a recreational centre. It is also observed in the study that natural landscapes in association with waterscapes are the most preferred recreational centres, as is the case either with Papikondalu or Kinnersani. It also needs to be mentioned that the sites that prominently appear as recreational centres are also the centres which are relatively well known either from historical or religious point of view.

Table 11.6 : Photographs - People Perception (Number of Respondants : 200)

Selected	*Poor*		*Average*		*Good*		*Very Good*		*Excellent*		*Total*
Toposheet Grid Reference	*Number of Respondants 1*	*Weightage*	*Number of Respondants 2*	*Weightage*	*Number of Respondants 3*	*Weightage*	*Number of Respondants 4*	*Weightage*	*Number of Respondants 5*	*Weightage*	*perception Value*
65 B/11 3/5	38	0.19	87	0.87	64	0.96	7	0.18	4	0.10	2.30
65 B/16 2/4	–	–	–	–	11	0.17	73	1.46	116	2.90	4.53
65 C/6 2/3	–	–	–	–	24	0.36	85	1.70	91	2.28	4.34
65 C/6 3/3	–	–	–	–	26	0.39	86	1.72	88	2.20	4.31
65 C/9 3/5	14	0.07	26	0.26	71	1.07	57	1.14	32	0.80	3.34
65 C/9 5/1	46	0.23	71	0.71	47	0.71	22	0.44	14	0.35	2.44
65 C/9 5/2	–	–	33	0.33	78	1.17	51	1.02	38	0.95	3.47
65 C/9 5/5	–	–	36	0.36	64	0.96	59	1.18	41	1.03	3.89
65 C/10 4/2	–	–	–	–	–	–	18	0.36	182	4.55	4.19
65 C/10 5/1	–	–	–	–	–	–	24	0.48	176	4.10	4.58
65 C/10 5/4	–	–	–	–	9	0.14	57	1.14	134	3.35	4.63
65 C/10 5/5	–	–	15	0.15	32	0.48	86	1.72	67	3.68	4.17
65 C/11 1/5	21	0.11	29	0.29	83	1.25	49	0.98	18	0.45	3.08
65 C/11 3/5	–	–	44	0.44	62	0.93	58	1.16	36	0.90	3.43
65 C/11 4/1	55	0.28	82	0.82	39	0.59	24	0.48	–	–	2.17
65 C/13 2/2	–	–	–	–	37	0.56	89	1.78	74	1.85	4.19
65 C/13 2/4	–	–	13	0.13	83	1.25	68	1.36	36	0.90	3.64
65 C/13 3/4	–	–	–	–	64	0.96	77	1.54	59	1.48	3.98

(Contd...)

Table 11.6 *(Contd.)*

65 C/13 4/1	–	–	–	–	–	–	38	0.76	162	4.05	4.81
65 C/13 4/4	–	–	–	–	–	–	32	0.64	168	4.20	4.84
65 C/14 1/4	20	0.10	38	0.38	53	0.80	53	1.36	41	1.03	3.67
65 C/14 4/2	–	–	–	–	–	–	38	0.76	162	4.05	4.81
65 C/14 5/5	15	0.08	79	0.79	64	0.96	42	0.84	–	–	2.67
65 C/15 5/1	–	–	22	0.22	83	1.25	57	1.14	38	0.95	3.56
65 G/2 1/5	33	0.17	62	0.62	48	0.72	43	0.84	14	0.35	2.70
65 G/2 2/5	11	0.06	19	0.19	64	0.96	78	1.56	28	0.70	3.47
65 G/2 4/1	21	0.11	49	0.49	63	0.95	46	0.92	21	0.53	3.00
65 G/2 4/2	14	0.07	36	0.36	79	1.19	54	1.08	17	0.43	3.13
65 G/2 5/3	10	0.05	19	0.19	86	1.29	62	1.24	23	0.58	3.35
65 G/3 1/1	12	0.06	26	0.26	61	0.92	65	1.30	36	0.90	3.44
65 G/3 1/3	22	0.11	21	0.21	76	1.14	52	1.04	29	0.70	3.20
65 G/3 3/1	23	0.12	31	0.31	63	0.95	45	0.90	38	0.95	3.23
65 G/3 5/2	35	0.18	54	0.54	67	1.01	26	0.52	18	0.45	2.70
65 G/6 1/4 &1/5	–	–	–	–	7	0.11	61	1.22	132	3.30	4.63
65 G/6 2/5	–	–	–	–	2	0.03	52	1.04	146	3.65	4.72
65 G/6 3/5	–	–	–	–	19	0.29	72	1.44	109	2.73	4.46
65 G/6 4/3	29	0.15	98	0.98	42	0.63	25	0.50	6	0.15	2.41
65 G/7 1/1	–	–	–	–	47	0.71	89	1.78	64	1.60	3.38
65 G/7 3/1	–	–	–	–	25	0.38	89	1.78	86	2.15	4.31
65 G/7 5/1	–	–	–	–	–	–	–	–	200	5.00	5.00
65 O/16 3/1 & 4/1	–	–	7	0.07	51	0.77	74	1.48	68	1.70	4.02

$$\text{WEIGHTAGE} = \frac{\textbf{Number of Respondants *PERCEPTION WEIGHTAGE}}{\textbf{Total Number of Respondants}}$$

PERCEPTION = Poor + Average + Good + Very Good = Excellent

Table 11.7 : Ground Water-Quality Chemical Analysis

Village	*pH*	*Turb (NTU)*	*Conductivity (uhos/cm)*	*T.D.S. (mg/l)*	*Total alkalinity (m g/l)*	*Total Hardness as $CaCO_3$ (mg/l)*	*Ca(mg/l) as Ca*	*Mg(mg/l) as Mg*	*Cl (mg/l) as Cl*	*SO_4(mg/l) as SO_4*	*NO_3(mg/l) as NO_3*	*NO_2(mg/l) as NO_2*
Amora	7.66	10	1900	1140	362	1200	100	123	105	750	11.5	0.15
Anta	7.2	8.8	1620	880	486	536	56.8	45	152	105	37.4	0.2
Ballabhpura	7.3	7.2	1170	902	454	676	50.8	150	64	95	25	0.05
Bamori	7.8	12.1	935	560	250	464	102.4	69	52	74	11	ND
Bhonra	7.27	8.5	1828	1080	274	470	107.2	100	57	840	17	ND
Cheensa	8.5	6.1	1400	900	380	202	23.2	125	35	105	19	0.05
Gadepan	7.44	10	1820	1170	290	600	180	175	275	221	75	0.1
Gulabpura	7.43	9.2	1120	960	354	1044	194	205	290	45	32	ND
Kakrawada	7.85	8.5	1200	1000	526	506	42	50	54	35	57	ND
Khera-Kheri	7.82	6.2	3005	2010	244	1700	224	500	57	600	60.5	0.07
Kherli Mahadeet	7.48	9.2	1000	730	220	1020	59	29	49	120	20.2	0.05
Nagda	7.4	6.4	992	605	412	1029	75	225	105	327	26.1	0.1
Palaitha	7.53	9.6	985	600	386	256	36	48	57	70	46	0.1
Simlia	7.63	10.5	1600	1200	256	798	54	29	95	615	170.1	0.18
Surela	8.2	6.8	1608	920	544	144	39	74	50	218	22.2	0.2

REFERENCES

Andhra Pradesh Year Book - 1999, Data News Features Publications, Hyderabad.

Andhra Pradesh Tourist Places - Published work of Andhra Pradesh Tourism Department.

Appleton, J. (1975), Lanscape evaluation: The theoretical vacuum, Transactions of the Institute of British Geographers, no. 66, 120-123.

Chubb and Banuman (1977). *Recreation Geography* (pp. 215-230, 317-319) London: Longman.

Cooke, R.U., and Doornkamp, K.C. (1974), *Geomorphology in Environmental Management-Introductionk*, Clarendan Press, Oxford.

Fines (1968), *Geomorphology in Environmental Management*—Introduction. Clarendan Press, Oxford.

Isachenko (1973), *Recreation Geography* (pp. 215-230, 317-319). London: Longman.

Leopold (1969), *Geomorphology in Environmental Management—Introduction.* Clarendan Press, Oxford.

Linton (1968, 73), *Geomorphology in Environmental Management—Introduction,* Clarendan Press, Oxford.

Padmajas (1998), "Aesthetic Evaluation of Landscape for Urban Recreation Andhra Pradesh", CSIR- Oxford.

Shafter (1969, 1970), *Recreation Geography* (pp. 215-230, 317-319). London: Longman.

Smith, R.J, (1974). The economics of outdoor recreation. In P. Lavery (Ed), *Recreation Geography* (pp. 215-230, 317-319). Newton Abbot: David and Charles.

Smith, S.L.J. (1983), *Recreation Geography* (pp. 215-230, 317-319). London: Longman.

Smith, S.L.J. (1989), The spatial analysis of recreation and leisure. In E.L. Jackson & T.L. Burton (Eds), *Understanding leisure and recreation: mapping the past, charting the future.*

Smith & Small (1980), *Recreation Geography* (pp. 215-230, 317-319) London: Longman.

Spate, O.H.K, Learmonth, A.T.A. and Farmer, B.H. (1967). *India, Pakistan and Ceylon—The regions*, Methuen & Co. Ltd. London.

12

Perspective of Land Resource Development in Madhya Pradesh and Chhatisgarh

Problems and Prospects

S.K. SHARMA

The contemporary meaning of development is the perceived progress of human being without undue stress and pressure on environment and fellow members of the society. Such development hinges on myriad of factors, land resource is one of them. For the present purpose land resource is restricted to the agricultural land only. Land is the only true resource, because, being non-transportable in nature, it must be used for development of the area of its occurrence. All other natural resources are transportable and need not to be used for the development of the region of their occurrences. Therefore, planning for development in developing regions is mainly concerned with the evaluation of land resources (Young, 1975, 9). Evaluation and development of land and water resources has been earmarked as essential preliminary to development, particularly to the agricultural development planning (Robertson, 1970, 148). It is such vital resource-complex that the proper and profitable use of land, both present and future, is of basic importance to any nation (Stamp, 1942). Proper and optimum utilization of land is of even more critical value for regions solely dependent on agriculture. Such as in Madhya Pradesh, agriculture provides employment to more than three-fourths (75.3%) of main workers and contributes more than one-third (37.1%) of the gross value of the net state products. Naturally the fate of the ruralites depends on the improvement of the conditions. Further, increasing pressure of population compels farmer to ignore the natural limits of land uses; and farming is extended to physically unsuitable lands, creating several environmental crisis. Therefore, it is pretinent to see the land utilization in perspective of population pressure on it, rather than viewing land as an independent variable of development. Present paper is an attempt to see the spatial pattern of the development of land resource in the light of the pressure of population on land resource in Madhya Pradesh and Chhattisgarh combined States.

Contemporary Situation of Land Resource Development

Agricultural land provides base for the economy and society of this state. It provides employment to 7.60 per cent of total workers (2001) and contributes more than one-third (37.1%) of the total value of domestic products (1991-92). In 1999-2000, about 19.9 million hectares are net sown, which is 44.86 per cent of the total area. Nearly 6.31 million hectares of net sown area is sown more than once, bringing the intensity of cropping to 131.7 per cent. Thus in respect of net sown area this state is above the national average (43.3% in 1997-98) but behind the states of northern plains and even Maharashtra (59.5%) and Rajasthan (45.7%). More than 9603 thousand holdings were in operation in 1995-96. Thus average size of operational holdings is 2.6 hectares but availability of agricultural land per rural person is only 0.327 hectare. With small per capita land area, agricultural productivity is also low. Per hectare yield of all food-grains is1211 kg in the state as against 1704 kg in the country in 1999-2000. This low productivity can be ascribed to small use of yield-raising inputs. For instance, only 33.9 per cent of net sown is net irrigated as against the national average of 38.4 per cent (1997-98). The use of the high yielding variety of seeds is very limited. Out of the total area under five cereals only one-third (48.2%) is sown with high yielding varieties of seeds in (1999-2000) the state as against the national average of 77.0 per cent (1995-96). Further, only 45.98 kg of all types of fertilizers are used per hectares of gross cropped area in (1999-2000) the state, which is less than two-thirds of the national average of 75 kg (1994-95).

Land Utilization

In 1999-2000, 19.90 million hectares are net sown, which is 44.86 per cent of the total area of the state. Another 3.3 per cent area is lying fallow, which was sometimes under cultivation. This brings the total cultivable area to 48.2 per cent of the total geographical area. Non-agricultural uses of land claim about 5.7 per cent of area. Thus, slightly more than half (53.9%) of the land could be put to some use.

Net Sown Area

Expansion of the net sown area has been comparatively rapid in the later half of this century. Total net sown area was 139.69 lakh hectares in 1950-51 which rose to 198.98 lakh hectares in 1999-2000, recording 45.6 per cent increase between 1950-51 and 1999-2000. Thus, 5.9 million hectares were added to net sown area. It is worth-mentioning that cultivation has been practiced since long ago and there was no good land left for colonization. Consequently, farming has been extended on marginal lands and on slopes of hills. It is evident from

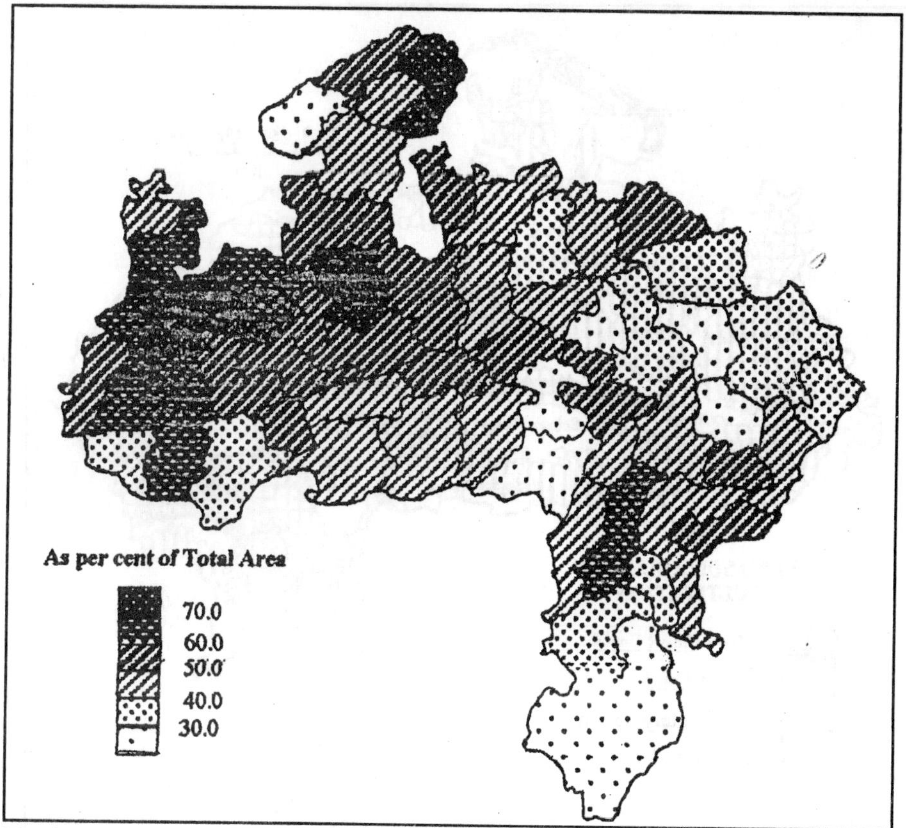

Fig. 12.1 : Madhya Pradesh—Net Sown Area 1999-2000

the fact that high and very high growth in net sown area occurred in regions not suitable for cultivation.

Proportion of net sown area varies from only 18.7 per cent in Koriya and 19.1 per cent in Dantewara districts (both in present Chhattisgarh) to 75.8 per cent in Bhind and 78.8 per cent in Ujjain district (both in present Madhya Pradesh) in 1999-2000 (Fig. 12.1). On block level variation is even more evident. It ranges between 9.3 per cent in Bharatpur (Surguja district) and 12.5 per cent in Kusumi (Sidhi) blocks and 94.6 per cent in Pansemal (Khargone) and 95.6 per cent in Shujalpur (Shujalpur district) blocks in 1993-94. Distribution of the development blocks according to the proportion of net sown area to total rural area is presented in Table 12.1 below.

Table 12.1 : M.P.—Distribution of the Development Blocks according to the Percentage net sown area in 1993-94

% of Net sown area	*Below 20*	*20 - 42*	*40 - 60*	*60 - 80*	*Above 80*	*All*
Number of blocks	10	64	149	198	37	458
% of total blocks	2.2	14.0	32.5	43.2	8.1	100.0

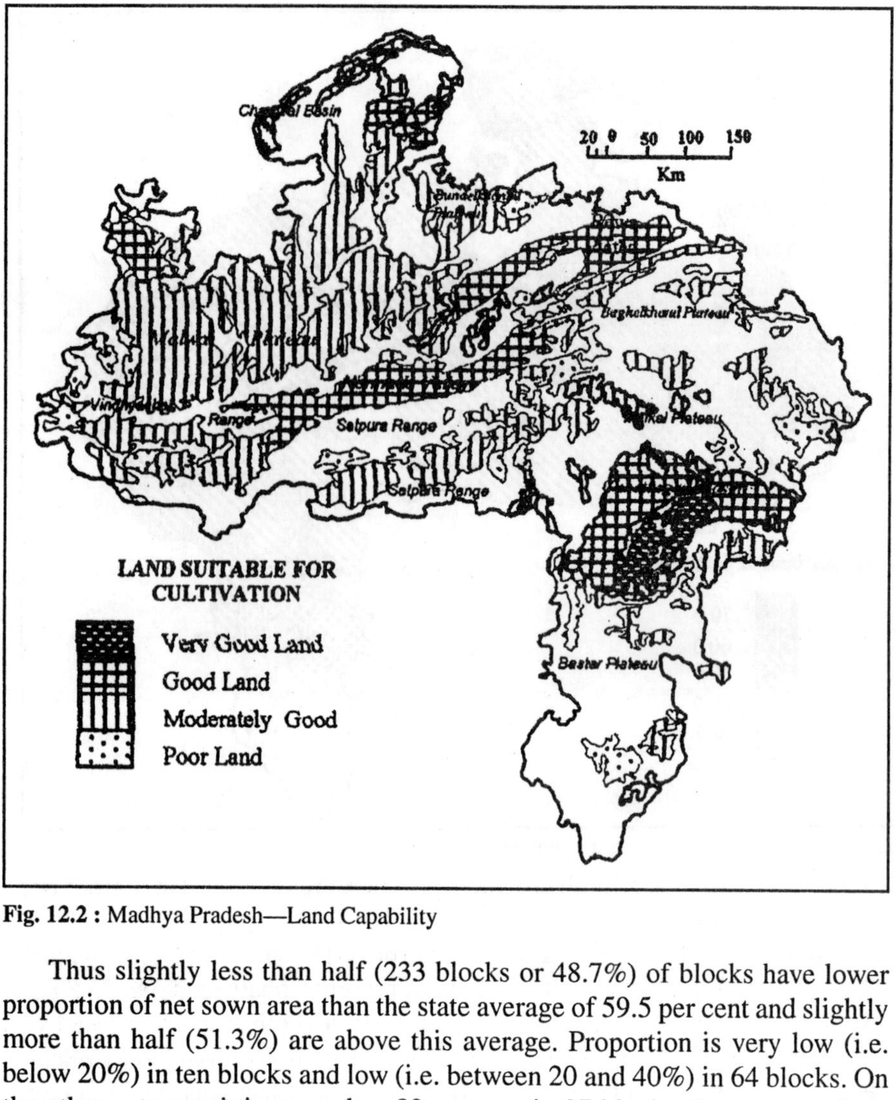

Fig. 12.2 : Madhya Pradesh—Land Capability

Thus slightly less than half (233 blocks or 48.7%) of blocks have lower proportion of net sown area than the state average of 59.5 per cent and slightly more than half (51.3%) are above this average. Proportion is very low (i.e. below 20%) in ten blocks and low (i.e. between 20 and 40%) in 64 blocks. On the other extreme, it is more than 80 per cent in 37 blocks. Proportion of net sown area is high and very high in the Malwa plateau, Sindh-Pahuj plain, Nimar plain, Narmada valley, Rewa plateau and Chhattisgarh basin. Contrary to them, proportion of net sown area is low and very low in the Bastar plateau, Baghelkhand plateau, Satpura region and Vindhyachal-Bhander hilly area.

Capability of Land and Distribution of Net Sown Area

The nature of terrain and characteristics of soils are the principal determinants of the land utilization. These elements along with some climatic characteristics determine the suitability of land for general use. Land must be used in accordance with its capability to avoid its deterioration and to adopt appropriate conservation measures. National Atlas and Thematic Mapping Organization, Kolkata has prepared land capability maps of India on two million scale. The net sown area map of the state is compared with these maps (Fig. 12.2) which reveals that proportion of net sown area is high (between 60 and 80%) and very high (above 80%) on very good to moderately good lands in terms of land capability. Such regions are the Chhattisgarh basin, Narmada valley, Rewa plateau, part of the Chambal basin, Malwa plateau, Nimar plain and Betul-Chhindwara-Lakhanadon plateaus in the Satpura region. At the same time, some of the poor quality lands also support high proportion of net sown area while they are capable of supporting pastures and forestry. Among such regions mention be made of the Dhar-Jhabua hilly tract, western Satpura, part of the Bundelkhand uplands and Raigarh hills.

Proportion of net sown declines with the increasing severity of limitations. Such lands are physically unsuitable for cultivation. Most of the lands of the Bastar plateau, Baghelkhand plateau, Satpura region and Vindhyanchal-Bhander hills are unsuitable for crop farming because of hilly, dissected and sloppy land and shallow and infertile soils. Such lands are also widespread on parts of the Madhya Bharat plateau and Bundelkhand uplands. Lands of these regions are classed as unsuitable for cultivation. Most of the area is bare and barren and suffer from erosion hazards. They are deemed suitable for pastures, grazing and forestry. Despite these facts, they are cultivated but proportion of net sown area ranges from low to moderate (40 to 60%). In fact, excessive dependency on agriculture led not only to the excessive utilization of land resources but also to extend physical limits of farming. There is competition for grabbing lands in every village, and farming has encroached on forests, fallowlands, village common lands and on land physically unsuitable for cultivation. Expansion of cultivation on these marginal lands has caused more serious damage to forests. Removal of forest cover has accelerated soil erosion, incidence of flood, development of gullies and ravines and excessive silting of dams and reservoirs (Sharma, 1987, 240-243).

Following conclusions can be drawn from the analysis of the distribution of net sown area in relation to land capability:

1. The spatial pattern as reflected by the block-wise data is directly related with the capacity of lands.
2. In most cases, capability of land has been over-used. Very good and moderately good lands are widely cultivated. Even non-cultivable lands are cropped ignoring their suitability.

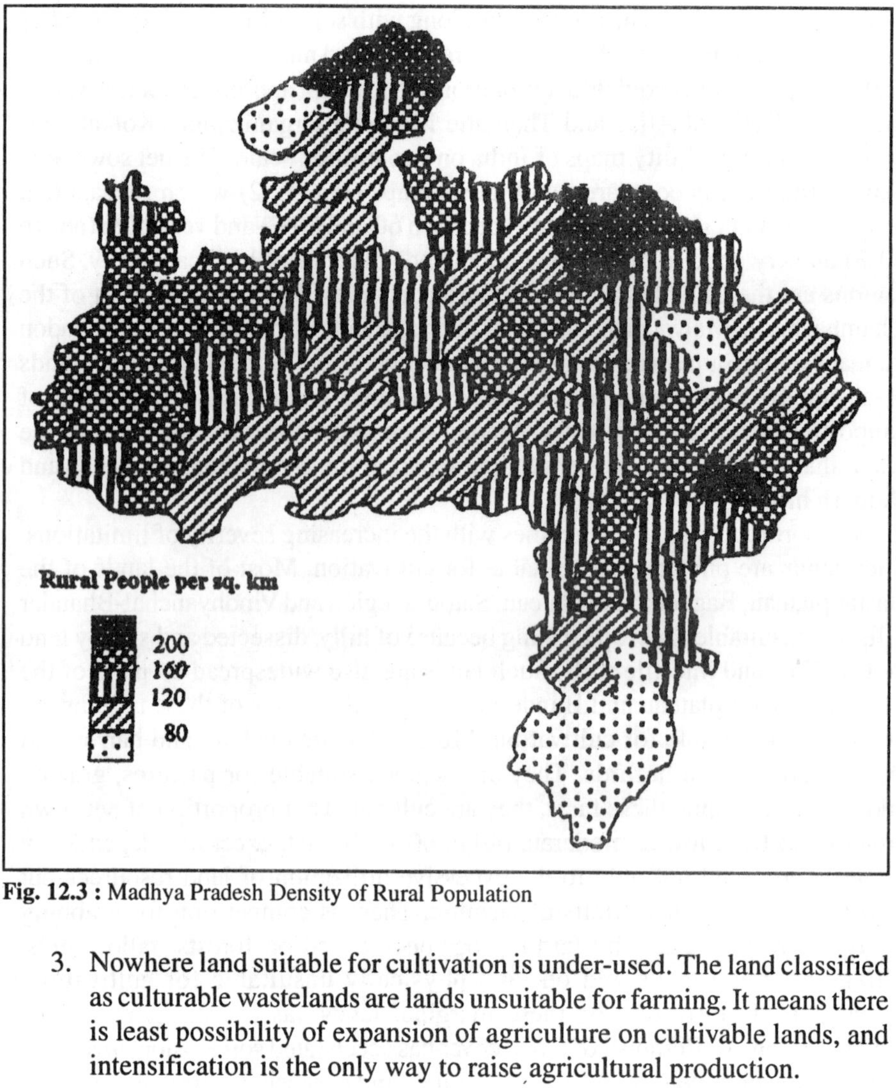

Fig. 12.3 : Madhya Pradesh Density of Rural Population

3. Nowhere land suitable for cultivation is under-used. The land classified as culturable wastelands are lands unsuitable for farming. It means there is least possibility of expansion of agriculture on cultivable lands, and intensification is the only way to raise agricultural production.

Population Pressure and Net Sown Area

Along with the physical character of land, demographic attributes of the region also go long way in influencing the development of land resources. The physical environment provides potentialities of and puts limitations to the land development. Among this group of factors only pressure of rural population is selected for the present purpose. It is more meaningful to assess the net sown area in relation with the rural population. It can be done in two ways: one by analyzing the density of rural population on net sown area, known as the nutritional density; and the other by calculating availability of net sown area per rural person. Total 60.9 million rural people are living on 4,35,538 sq. km of rural area of the state in 2001. Thus on average, there are 140 rural persons per sq. km. It ranges from only 43 persons in Dantewara and 69 in Bastar and Koriya districts to 262 in Janjgir and 263 in Rewa district. On block level, it ranges from 6 persons in Orchha block (Bastar district) and 26 persons in Bharatpur block (Surguja district) to 377 persons in Dharsiba block (Rajpur district) and 381 persons in Katghora block (Bilaspur) in 1991. From comparison of rural density map (Fig. 12.3) with that of net sown area (Fig 12.1) three types of situations are witnessed.

Firstly, densely populated rural area usually have higher proportion of net sown area. Secondly, not all widely cultivated blocks have high rural density. Major portion of the Malwa plateau, middle Narmada valley and Betul Lakhanadon plateaus are good examples of low rural population density despite of high proportion of net sown area.

Thirdly, contrary to it, areas of low rural density usually have low and very low proportion of net sown area. Consequently, even low density exerts high pressure on cropped area in these regions and land physically not suitable for cultivation is brought under cultivation. Because of this fact, growth of net sown area during last five decades has been very high in these areas.

Nutritional Desity

Nutritional density is more realistic expression of pressure of rural population on cultivated land. It is ratio between rural population and net sown area. Average nutritional density for the state, based on population data of 2001 and net sown area in 1999-2000, is 306 rural persons per sq. km of net sown area. On district level, nutritional density ranges from only 179 rural people per sq km of net sown area in Vidisha district to 481 persons in Kobra district of Chhattisgarh in 2001. On block level, it ranges from only 120 persons in Sukma block of Bastar district to 810 persons in Katghora block of Bilaspur district in 1991. Total 207 development blocks out of 458 have lower density than 260 persons per sq. km. of net sown area (state average in 1991).

Spatial pattern of nutritional density is quite different from those of arithmetic density and rural density. Salient features of this distribution are as follows :-

1. All those areas, which have high rural density and are basically dependent on agriculture also, have high nutritional density. Such regions are the crores of the Chhattisgarh basin, Rewa plateau, Wainganga basin, and the lower Chambal valley. It is also higher on parts of the Nimar plain. Proportion of net sown area is high in these areas.

2. Besides them, in low rural density areas, where not sown area is very limited due to physical limitations, nutritional density is higher than the state average. Among such regions mention be made of the Banghelkhand plateau and Bundelkhand uplands. Proportion of net sown area is low and very low in these regions; and therefore, small population is too much on cropped land. Similar situation exists in 15 blocks of Bastar district which happens to be least populated tract of the state.
3. Contrary to it, in areas where arithmetic density is enhanced due to industrialization and urbanization, nutritional density is low; such as in Bhopal-Vidisha tract, Betul-Chhindwara plateau, Indore-Ujjain tract and Ratlam-Mand-saur area. In fact, in most of these areas rural densiry is moderate, not low. At the same time, proportion of net sown area is high. Consequently, pressure of rural population on net sown area is low and very low. The Malwa plateau, Narmada valley, central Satpura region and Sagar-Damoh plateau are such areas.

Table 12.2 : M.P. Distribution of Districts according to Nutritional Density, 2001

Density Class	*<200*	*200-250*	*250-300*	*300-350*	*350-400*	*>400*	*All*
No. of districts	01	09	24	08	09	10	61
% of total districts	01.64	14.75	39.34	13.11	14.75	16.39	100.00

It can be summed up that the pressure of rural population on agricultural land is comparatively high in the eastern half of the state and also in the lower Chambal valley but it is low in the western half. This scenario has two important effects on utilization of land. Firstly, it determines per capita availability of land and size of the operational holdings and secondly, it goes long way in influencing the use of modern inputs in agricultural development.

Per Capita Sown Area

Per capita availability of net sown area would clear the situation even more clearly. On average, only 0.327 hectare of net sown area per rural person is available in the state. It ranges from 0.21 ha in Korba district to 0.559 ha in Vidisha distirct in 2001. On block level it varies from 0.123 ha. in Katghora block (Bilaspur) to 0.833 ha. in Sukma block of Bastar district. It is the function of the nutritional density, and therefore, blocks of eastern and northern parts of the state have low and very low per capita net sown area because of high and very high nutritional density. Situation improves westwards. Even then the average situation is very grim in the state as a whole. If we presume the average size of a family with five persons, then per family net cropped area arrives at 1.94 hectares. This much land is insufficient for sustaining an average family along with their livestock.

Social Structure and Agricultural Land Use

Use of land in general and of agricultural land in particular depends on the socio-economic status as well as characteristics of farmers. Adoption of modern methods, techniques and inputs varies not only spatially and temporally but it also varies from community to community and from person to person. Social structure is the third dimension of the reality. In fact, the capacity of adoption hinges upon the size of holding of the farmer and his social status donated by caste. Even two characteristics of farmers have resulted in unequal accessibility of farmers to the new scientific knowledge and technology. It has manifested in differential level of utilization of land resources.

Size of the Operational Holdings

This situation is corroborated by the distribution of the size of operational holdings. Proportion of marginal (less than 1 ha) and small (1-2ha) ranges from 39.9 per cent of all holdings in Dewas district in the west to 79.5 per cent in Bilaspur district in the east. Spatial pattern clearly shows that more than 60 per cent of the operational holdings in the eastern part is marginal and small in size while on major portion of Malwa and Narmada valley these holdings constitute less than 50 per cent of the total holdings. Such distribution pattern of the per capita net sown area as well as size of the holdings has bearings on adoption of agricultural innovations and also on agricultural productivity.

Table 12.3 : Variation in adoption of Agricultural Innovations by Size of Holdings (Adopters as percentage to total households of size-class)

Innovations	*Chhindwara-Betul*				*Chhattishgarh Basin*				*Tons Basin*			
	ST	*SC*	*BC*	*OC*	*ST*	*SC*	*BC*	*OC*	*ST*	*SC*	*BC*	*OC*
Irrigation	38.7	82.5	100.0	63.0	44.5	64.6	99.7	53.3	59.3	88.5	100.0	81.8
HYVs	80.6	80.0	81.3	79.8	11.5	37.5	50.0	25.2	56.0	77.0	78.1	66.2
Fertilizers	57.4	63.6	63.3	61.6	88.5	98.0	100.0	93.9	62.6	98.4	100.0	85.1

Source : Sharma 1998, p. 168.

Adoption of modern agriculture inputs has proved panacea for raising production. Being capital intensive and scientific in nature, the use of modern inputs depends on the capacity and capability of the farmer. The capacity of the farmer, on the other hand, depends mainly on the size of holdings. In fact size of holdings is necessary qualification for most of the opportunities including the access to the institutions and organisations involved in the process of diffusion of innovations and to the entire infrastructure. Therefore, the size of holding creates disparity in level of adoption. It has been on meso (Sharma, 1998) and household level (Sharma, 1998). Further, adoption and size of holdings have direct relation with the productivity (Sharma, 1999).

Surveys conducted in this connection on household level in the Chhattisgarh basin, Betul-Chhindwara plateau and Tons basin of this State reveal this fact. Situation of adoption of certain innovations in these regions is presented in Table12.3 in which adopters of each variable are converted as percentage of total sample households to the size class or the community.

Farmers have perceived the significance of irrigation and are trying their best to irrigate their fields. It is evident from the above table that more than half of the households of the Chhattisgarh basin, two-thirds of the Betul-Chhindwara plateau and more than three-fourths of the Tons basin practice irrigation. This proportion varies with the size of holdings. As a rule, proportion of irrigating households is lower in case of marginal and small holdings than their averages and much lower than the proportion of medium and large holdings. This

Table 12.4 : Variation in adoption of agricultural Innovations by Communities (Users as per cent of total household of the community)

Innovations	*Chhindwara - Betul*				*Chhattishgarh Basin*				*Tons Basin*			
	ST	*SC*	*BC*	*OC*	*ST*	*SC*	*BC*	*OC*	*ST*	*SC*	*BC*	*OC*
Irrigation	49.7	44.4	73.9	82.9	43.9	24.3	67.9	48.4	39.1	83.3	75.3	86.9
HYVs	60.8	66.0	93.6	100.0	26.3	8.1	27.7	32.3	47.8	54.2	56.5	71.5
Fertilizers	33.4	39.3	79.3	82.6	84.2	52.6	100.0	100.0	39.1	66.7	87.1	89.7
Land treatment	2.3	6.7	23.4	22.9	1.8	2.7	5.8	3.2	4.3	4.2	2.4	4.1
Seed treatment	4.1	8.9	28.0	41.5	1.8	2.7	8.8	3.2	4.3	8.3	3.5	10.3
Pesticides	10.5	15.6	30.3	53.7	26.3	45.9	52.6	74.2	00.0	00.0	2.4	7.2
Tjresjer	7.6	35.6	67.4	70.7	00.0	00.0	0.7	00.0	na	na	na	na

Source : Sharma, 1998.

proportion also varies with the community (Table 12. 4). Scheduled castes (SC) are at bottom while backward communities are at top in this respect.

Under the scintific management of farming certain measures have been used to maintain and enhance the soil fertility. With the advent of green revolution, use of chemical fertilizers is popularized and they are widely used in regions under consideration. There is clear cut influence of size of holding and social status of the farmers on use of fertilizers. Marginal farmers and scheduled population are trailing far behind in use of fertilizers. Along with the yield-raising technology, certain protectional measures also have been propagated. These protectional measures include land and seed treatments, rat control etc. The use of these practices is also positively related with the size of holdings and community. These practices are usually limited to semi-medium, medium and large holdings. Further, it seems that these practices are exogenous to Scheduled Tribes and, to some extent, to Scheduled Caste also. They are incapable of accepting these methods radily. Mechanization is also associated with the socio-economic capacity of farmers. Example can be cited of the thresher, use of which is size-specific and caste-specific. It can, therefore, be inferred that the size of the operational holdings is the kingpin of the entire socio-economic structure and manifests itself in the use of agricultural development measures. At the same time, place of farmer in social arena goes long way in influencing the adoption.

Agricultural Productivity

The problem of tiny piece of land of the rural people in multiplied by its low productivity. Agricultural productivity is still low in this state in comparison to that of the country. It is evident from the fact that per hectare production of all food-grains is 1028 kg in this state as against 1547 kg in the country in 1995-96. On this basis per capita production thus arrives at (0.338 ha × 1028 kg) 399 kg and per family it comes to (1.94 ha × 1028 kg) 1994 kg. All major crops have lower yield rates than that of the country. Consequently, this state could produce only 9.5 per cent of food-grains on 14.3 per cent of total area under them of the country.

Agricultural productivity, measured in terms of value of crops per hectare, ranges from only Rs. 3303 in Mandla district to Rs. 8219 in Narsingpur district with state average of Rs. 5708 in 1992-93 at current prices. Twenty-two out of the forty-five districts have lower value than the state average, and fourteen of them have low and very low productivity. They form a continuous belt, extending from Seoni to Panna, covering Seoni, Mandla, Shahdol, Sidhi, Rewa, Satna, Panna and Damoh districts. Jhabua, Rajgarh, Betul, Rajandgaon, Surguja, and Bastar are out of this belt. Contrary to them, twenty-three districts with higher productivity form four extensive tracts. Except Chhattisgarh basin, other three tracts are in the western part. They cover the lower Chambal basin, core of the Malwa region and Narmada valley and adjoining Chhindwara plateau. It can safely be concluded that productivity is usually lower in areas of smaller holdings measured in terms of per capita availability of net sown area or the size of the operational holdings. Contrary to it, it is higher in areas of large holdings. This situation goes long way in determining per worker productivity.

Productivity per Agricultural Worker

Even more realistic picture of the land resource development would emerge from the analysis of the agricultural productivity per agricultural worker. It ranges from Rs. 1364 in Bhopal district and Rs. 2859 in Mandla district to Rs. 15278 in Hoshangabad district. Spatial distribution of productivity per worker reveals several interesting facts. For example, worker productivity is also high in areas where land productivity is high. At the same time, worker productivity is also low in eastern districts of low per hectare productivity. But low productive districts of eastern Malwa and southern Madhya Bharat plateau present high productivity per agricultural worker. Thus larger availability of net sown area and bigger size of holdings yield higher per worker productivity in the western part of the state. But heavy pressure of population causes smaller holdings as well as smaller per capita available net sown area, both of which result in low and very low productivity per agricultural worker in the eastern part.

Intensity of Cropping

Anatomical analysis of agricultural productivity reveals that the only way left for raising agricultural production is the intensification of cropping. Intensity of cropping means raising several crops from the same land in a year and can be calculated by converting the gross cropped area as percentage of net sown area. Thus it is proportion of area sown more than once. The area sown more than once increased from 15.85 lakh hectares in 1951-52 to 63.09 lakh hectares in 1999-2000. Thus intensity of cropping increased from 111.2 per cent in 1951-52 to 131.7 per cent in 1999-2000. Increase in area sown more than once was nearly four times during last five decades.

Distribution of area sown more than once is very uneven. It ranges from only 100.3 per cent in Konta and 100.4% in Usoor blocks (both in Bastar district) to 171.4 per cent in Dhamtari and 177.3 per cent in Kurund blocks (both in Raipur district). Intensity is high in five areas while it still low over extensive areas of the state. Areas of high intensity of cropping are : (i) the crore of the Chhattisgarh basin, (ii) Wainganga basin, (iii) Rewa division, (iv) Bundelkhand region, and (v) Mandsaur region of the Malwa plateau. With the exception of Rewa division, high intensity is closely associated with the provisions of irrigation. Use of other yield-raising inputs follows irrigation. It leads to the conclusion that there is vast scope of increasing intensity of cropping in areas with moderately good and good lands by increasing the irrigation facilities, which is also essential for popularizing modern farm technologies and for raising agricultural productivity.

Conclusions

The preceding analysis suggests that the density of rural population declines with the declining proportion of net sown area. But the rural density correlates rural population with rural area and not with the net cropped area. Further, the ratio of rural population and net cropped area is not uniform everywhere. Pressure of rural population on net cropped area is high and very high in areas of low rural density. Consequently, per capita net cropped area is low and very low.

Such situation exists in the eastern part of the state where cultivability of land is limited by the physical limitations. Nevertheless, farming is still the principal means of livelihood and proportion of main workers engaged in agricultural activities is comparatively high in this part. Low land productivity and larger number of agricultural workers results in low productivity per worker. This situation suggets the general conclusion derived by several studies that the density of rural population is directly related to the productivity of land (Nityanand, 1966, Ayyar and Dube, 1968, Mehta, 1973). In the poverty-ridden society the survival algorithm is operative rather than the principle of productivity. People still live in suppressing poverty in low productivity areas. Large number of ruralites constantly searches outlet for emigration for profitable jobs. The contemporary land resource development is incapable of improving the fate of the people unless it is integrated with the development (not exploitation) of other resources within the region.

REFERENCES

Ayyar, N.P. and R. Dube (1968) Some aspects of rural population of Madhya Pradesh, *Geographical Outlook,* Vol. 5, p. 53.

Mehta, B.C. (1973) Spatial distribution of population in Rajasthan, *National Geographical Journal of India,* Vol. 19, parts 3-4, pp. 149-157.

Nityanand (1966) Distribution and spatial arrangement of rural population in East Rajasthan, India, *Ann. Asso. Am. Geogrs.* Vol. 56, p. 205.

Robertson, V.C. (1970) Land and water resource planning in developing countries, *Outlook on Agriculture,* Vol. 6, No. 4, pp. 148-157.

Sharma, S.K. (1987) Dilemma of Resource Development and Environmental Crises in Madhya Pradesh, pp. 236-254 in H.S. Sharma & M.L. Sharma, eds. *Environmental Design and Development.* Jodhpur. Scientific Publishers.

Sharma, S.K. (1988) Adoption of Agricultural Development measures in relation to Size and tenancy of Operational holdings in Madhya Pradesh, *Journal of India Regional Science,* Vol. 20, No. 2, pp. 27-39.

Sharma, S.K. (1998) Reflection of Social Inequality on Adoption of Development Measures : An Example of Agricultural Innovations, pp. 165-77 in Y.G. Joshi & D.K. Verma, eds. *Social Environment for Sustainable Development.* Jaipur : Rawat Publications.

Sharma, S.K. (1999) *Adoption of Agricultural Innovations and their impact on Agricultural Productivity in Madhya Pradesh,* paper contributed to the National Seminar on Geographical Perspective on Fifty Years of Indian Agriculture held at the Department of Geography, AMU, Aligarh, 17-19 March 1999, being published in the Proceedings of the Seminar.

Young, A., 1975, Rural land evaluation, pp. 382-423 in *Trophical Soils and Soil Surveys.* Cambridge University Press.

13

Settlements Distribution and Their Correlation with Morphometric Attributes in the Parbati Basin, South-East Rajasthan

L.C. AGRAWAL

Introduction

The study of distribution settlements of and their correlation with morphometric attributes is one of the applied aspects of geographical studies. To advance the applied aspects of geography for the welfare of the human being. It is now defined as the science which provides the base to understand the earth as a home of man (James, 1972). Within this frame, a new trend is geomorphological attributes to the human phenomena as such it gives a relevancy for the physico-cultural planning. Such a physical and cultural unity concerning its processes, impact and involving structure will certainly represent the real world situations which has been long advocated by Sauer (1965).

The present study is limited within the correlative study of the impact of geomorphological attributes upon the spatial characteristics of the settlements.

Distribution of Settlements

The spatial distribution of human settlements are mainly governed by climatic conditions, relief, slope, water frontage and socio-economic factors. But in the Basin, topography, water bodies, river channels and climate exert their influence on the distribution of settlement more than other factors. Due to the control of the above mentioned factors to a different land of their intensity, the settlement distribution in the whole area is uneven (Fig. 13.2).

In the present case study the distribution of settlements has been analysed by quadrate analysis. This analysis helps in finding out the distribution pattern of settlements at micro level.

The degree of concentration has been worked out. The whole basin has been divided into three categories as follows:

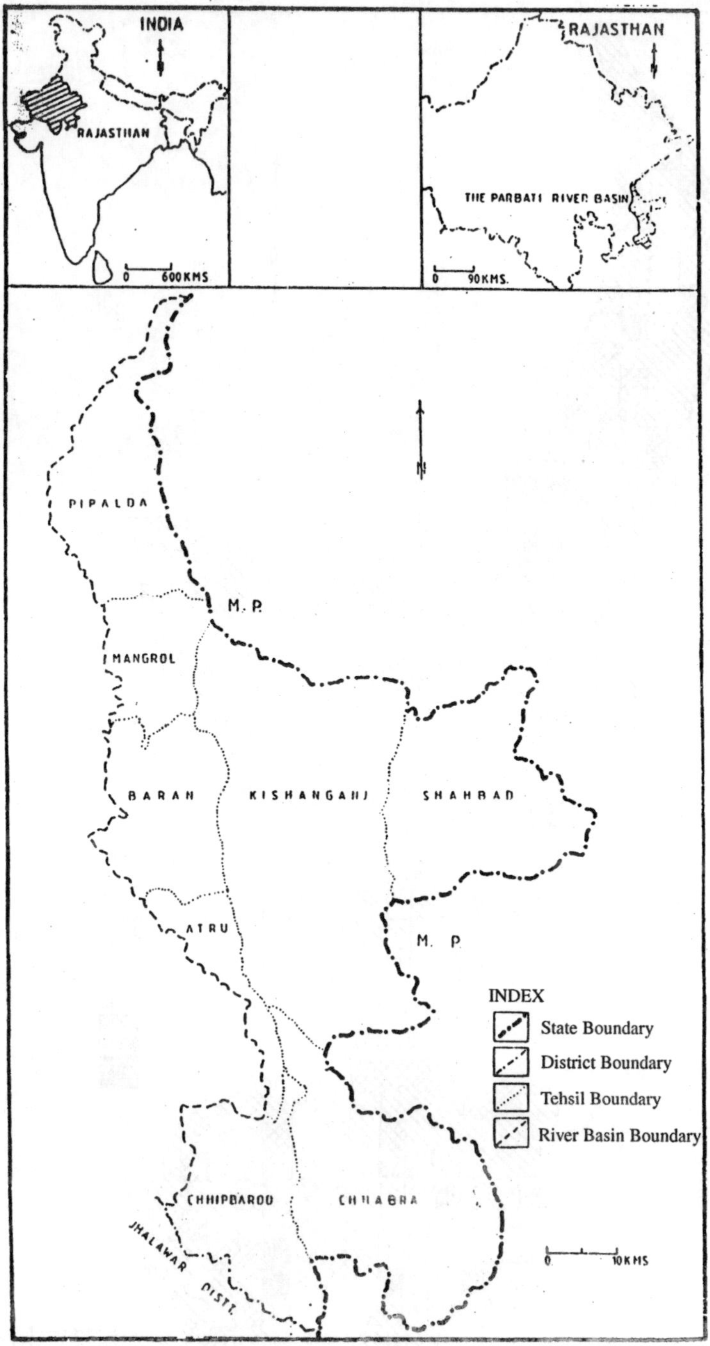

Fig. 13.1 : Key Map—the Parbati River Basin

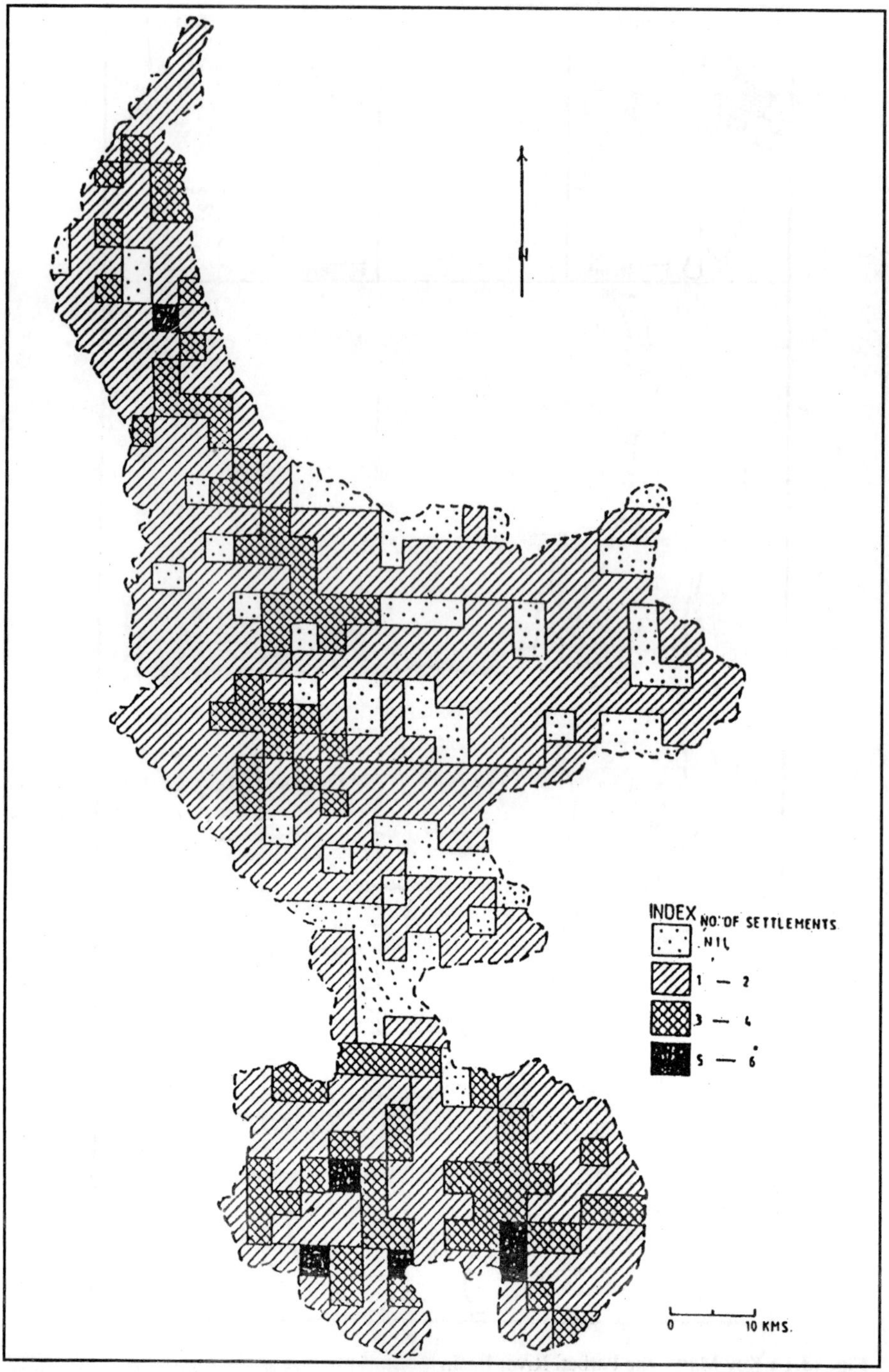

Fig. 13.2 : The Parbati River Basin—Concentration of Settlements

(i) Maximum concentration,
(ii) Moderate concentration, and
(iii) Minimum concentration.

The map (Fig. 13.2) reveals that the maximum concentration of settlements has been observed in the parts of Chhabra and Chhipabarod tehsils. In these areas fertile black soil is found. The concentration of water bodies is maximum. The moderate concentration of settlements distribution is found in Pipalda, Mangrol and some parts of Kisanganj, Baran, Chhipabarod, Chhabra and tehsils. Since these are the settlements which have been noticed along the course of the river channels. This region represents undulating nature of topography. The minimum concentration of settlements has been noticed in all the remaining tehsils. This category of settlements distribution covers maximum area of the Basin. This region is characterised by hilly terrains and Shahbad hills, Chhabra and Chhipabarod hills, Pedipla in and upland topography. Due to the dearth of open space the habitations have been observed in valley sides and along the course of river channels.

Correlation of Settlements Distribution with Orphometric Attributes

After the qualitative analysis of settlement distribution, the interpretation now switches on the quantitative analysis of the settlement distribution in relation with the slope and relief. In the Parbati Basin, the settlement distribution is mostly affected by above mentioned two morphometric attributes. In general it can be inferred that the low to moderate altitudinal zones and moderate relief zones in the study shares much settlements whereas the high altitudinal zones and high to very high relief zones do not enjoy much number of settlements.

How far and how much settlement distribution in the study area is affected with the morphometric attributes will be discussed here with the help of entropy analysis and village density pattern.

Slope vs. Settlement Distribution

A comparative analysis has been done to find out the relationship between degree of uniformity in the settlement distribution and slope conditions, as given in Table 13.1.

It may be inferred from Table 13.1 that the degree of uniformity in the distribution of settlements is largely governed by the physical configuration of the area as the rank correlation coefficient between uniformity and slop is + 0.62 which show that the higher is the proportion of area above 2°slope, the better is the degree of uniformity in settlement distribution.

Table 13.1 : Showing Ranking of Tehsils for Different Variables

Sl. No.	*Tehsil*	*Uniformity %*	*% Area more than 2°slope*
1.	Pipalda	3.07 (8)	Nil (7.5)
2.	Mangrol	56.42 (2)	4.57 (5)
3.	Baran	36.49 (5)	Nil (7.5)
4.	Shahbad	69.30 (1)	34.14 (1)
5.	Atru	36.30 (4)	1.46 (6)
6.	Chhipabarod	18.71 (6)	34.06 (2)
7.	Kishanganj	41.90 (3)	20.96 (4)
8.	Chhabra	17.75 (7)	31.94 (3)

Notes : Figures within brackets shows the ranks.

It can also be seen that Pipalda tehsil with a ravine affected topography and having nil of its are above 2° slope has the least uniformity. The Shahbad and Mangrol tehsil with only 30.44 per cent of its area being above 2°slope have maximum uniformity in the settlement distribution.

Relief vs. Settlement Distribution

Table 13. 2 depicts relationship between relief and settlement density. It is clear from the above table that highest settlement density (4.88 settlement/square, mile) belongs to above 400 metres relief, while lowest in 300-400 metres relief group. It is because in the higher relief regions clustering of settlement pattern observed, while in upland and plain region uniform settlement pattern noticed.

Table 13.2 : Relief and Settlement Density

S. No.	*Relief (Metres)*	*Area (Square mile)*	*Settlement density*
1.	< 200	20	2.50
2.	200-300	720	2.63
3.	300-400	840	2.31
4.	> 400	200	4.88

Negative Areas for the Growth of Settlements

In the study area, there are some stretches which are not suitable for the growth of settlement as the physical factors do not allow them. The areas which are not suitable for the growth of settlements correspond with the area of ravine affected, steep to very steep slop, high to very high relative relief and high altitude. About 50 per cent of such localities are marked in the Pipalda, Shahbad, Atru and Baran thesis. Apart from these zone some other pockets are also in the area which are not suitable for the settlement growth. These zones also correspond with rugged terrain, steep to very high relative relief.

Acknowledgements

The author is thankful to his teacher and supervisor, Professor H.S. Sharma, Director U.G.C. Academi Staff Collage, U.O.R., Jaipur for his suggestions and guidance throughout the preparation of this paper.

REFERENCES

Inderpal and Mishra, R.N. (1984): Entropy Analysis in distribution of Tribal settlements. A case stikly of District Banswara, *Modern Geographical Trends* Ed. P. Pandey pp. 585-597.

Kharwal, S.C. (1969) : *Landforms and settlements in Nainital and its Environs.* Unpublished Ph.D. thesis, B.H.U., Varanasi.

Kharwals, S.C. and Singh, S.B. (1988) : *Settlement distribution and their correlation with morphometric attributes in Simla hills.* Transactions of the Institute of Indian Geographers, Vol. 10, No. 2, July, 1988, pp. 47-52.

Asthana, V.K. (1968): *Landforms and settlements in Almora and its Environs.* Unpublished Ph.D. thesis, B.H.U. Varanasi.

SECTION-III

ENVIRONMENTAL CHANGE

14

Understanding Future Dryland Environmental Changes from Past Dynamics

An Overview

S.K. WADHAWAN AND A.K. SINGHVI

Introduction

Deserts and drylands cover almost 50 per cent of the global land area and pose challenges for their human populations. Desertification has been recognized as a globally important environmental issue. Problems of water resource availability and food production are bound to increase under the dual impacts of enhanced population pressure and the future greenhouse warming induced climatic changes. Despite their great extent and growing population, deserts and drylands are marked by their relative scientific neglect. It is imperative to improve understanding of spatial complexity of dryland systems and change causing linkages to allow better, region specific predictions of future changes and their remedial measures for sustainable utilization of the available resources. Understanding the nature, magnitude, causes and timing of past environmental changes in drylands provides critical data inputs into models and predictions of the future changes that will significantly affect societies.

International Geological Correlation Programme (IGCP) Project No. 413 of the International Union of Geological Sciences (IUGS) worked within the mandate of the United Nations and is considered the most important activity of the IUGS and the United Nations Educational, Scientific and Cultural Organization (UNESCO). IGCP 413 commenced in 1998 on the focal theme of 'Understanding future dryland from past dynamics' and completed its tenure in 2003. Aims and the work-plan of the IGCP 413 were an enhanced understanding of the of the processes controlling the timing and amplitude of dryland environmental changes, together with nature of these changes and an improved recognition of likely spatial and temporal complexities of future changes in drylands. Their impacts on societies by highlighting regional differences and hemispheric contrasts in the timing of desert fluctuations, the operation of different processes and understanding past human relationships

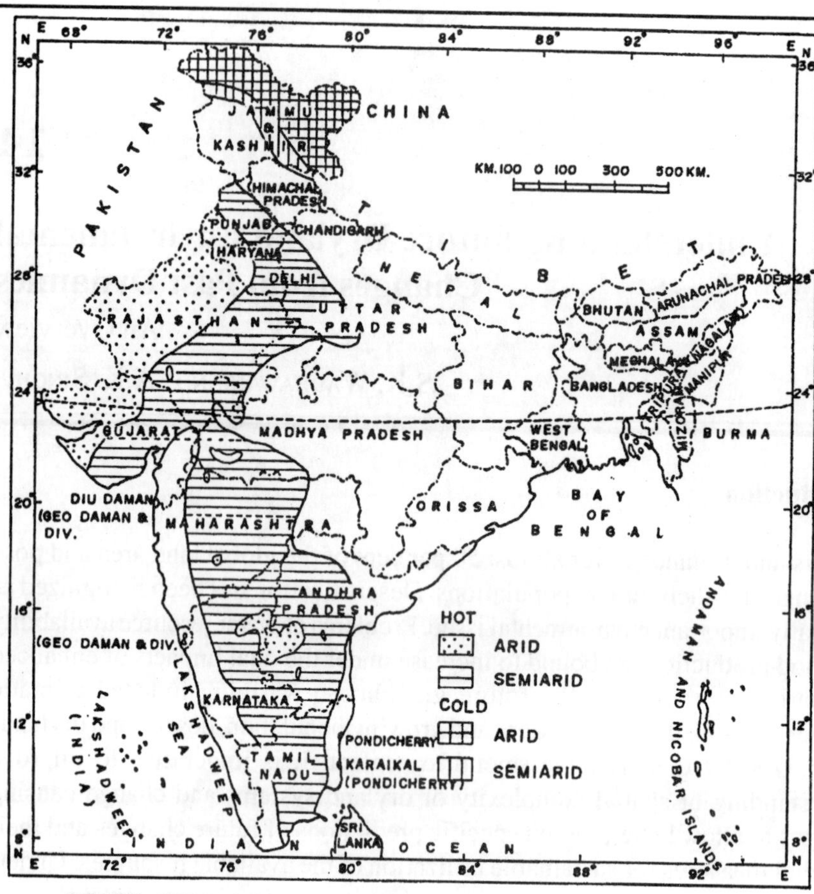

STATE-WISE AREA OF ARID AND SEMIARID ZONES IN INDIA

REGION	STATE	AREA IN SQ KM		PERCENTAGE OF AREA UNDER EACH STATE	
		ARID	SEMIARID	ARID	SEMIARID
	COLD				
NORTH	JAMMU & KASHMIR	70300	13780	–	–
	HOT				
	UTTAR PRADESH	–	64230	–	7
	MADHYA PRADESH	–	59470	–	6
NORTH-WEST	RAJASTHAN	196150	121020	61	13
	GUJARAT	62180	90520	19·6	9
	PUNJAB AND HARYANA	27350	58650	9	6
SOUTH	MAHARASHTRA	1290	189580	0·4	19
	MYSORE	8570	139360	3	15
	ANDHRA PRADESH	21560	138670	7	15
	TAMIL NADU	–	95250	–	10
	TOTAL	387300	970550	11·80%	29·57%

TOTAL AREA OF INDIA – 3282016 SQ. KM.

Fig. 14.1 : Arid and Semi-arid Zone in India

with environmental changes. This also entailed identification of the role of high magnitude events in dryland landscape development, enhanced understanding of the stability/instability of drylands.

IGCP-413 Project had focused on collating and synthesizing data from different parts of the Thar Desert and dryland environments in India. These scientific group activities helped to infer new results on the critical geomorphic thresholds, rates of dune construction and its mobility nature and magnitude of aeolian dynamics and response/reaction periods (leads or lags) of geomorphic processes to climatic perturbations.

The Project was motivated by a variety of factors such as:

(1) The importance of drylands and deserts in influencing atmospheric circulation pattern *via* the changes in the surface *albedo*,
(2) Location of a large human population in the desert margin region,
(3) Promise of geological records of climatic change with minimal biogenic influences, and
(4) Recent developments of new techniques for absolute dating of continental sediments such as Thermoluminescence (TL), Optically Stimulated Luminescence (OSL), Infra-Red Stimulated Luminescence (IRSL), and Electron Spin Resonance (ESR) dating, etc.

Dryland environments in India extend nation-wide (Fig. 14.1). The hot dryland environment of the Thar Desert covers major parts of Rajasthan, Gujarat and Haryana in western India. The Ladakh region in the Trans-Himalayas is the typical cold desert in India whereas the semiarid tracks traverse through parts of Maharashtra, Madhya Pradesh, Andhra Pradesh, Karnataka and Tamil Nadu in Peninsular India. The need for carrying out collaborative studies to work out the short-term and long-term changes in dryland environments was suggested and considerable progress has been made in collation, creation and synthesis of a large data base during the tenure of this Project. In the deserts and drylands including the drought prone areas, human populations are set to face many problems in response to the pressures from increased population of humans and livestock with increasing demands on the available meagre water resources. Mining of rock and mineral resources in dryland environments is bound to have its impacts on the rapidly changing land use scenario. Climatic variability both within a desert and between dryland settings significantly affects the operation of geomorphological and hydrological processes and their response times. IGCP-413 had served as a focal point for research into dryland environmental change and has contributed to the initiation of major dryland environmental research programmes. It has also intensified ongoing programmes by strengthening them in respect of both extending chronometric basis and

Table 14.1 : Phases of Quaternary climatic change in Western India

Sl. No.	*Period*	*Thar*	*Luni-Mahi-Sabarmati*	*Narmada-Tapi-Godavari-Bhima*	*Palaeoclimatic Inference*
1.	Neogene (Reliect features)	Hundreds of metres of gravels filling grabens, gravels with inverted relief. Calcretized and Ferricretized regolith.	"Super-Fluvial" systems represented by hundreds of metres of fluvial sediments filling grabens	Dedunational surfaces, high level gravels, weathered regolith and anomalous fills in source regions of major rivers	Climate was more humid than today. Tectonic forces are a factor in the formation of these units.
2.	Pre Late Quaternary	Dune and plays facies are known which pre-date the Late Quaternary but absolute dating is so far scanty.	Gravels containing Palaeolithic artefacts in the Sabarmati and Type II sequence from the Luni	Many alluvial fills with Acheulian artifacts	These units primarily reflect Quaternary climatic change, which could be either relatively humid or arid.
3.	-50.30 kyr	Shergarh Trijunction shows evidence for reduced aridity	A "red soil" horizon has been identified and correlated from the three basins	Not well documented due to lack of dating inputs, but many aggradational records begin close to 25-30 kyr, implying incision in this period	Relatively humid phase.
4.	-25 kyr to 1 kyr	Dune system is relatively inactive, some playa show ephemeral shallow phases	—		Major phase of Late Pleistocene aridity.
5.	14 kyr	Both dune and playa systems become active with re-establishment of monsoon	Major phase of Luni activity near Balotra begins at 14 kyr	A major change in fluvial behaviour documented from Bhima, Godavari and Narmada systems	14 kyr appears to be a very significant period for a rapid transition from glacial to post-glacial climate.
6.	14-8 kyr	11-14 kyr is the major phase of dune building while lake levels fluctuated from 14 to 8 kyr	The major activity of the Luni 14 to 11 kyr on the basis of luminescence dating	Short aggradational phases from Bhima tributaries Major aggradational phase on Sindhphana. Major incisional phase of Narmada, Bhima.	A period of rapid and frequent climate changes.

(Contd...)

Table 14.1 *(Contd)*

7.	8-4 kyr	Major phase of lake filling. Pedogenesis of dunes.	Gastropod date from Early Holocene Luni gravels near Balotra	Floodplain development on Godavari seen at Nevasa, Dharangaon and Apegaon	Most humid phase of the Holocene, increasing aridity after 4 kyr when lakes dry in Rajasthan.
8.	2.7-3 kyr "Dark Age"	Reduced number of settlements, with some being covered sand during this period	Chalcolithic site covered with sand at Manawara on the Luni	Little geological data. A reduced number of human settlements	The archaeological data does document a period of reduced settlements but the links with climate are controversial. More independent geological data needed.
9.	1.9-2.2 kry "Roman Optimum"	Dune stability	No data	Increased floods on Narmada and flood damage to settlements	Archaeologically a period with more settlements and some evidence for good rainfall from palaeoflood studies.
10.	0.7 kyr "Medieval Warm Period"	Dune stability	No data	Increased floods on the Narmada	A period of increased numbers of settlements. Evidence for dune stability and larger floods, indicates "good" monsoon.
11.	.3 kyr "Little Ice Age"	Scanty data	Scanty data	Reduced floods on Narmada, aggradation in Godavari and Bhima rivers	Only palaeoflood data so far documents this phase.
12.	20th century little data	Flood of 1979 on Luni probably the largest in the last 4 kyr	Increased flood frequency on Tapi and Narmada	The present century shows increased floods onall the rivers studied, probably due to global warming.	—

elaboration on relevant societal aspect. IGCP-413 has also formulated joint investigations with researchers investigating current environmental changes in drylands in five continents while contributing to the establishment of a unique project that links several different IGCP projects together. Improved data inputs from dryland areas to Global and Regional Climate Models was attempted in order to contribute to better predictions of the impacts on future climatic changes on desert and peridesert societies.

Drylands in India support a large human and livestock population whose stability and sustainability depends on the climatic and environmental conditions. Adverse conditions in the past have witnessed large-scale migrations—an aspect with significant social and economic ramifications. This aspect becomes all the more crucial in view of an ever-increasing population pressure and the anticipated trends on human induced climatic changes. The research results of the IGCP-413 such as estimation of cyclic climatic changes in dryland environments, response of dryland societies to rapid landuse-landcover changes, etc., would be of direct relevance to society.

Characterisation of prevalent nature of harsh conditions of dryland environments and scientific research on understanding the operative forces in the inhospitable region are important for predicting the future scenario from the past dynamics. Drylands have been subject to increasing human and biotic pressures as the "last frontiers" on earth's extremely stressed desert ecological system. This has enforced rapid and drastic changes in land uses. Human induced changes could impart drastic degradation of desert vegetation through overgrazing and excessive soil erosion. Present intensified use of land resources by humans and livestock in Thar Desert in Rajasthan has resulted in steady depletion of pastoral lands, groundwater reserves and degradation of stable dunes.

Multi-disciplinary research from various desertic terrains over the past decade has demonstrated that the desert margins have fluctuated significantly during the Pleistocene and Holocene. Besides broad conformities in Quaternary climato-stratigraphic and geologic evolution, operative geomorphic processes in different deserts within the given hemispheric domains have had different response times to climatic perturbations. Different desert surfaces and climatic zones within the desert show marked differences in ecosystems characteristics and offer varying potential for human activities. Scenarios for the future impact of greenhouse gas induced global warming predicted that most drylands would experience increased mean temperatures and enhanced aridity. This is likely to affect operation of hydrological and geomorphological process domains or intensity of their operation thus leading to spread of desert margins at the expense of more productive lands and triggering rapid land use changes and related inevitable land degradation.

Dryland Environment of Western India : The Thar Desert

Three major dune-forming episodes are presently inferred in hot dryland environment of the Thar Desert. Massive and large-sized oldest dunes show crescentic, stabilized megabarchanoids or transverse draas formed by stronger intensity southerly to southeasterly winds. These draas were transformed into linear and parabolic dunes through aeolian scouring and superimposed accretion of second and subsequent generations. The youngest aeolian reactivation here is represented by selective deflation and depositional linear arrays of active small barchans, vegetation-induced aeolian accumulations and prograding sand sheet. Dune trends of younger-evolved forms suggest establishment of the southwesterly sand-moving wind circulation patterns during second-generation aeolian episode and its persistence of directional flows though with significantly lesser intensity into the contemporary wind regime in the area. Implications for future changes are discussed in view of contemporary rapid land use changes, mining for limestone and importing and distribution of surface water through a network of canals and other human interventions in this ecologically fragile dryland environment (Wadhawan, 2002).

A framework of Quaternary litho-stratigraphic classification has been established in different sectors of the Thar Desert. Fluvial and fluvio-lacustrine deposits form the base for aeolian deposition and control the configuration and disposition of older aeolian deposits (dune fields) in the Thar Desert. The earlier dominant fluvial regime of well-integrated drainage system could not re-establish itself once it got disrupted and disorganised by the advancing dunes. The subsequent wet interludes could create semiarid consequent drainage at best, flowing into inland depo-centres or the lacustrine sedimentation basins, e.g. the Sambhar and Didwana Lakes, Pokaran, Thob and Bap Ranns, etc. Although climatic change is responsible for the aeolian episodes, field studies and interpretation of subsurface geological data have now revealed that tectonically active Quaternary domain has influenced the successive burial of different sedimentation units of fluvial/fluvio-lacustrine and aeolian generation particularly in the northern and western parts of the Thar Desert. Isopach map of sub-surface Quaternary formations and palaeo-basin configuration show that the Ghaggar basin deepened to abnormal depth and hosted a thick (over 350 m) pile of continental sediments in its northern part. Northerly flowing, structurally controlled palaeo-drainage pattern for the early Quaternary fluvial regime is deduced. This contribution attempts synthesis and evaluation of the Quaternary litho-stratigraphy and dune hierarchy as established presently in the sub-basins confined to arid tracts in the Thar Desert. Geological issues and prospects bearing on the evolution of Thar Desert in India are analyzed (Wadhawan *et al.*, 1999).

It has been established that during the Last Glacial Maxima (LGM) the geomorphic activity (both aeolian and fluvial) was largely dormant in the Thar Desert. This accrues from the near total dependence of the aeolian and fluvial processes on the existence of an effective south-west monsoon system. New field-based investigations of geomorphic records from the region also indicate a climatic instability during the period 11-13 Ka, in terms of fluvial interludes during a predominantly aeolian regime. This is attributed to an enhanced winter precipitation during the Younger Dryas period, but will need further elucidation.

Deeper level aeolian sand sequences showing multiple episodes of accumulation from numerous excavated pits/sections in the core and margins of the Thar Desert have been dated (TL and ESR methods) to be more than 100 ka in age. This confirms the view that the Thar Desert came into existence due to a combination of geological and climatic processes and it does not owe its origin to anthropogenic causes. Recent efforts on dating of aeolian sequences and the geoarchaeological studies in the Thar Desert have shown relatively better cultural developments around 100 kyr, 30-40 kyr, 9-5 kyr, 2.5-1.3 kyr and possibly between 0.9 kyrs and 0.7 kyr (Kar *et al.*, 1998; Thomas *et al.*, 1999; Kar *et al.*, 2001). A corresponding decline in human occupancy of the dryland environment of the Thar Desert is inferred around 20-16 kyr, 4 - 2.5 kyr, 1.4 - 1 kyr and between 4th and 10th century A.D. —the so called Dark Ages in Europe (e.g., Deotare *et al.*, 1998; Mishra and Rajaguru, 2001). Chronological determinations also suggest that the peak of aeolian accretion or dune construction occurred at about 13 ka, which is about 8 ka later than the glacial epoch maxima (LGM) at 21.5 ka. This observation can be explained on the basis of close relationship of dune accretion conditions to a transitional climatic regime from glacial to interglacial and to the time when the SW monsoon was being re-established after the glacial epoch (DST, 2002). A fine resolution aeolian stratigraphy during the Holocene is also attempted implying an approximate 1500 years periodicity of intensification of aeolian activity separated by relative quiescence or non-accumulation and landscape stability. However, it is difficult to entirely rule out anthropogenic factors for aeolian remobilization in such cases. Change detection studies in respect of the mobility of transverse dunes during the past 2000 years indicated high migration rate at ~2ka, reduced migration rate during the medieval warming and at least 6-10 fold higher migration rate during the past 200 years. This is mainly due to human activity *via* the reduction of vegetation cover caused by uncontrolled grazing and land degradation.

Studies on Thar Desert margins also indicate a northward shift in monsoon during the Holocene, such that aeolian accretion in southern areas ceased at 10 ka and 6 ka ago at areas north of Ahmedabad. Aeolian accretion in the core region of the Thar continued till 800 to 600 years ago. Therefore, these data indicated spatial shifts in the monsoon gradient varying form south to north in contrast to the presently prevailing progressive decrease in intensity of rainfall from east to west. Results also indicated multiple episodes of aeolian activity during the Holocene, which had an approximate periodicity of 1500 years. Thus in this regional context the dunes and other records indicated significant climatic excursions even during the Holocene, which so far had been considered as a stable climatic period. Stable carbon and oxygen isotopes ratios of pedogenic carbonates along with luminescence dating indicated that during the Marine Oxygen Isotopic Stage (OIS) - 3 and - 1, the region experienced normal monsoon conditions and during the OIS - 2 and - 4, the region experienced an arid climate. Other evidence from the Thar Desert margin also indicated that the isotopic Stage-3 experienced a wetter climate. The final phase of natural aeolian accumulation appears to have ended at 600 - 800 years in the area bounded by annual rainfall ~200-350mm. This observation is significant for the land use planning of the region.

Based on the chronostratigraphy worked out for dune deposition in the Thar it is inferred that favourable climate for dune construction occurred during a transitional period between glacial and interglacial and over a restricted time window confirming episodic nature of such activities. It was established that geomorphological activities, both aeolian and fluvial, were largely dormant during the LGM in the Thar Desert. This accrues from the near total dependence of aeolian and fluvial processes on existence of an effective southwest monsoon system. Field investigations also indicated a climatic instability during the period 11 – 13 ka, in terms of fluvial interludes during a predominantly aeolian regime. This could be attributed to an enhanced winter precipitation as ascribed to the Younger Dryas period.

Relationships between Holocene monsoon changes and urban settlements/ expansion episodes were explored. There appears to be a gross correlation between Holocene urban settlements and occurrence of monsoon changes. For instance the major Chalcolithic Settlement phase (5-3 ka ago) and Early Historical (2.2 – 1.8 ka ago) and Medieval Period (1.1 – 0.7 ka ago) also coincided with and included the lake desiccation and dune building activity in the Thar desert. A better understanding of human responses to climatic changes and human contributions to environmental changes is being achieved by such studies. However, microliths belonging to the Mesolithic Culture have been reported to be the most widespread in the Thar Desert implying amelioration in the climate and increase in the food resources and human population during the Holocene Optimum, ca. 6 - 7 ka ago. Subsequently the settled mode of agro-pastoral living on the pattern of the river valley civilizations (Harappan Culture) have been recorded all along the Ghaggar river basin in the northern peripheral parts of the Thar Desert. However, neotectonic activities and shifting of the Himalayan rivers led to abandoning of the Kalibangan-Harappan settlements ca. 2nd and 3rd Millennium B.C. Sheila Mishra and Rajaguru (2001) applied geoarchaeological approaches for palaeoclimatic deductions in aridity affected western India. Human artifacts recovered from the middle reaches of the Luni river basin have been classified into archaeological assemblages related to the associated Quaternary fluvial formations and dated to Lower Palaeolithic (ca 200 kyr), 80 kyr and 11.4 kyr and Chalcolithic Culture older than 3 kyr respectively (Mishra *et al.*, 1999).

A synthesis of the geoarchaeological record from the Thar Desert and its margins is enabling reconstruction of socio-environmental relationship such as the artifacts and their host strata that could be used to reconstruct environmental changes during the Quaternary (Table 14.1). The presence of Acheulian artifacts in a playa shore context and of middle Paleolithic artifacts in fluvial contexts reflects a relatively humid environment during early to late-mid Pleistocene. Isotopic Stages-4 and -2 present sparse evidence of human occupation and of geomorphic processes and Stage-3 witnessed relative abundance of flake industry. Surprisingly, evidence of human occupation during early Holocene is also sparse indicating that the contemporary climate was not conducive for human occupation. Overall, it is seen that, in this region, the human occupation thrived only during the humid phases.

Work on the reconstruction of the limits of the Thar Desert at the Holocene Optimum (ca. 6 ka) and at 14 ka (when the monsoon was re-established) has been attempted. Clear signatures of the progressive northward stabilization of the dune activity was seen such that dune accretion ceased at 10 ka at the southern margins, at 6 ka north of Ahmedabad and 600 years in the present core of the Thar. This reflects a progressive northward shift in the dune accretion climate (i.e. monsoon) and provides a dataset that can be used to validate Global Climate Models. Change detection studies in respect of the mobility of transverse dunes during the past 2000 years indicated high migration rate at ~2ka, reduced migration rate during the medieval warming and at least 6-10 fold higher migration rate during the past 200 years. This is due mainly to human activity *via* the reduction of vegetation cover caused by uncontrolled grazing and land degradation.

Dryland Environment of Peninsular India

IGCP-413 research in India has also focused on the reconstruction of palaeoclimatic record from semi-arid regions of South India, the synthesis of geo-archaeological records to identify gaps in the record and to examine the man-environment relationships. Climatologically, a major part of the Southern Indian peninsula is semi-arid to arid with seasonality of rainfall and high potential evapotranspiration. Construction of ponds during the fifteenth century reflects enhanced aridity in the region that needed massive rainwater harvesting. The sediment characteristics of the pond fill sequences such as the magnetic susceptibility, granulometry and pollen assemblage changes along with the chronology are being examined to understand both the climate and human dimension of these sediments. Results so far indicate a progressive desiccation of the region for over 500 years. On a longer time scale, the alluvial sequences of the River Pennar indicate arid-ephemeral regimes 2400-2000 years ago, a more humid regime 1800-1000 years ago and drier conditions for the past 400 years. Given that the hydrology of these rivers depends primarily on the southwest monsoon, these inferences provide a surrogate for the southwest monsoon performance in the region. Studies so far reflect that the records though accord with global signatures of aridity, exact timing of the geomorphic process response differs. This implies that land use planning strategies in dry lands should take due cognizance of the response times of geomorphic processes and as a consequence the global climatic events are likely to be manifested somewhat asynchronously in the geomorphic settings. This is an aspect crucial also for the geological correlation.

Geoenvironmental Studies in Dryland Regions of India

Characterization of prevalent nature of harsh conditions of dryland environments and scientific research on understanding the operative forces in the inhospitable region are important for predicting the future scenario from the past dynamics. Drylands have been subject to increasing human and biotic pressures as the "last frontiers" on earth's extremely stressed desert ecological system. This has enforced rapid and drastic changes in land uses. Human induced changes could impart drastic degradation of desert vegetation through overgrazing and excessive soil erosion. Present intensified use of land resources by humans and livestock in Thar Desert in Rajasthan has resulted in steady depletion of pastoral lands, groundwater reserves and degradation of stable dunes.

Presently drylands in India are subject to intense human occupancy and utilization of available resources. With the development of infrastructure, advent of surface water availability through inter-basinal transfer of Himalayan water [e.g., Indira Gandhi Nahar Pariyojna (IGNP) in the Thar Desert] or exploitation of meagre groundwater resource, rapid changes in land use is recorded leading loss of biological productivity, land degradation and desertification. Geological Survey of India has carried out several geoenvironmental appraisals on regional scale as well as environmental impact assessment and seismo-tectonic studies on large scale thematic maps in aridity affected areas, such as Jodhpur, Jaisalmer and Bikaner and Ajmer districts in Rajasthan; Jamnagar and Kachchh districts and Sabarmati and West Banas basin in Gujarat; Beed and Ahmednagar districts in Maharashtra; Anantpur and Nalgonda districts in Andhra Pradesh; Chitradurga and Bellary districts in Karnataka; Sivaganga and Dharmapuri districts in Tamil Nadu, etc. Synoptic integrated and derivative maps depicting land-capability aspects and geoenvironmental resource potential besides delineation of natural and human induced hazards are proving to be useful for perspective land use planning, rehabilitation of natural disaster prone areas and sustainable utilization and conservation of available resources and hazards mitigation. These thematic geofactor maps are particularly useful and provide essential data base in ecologically stressed drylands.

The Cold Desert of Ladakh Region, Jammu and Kashmir

Quaternary geological studies and geofactors controls on land use—land cover aspects of cold dryland environments in Ladakh region, Jammu and Kashmir, India are being carried out by the National Working Group Members of GSI with feasibilities of involving collaborative multi-institutional work for a comprehensive understanding and base-line data generation. High incidence of silicosis is reported in Chushot area near Leh, which is the major track along Indus River forming a broad open valley exposed to high-speed winds and prone to deflation. In the Upper Indus river basin, the most characteristic deposits are those of lacustrine origin, occurring in linear palaeo-lake disposition owing to periodic damming of the Indus and its tributaries by the Quaternary glacial and glacio-fluvial deposits. Investigation of palaeo-lake deposits along the Indus river trunk channel show potential of distinguishing several episodes of water borne clay and wind borne sandy-silt horizons indicating alternating dominance and succession of dry and wet conditions in Ladakh. Aeolian deposits also occur as obstacle dunes of different generations that require absolute dating by luminescence methods.

Quaternary Aeolian and Tephra Deposits of Peninsular India

Considerable new data is generated on the volcanic tephra deposits and coastal and inland dune systems of eastern coast of Orissa (red dunes of Ganjam coast), Andhra Pradesh and Tamil Nadu. The coastal dune complexes in parts of Tamil Nadu have been assessed to be rich deposits of white glass sands with high silica content (87 to 98 %). Volcanic ash beds are used as marker horizons on the basis of similarity in morphology of glass shards and geochemical characteristics. Considering the significance of wind-borne tephra deposits as marker horizon within the Quaternary sequences, geological investigations are carried out for inter-basinal correlation based on the volcanic ash bed occurrences amongst the alluvium of major rivers in Peninsular India. These tephra deposits have been correlated with the Youngest Toba Tephra (YTT) and dated as ca.

70,000 years B.P. Ash beds associated with alluvium of Baneta Formation of the Narmada basin is correlated to the Brown Silt Formation of the Purna basin and Belora Formation of the Wardha basin. Therefore, interregional correlation of Quaternary stratigraphy has been established and assigned middle to upper Pleistocene age by geoscientists of the Geological Survey of India.

Application of modern tools and use of the remotely sensed data products with improved spatial, spectral and temporal resolution and data integration and analyses in GIS environment by IGCP-413 NWG Members at Space Application Centre, Ahmedabad and RRSSC Jodhpur have been proving fruitful in delineating palaeochannels in different parts of Thar Desert as groundwater prospective zones and suitable sites for artificial recharge. Similarly the relationships between palaeo-drainage courses and the archaeological sites in parts of Saraswati river basin of northern and western Thar Desert are being carried out for improved understanding of human adaptation to dryland environmental changes through prehistoric and historic times. Further studies on palaeo-hydrological records and nature of bedrock weathering in the arid tracks of Jaisalmer are being pursued to interpret history of climatic changes, nature and magnitude of past flash floods and their societal impacts.

Conclusion

Drylands are essentially moisture deficient environments. Hydrological resources and hydrological changes, at a range of time scales, are therefore the principal controls on many dryland environmental systems, be they ecological or geomorphological systems. Most GCM scenario forecast an expansion in extent of drylands and intensification of conditions in existing drylands during the 21st century as a function of anthropogenically enhanced global warming. Even where dryland precipitation is forecast to increase rather than decrease, wetting effects are commonly more than offset by temperature increases that are predicted to decrease effective precipitation and soil moisture levels. IGCP-413 placed considerable emphasis on the understanding of contemporary environmental process in drylands, and on understanding the roles in and responses to environmental changes of dryland people. Considering the significance of both dust and aeolian sand deposits in dryland geomorphic systems advances in and wide application of luminescence dating is increasing the chronometric control on understanding of the timing of arid and humid events in drylands, desert and peri-desert locations. Based on such enhanced understanding of palaeo-environmental reconstructions, it is recognised that the timing of major millennium scale dryland environmental changes was not synchronous from place to place owing to the roles of different climatic forcing factors that operate from decadal (e.g. ENSO events) to multi-millennial (e.g.

orbital forcing and high latitude forcing) time scales. For instance it is inferred that arid phases during the last glacial cycle were probably out of phase north and south of the equator due to significance of insolation changes in driving tropical and subtropical climatic changes. Similarly, studies in Indian and Arabian deserts within the same hemisphere also proved that due to differential timing of the monsoon re-establishment after the last glacial maximum, environmental changes were not necessarily synchronous within the latitudinal belts.

Environmental sustainability and future of Thar Desert is affected by growing human population and increase in demand for urban land and rapidly chaging land uses, particularly overgrazing and agricultural activities related to intensive cultivation and destabilization of dunes. Soil erosion has become the most destructive, pervasive and enduring element in dryland change and consequent degradation. Thar Desert and its ecologically fragile peripheral tracks are particularly sensitive and prone to intensification of desertification. Therefore, it is imperative to monitor and adopt geoscientific, integrated approaches of alternate land use for optimum utilization of available dryland resources.

REFERENCES

Bajpai, V.N., Saha Roy, A. and Tandon, S.K. Subsurface sediment accumulation patterns and their relationship with tectonic lineaments in semi-arid Luni river basin, Rajasthan, western India. *J. Arid Environments* 48, 603-621, 2001.

Bakliwal, P. C. and S. K. Wadhawan . Geological Evolution of Thar Desert In India—Issues and Prospects, *Proc. Indian Natn. Sci. Acad.*, 69 A (2), pp. 151-165, New Delhi, 2003.

Deotare, B.C., Kajale, M.D., Kshirsagar, A.A. and Rajaguru, S.N. Geoarchaeological and Palaeoenvironmental studies around Bap-Malar Playa, District Jodhpur, Rajasthan. *Current Science* 75 (3), 316-320, 1998.

Deotare, B.C., Kajale, M.D., Kusumgar, S. and Rajaguru, S.N. Late Holocene environment and culture at Bari Bavri, Western Rajasthan. *Man and Environment* XXIV (1), 27-32, 1999.

Gautam, A.M., 1998, Studies of Drainage Anomalies of Himachal Pradesh and Jammu and Kashmir, *Records Geo. Surv. Ind.* 131 (8).

Gautam., A.M. 2000, Updating and synthesis of Geological Information of Ladakh Region *vis-à-vis* interpretation of Remote Sensing Data, *Rec. Geol. Surv. Ind.*,133 (8).

Gautam, A.M., 2001, Tectonic Fabric and Neotectonic Manifestations in Parts of Western Himalaya—a Remote Sensing Based Study. Proc. Nat. Symp. Role of E. Sci. in *Integrated Development and Related Societal Issues*, GSI Spl.Pub. No. 65 (III), 2001: 209-213

Jain, M. and Singhvi, A.K. Limits to depletion of blue-green light stimulated luminescence in feldspars: implication for dating. *Radiation Measurements* 33, 883-892, 2001.

Jain, M., Tandon, S.K., Bhatt, S.C. and Singhvi, A.K. Alluvial and aeolian sequences along the River Luni, Barmer district: Physical stratigraphy and feasibility studies towards Luminescence dating. *Geol. Soc. India. Memoir.* 42, 273, 1999.

Juyal, N., Singhvi, A.K. Kar, A., and Rajaguru, S.N. Chronostratigraphic evidences for episodes of Desertification since the Last Glacial Epoch the southern margin of Thar Desert, India. In A.S. Alsharhan, W.W. Wood, A.S. Goudie A. Fowler and E.M. Abdellatif (Eds.) *Desertification in the Third Millennium.* 2003. pp. 123-128. Swets & Zeitlinger Pub. Lisse, The Netherlands.

Kale, V.S., Singhvi, A.K., Mishra, P.K. and Banerjee, D. Sedimentary Record and luminescence chronology of late Holocene Paleofloods in the Luni valley, Thar Desert, northwest India. *Catena* 40, 337-358, 2000.

Kailath, A.J.,.Rao, T.K.G., Dhir, R.P., Nambi, K.S.V., Gogate, V.D. and Singhvi, A.K., Electron Spin Resonance Characterization of Calcretes from Thar Desert for dating applications. *Radiation. Measurements* 32, 371-383.

Kar, A., Felix, C., Rajaguru, S.N. and Singhvi, A.K Late Holocene growth and mobility of a transverse dune in Thar Desert. *Journal of Arid Environments* 35, 175, 1998.

Kar, A., Singhvi, A.K., Rajaguru, S.N., Juyal, N., Banerjee, D., Thomas, J.V. and Dhir, R.P. Reconstruction of Late Quaternary Environment of the Lower Luni plains, Thar Desert, India. *Jour. Quaternary Science* 16 61-68, 2001.

Mishra, Sheila, Jain, M., Tandon, S.K., Singhvi, A.K., Joglekar, P.P., Bhatt, C., Kshirsagar, A., Naik, S. and Mukerjee-Deshpande, A. Prehistoric cultures and late quaternary environments in the Luni basin around Balotra. *Man and Environment* 24, 39-49, 1999.

Mishra, Sheila and Rajaguru, S.N. Late Quaternary Paleoclimates of western India: A geoarcheological approach. *Mausam*, 52, 285-296, 2001.

Narula, P.L., Shome, S.K., Nandi, D.R. 1989. *Neotectonic Activity in the Himalaya, Geology and Tectonics of Himalaya*, Geological Survey of India Special Publication No. 26.

Proceedings of the 3rd Meeting of the IGCP 413 - Volume of Abstracts/papers incorporating 17 contributions by the NWG Members and delegates to the 3rd Meeting of the IGCP 413 held in GSI, New Delhi, April 5-6, 2002.

Proceedings of the 4th Meeting of the IGCP 413 - Volume of Abstracts/papers incorporating 29 contributions by the NWG Members and delegates to the 4th Meeting of the IGCP 413 held in Anna University, Chennai, February 3-8, 2003.

Shankar, Ravi, Kumar Gopendra, Saxena, S.P.; 1989, Stratigraphy and Sedimentation in Himalaya,: A reappraisal. In: *Geology and Tectonics of Himalaya,* Special Publication, 26, Geological Survey of India.

Singhvi, A.K. and Krbetschek, M. Luminescence dating a renewed perspective for arid zone sediments. *Annals of Arid Zone* 35, 249-279, 1996.

Singhvi. A.K. and Lang, A. Improvements in Infra-red Stimulated Dating of Poorly Bleached Sediments - the 'Differential' Partial Bleach Method. *Ancient TL*, 16, 63, 1998.

Srivatsava, P.K., Juyal, N., Singhvi, A.K. and Wasson, R.J. Luminescence chronology of River adjustment and incision of Quaternary Sediments in the Alluvial Plain of Sabarmati River, North Gujarat, India, *Geomorphology* 36, 217-229, 2001.

Thapa, B.D, Som Nath, Shali, A.K., 1999. Quaternary geology and geomorphology of Demchok-Hanle-Kidmang belt, Upper Indus basin, Leh Distt., Jammu & Kashmir, Records Geological Survey of India, 132 (8).

Thomas, J.V., Kar, A., Kailath, A.J., Juyal, N., Rajaguru, S.N. and Singhvi, A.K. 1999. Late Pleistocene-Holocene History of Aeolian Accumulation in the Thar Desert, India, *Zeitschrift f. Geomorphology* Suppl. Bd. 116, 181-195, 1999.

Wadhawan, S.K. Geoenvironmental appraisal of Ajmer District, Rajasthan with special emphasis on Desertification beyond Thar Desert margins and Aravalli Hills gaps. 2002. *Geological Survey of India Report* for F.S. 1998-2000, Jaipur.

Wadhawan, S. K. Geoenvironmental appraisal of Jodhpur District, Rajasthan, 1996. *Geological Survey of India Report* for F.S. 1993-1994. Jaipur

Wadhawan, S.K., B.K. Sareen, N.K. Pal and K.S. Raghav, Final report on geological and environmental evaluation of Thar Desert, Rajasthan and Gujarat, India. 1999. *Geological Survey of India Report* for F.S. 1994-1996, pp. 199, 1999, Jaipur.

15

Geomorphological Indicators of Climate Change

SAVINDRA SINGH

Introduction

Variability, in both time and space, is an inherent feature of climate, as the atmosphere is always in the state of turmoil and instability leading to variations in weather and climatic conditions. The climatic change, thus, is defined as variations and shifts in weather conditions over space and time of different scales and magnitude resulting into change of climatic type for example from warm and moist climate to warm and dry climate, from warm and moist climate to cool and moist climate (as happened during Carboniferous period in India) etc. In fact, climatic change refers to drastic or secular changes in heat balance of the earth-atmosphere system, moisture, cloudiness and precipitation caused by either external factors such as variations in orbital characteristics of the earth, solar variability (fluctuations in radiation from the photosphere of the sun), tectonic processes (mainly plate tectonics and displacement of continents and ocean basins), vulcanicity, changes in atmospheric composition in terms of concentration of atmospheric aerosols and carbon dioxide contents etc. or by internal factors such as exchanges of energy between the atmosphere, hydrosphere, lithosphere and cryosphere (ice covered surfaces of both lithosphere and hydrosphere) or by both, at local, regional and global levels. The climochronology (history of palaeoclimates) reveals the fact that climates have changed in the geological past and hence it is opined that 'the world's climates have changed in the past, are changing now, and there is every reason to expect that they will change in future' (Hobbs, 1980). If the change is law of nature, the change in climates is a reality because the climate of a region is not fixed and static rather it goes on changing. Sometimes, the changes are cyclic and rhythemic, such change is called climatic cycle. It may be mentioned that James Hutton (a Scottish geologist) while propounding the concept of uniformitarianism, postulated the concept of cyclic nature of earth's history. The example of occurrences of ice ages during (1) pre-Cambrian period (850-600 million years before present, mbp = million years before present),

(2) Ordovician period (450-430 mbp), (3) Carboniferous-Permian periods (300 mbp), and (4) Pleistocene period (2-3 mbp) validates *the concept of cyclic nature of climatic changes.*

The climatic changes are supposed to be quick and rapid rather than slow and gradual but this may not be always true as climate changes both gradually and rapidly, partly and drastically. For example, the climatic change, which occurred during Jurassic period leading to mass extinction of dinosaurs due to sudden onset of cold climate, was rapid and instantaneous. In fact, the rate of climatic changes depends on the nature of causal factors. The rapid Jurassic climatic change is related to sudden collision of the earth and a giant meteor and consequent release of enormous amount of dusts in the atmosphere. One cannot infer the nature of long-period climate on the basis of present-day climatic conditions. It is also an observed fact that the 'cool periods of earth history are periods of greater than normal climatic instability' (Hobbs, 1980).

The human society with present-day weather conditions is seized with the problems of possible climatic changes in near future. The most significant global environmental problem faced by the world community is *global environmental changes* (GEC) leading to probable climatic changes consequent upon global warming resulting from a host of causal factors, namely ozone depletion, increase in the emission of greenhouse gases at alarming rate, deforestation etc. The probable net result of global warming would be climatic changes at local, regional, and global levels. The international communities are scared of catastrophic adverse effects of future climatic changes on different spheres of man and nature, e.g. deglaciation and sea level changes, submergence of island nations and major coastal lowlands, atmospheric dynamics including evaporation and precipitation, global radiation balance, photosynthesis and ecological productivity, plant and animal communities, human health and many more. It may be summarized that climatic change is a reality, it has changed in the past, it is changing at the present, and it will change in future. The change of climate may be slow and gradual, rapid and catastrophic, periodic, semi-periodic or non-periodic, short-term or long-term, may be at local, regional and global scales, it may be due to natural factors or anthropogenic factors. It is, thus, necessary to discuss various aspects of climatic changes, namely scales of climatic changes (both spatial and temporal scales) and indicators of climatic changes (i.e. evidences of climatic changes). The climatic variations and changes are viewed in terms of temporal and spatial scales depending on the purpose of studies. The temporal scales of climatic changes range from a very micro-scale involving 10-day period to macro-temporal scale involving thousands to millions of years. 'The variability may be periodic (A), quasi-periodic (B), or non-periodic (C), or alternatively it may show a progressive trend' (Barry and Chorley, 2002). It may be mentioned that spatial and temporal scales of climatic changes are correlateds, e.g. as the temporal scale of change becomes shorter, the area also becomes smaller. In other words, the changes may be more perceptible and pronounced in localized area during short period or recorded climatic data, but if we consider climatic data at regional level involving large countries or continents, and at global level, the local level changes are overshadowed. Thus, the climatic changes may be viewed in terms of three temporal scales, namely (i) macro-temporal scale (millions of years), (ii) meso-temporal scale (thousands of years), and (iii) micro-temporal scale (hundreds of years). These three temporal scales correspond to global, regional, and local spatial scales. At regional and local spatial scales the climatic variability can be looked upon at even shorter temporal scales which may range from very micro-temporal scale to increasing time span, e.g. (1) 10 days time scale, (2) 10-100 days time scale, (3) 100-1000 days time scale, (4) 1000-10000 days time scale etc. It may be remembered that such changes may be detected only if the instrumented data are very accurate and regular. The climate changes which occurred in the past, say before the industrial revolution, cannot be viewed in such a micro-temporal scales.

Generally, climatic changes are considered at two levels, e.g. (1) short-term changes, and (2) long-term changes. Short-term climatic changes involve the consideration of changes in the energy balance of the earth-atmosphere system leading to periodic changes in weather and climate. This temporal scale involves time span ranging from few years to thousands of years. Short-term climatic changes are either inter-annual or they may persist for many years and are generally caused by anthropogenic factors. On the other hand, long-term climatic changes persist for thousands to millions of years and are exceedingly slow. These are always caused by natural factors.

Incidators of Climatic Changes

The evidences of climatic changes in the past, which are utilized in the reconstruction of climochronology (history of palaeoclimates) are called indicators of past climatic changes. The riddle of reconstruction of palaeoclimates is a fascinating puzzle, the solution of which basically depends on proxy data (which include alternative sources of data as substitutes for real data) and logical deductions by the investigators. It may be mentioned that recorded data of weather and climate are available only since industrial revolution, more precisely for the last 100 years or so, and hence proxy data for recorded data are the only indicators to decipher the pages of past climatic history and changes on the basis of Huttonian theory of uniformitiarianism (postulated by Scottish geologist, James Hutton in the year 1785). This theory was based on two concepts of Hutton, namely (i) present is key to the past, and

(ii) no vestige of a beginning, no prospect of an end. The principle of uniformitarianism states that 'the same physical processes and laws that operate today, operated throughout geological time, although not necessarily always with the same intensity as now (Hutton postulated same intensity of processes throughout geological time).

The indicators (clues) of palaeoclimaties may be classified on the basis of basic sources as follows :

(1) Biological Indicators

(A) *Floral Indicators*
- (i) fossil remains of plants
- (ii) fossilized pollens
- (iii) oxygen isotopes
- (iv) tree ring growth (dendrochronology)

(B) *Faunal indicators (evidences)*
- (i) fossilized animal remains
- (ii) distribution and dispersals of animals

(2) Geological Indicators

(A) *Terrigeneous ancient deposits*
- (i) lacustrine deposits (varves)
- (ii) evaporite deposits
- (iii) sedimentary deposits (coals)

(B) *Marine deposits*
- (i) sea floor deposits

(C) *Pedogenic indicators*

(3) Cryogenic Indicators

(A) *Glaciation*
- (i) ice ages
- (ii) glaciers and glaciation
- (iii) ice sheets and ice cores

(B) *Periglacial evidences*

(4) Tectonic Indicators

(A) *Plate tectonics*
- (i) pole wandering and continental drift

(ii) palaeomagnetism and sea floor spreading

(B) *Sea level changes*

(5) Geomorphological Indicators

(A) *Morphological Features*

(i) erratics
(ii) sand dunes
(iii) river terraces
(iv) duricrusts
(v) pediments
(vi) tors

(B) *Geomorphological Processes*

(6) Historical Records

(i) flood records
(ii) drought records
(iii) migration of man and animals

In a broader sense geomorphological indicators of palaeoclimatic changes include geological indicators, cryogenic indicators, tectonic indicators, morphological indicators, dendrochronology, sea level fluctuations etc. These indicators are briefly examined below:

Geological Indicators

Geological indicators (evidences) of palaeoclimates include lithological characteristics of mainly sedimentary deposits such as lacustrine deposits (varves), evaporites, limestones and coal seams, marine deposits (sea floor deposits), soil profiles etc. *Varves* are the alternate sequences of layers of fine silts and clays deposited at the floors of lakes and large ponds in such an area which is characterized by alternate freezing and thawing during winter and summer seasons respectively. Thus, the lithological sequences of lacustrine deposits in the aforesaid condition denote periglacial climate. The analysis of annual layers of silt and clay provides an idea of seasonal changes in climate. It may be mentioned that when the surface freezes during winter season, very fine suspended clay particles are deposited in the lakes and ponds but during summer season, when frozen surface thaws (melts), silts with melt water are deposited in the lakes and ponds. Such type of annual banded alternate layers of clay and silt denote periglacial climate.

Evaporite deposits, represented by salt deposits, occur when climate is characterized by high temperature and aridity wherein evaporation exceeds precipitation. In such circumstances water is evaporated and salt contents are left behind. It may be mentioned that this happens only when water on land is saline and thus salt is in solution form. So, the massive salt rocks deposited on land surfaces, such as in South-West USA, Germany, Central Asia, North-West India (mainly Rajasthan and Gujarat), denote hot and arid climatic environment.

Limestones ($CaCO_3$) containing calcium carbonates are deposited in tropical warm oceanic water and hence the occurrence of limestones in the regions having cold climates denotes the fact that the concerned region was under tropical warm climate at the time of limestone deposition.

Duricrusts are indurated hardened surfaces of different kinds such as *laterites, silcretes, calcretes, alcretes, ferricretes* etc. depending on the dominance of constituent minerals. Normally, lateritic crusts are supposed to have been formed in hot and humid climate of tropical and subtropical areas and therefore, these are indicative of hot and humid climate. Lateritic crusts are predominantly found in Chhotanagpur high lands (Patlands of Ranchi and Palamau plateaus of Jharkhand); and over many areas of Deccan plateau {e.g. Mahabaleshwar and Panchgani plateaus of Maharashtra). The presence of lateritic crusts in certain parts of Europe (e.g. U.K. and Germany) clearly demonstrates the fact that these are not the result of the present temperate climate. 'Such crusts are often interpreted as of Tertiary age, or as having been under continuous formation since the end of the Mesozoic. Exposures of silcretes and calcretes similarly are often related to past rather than present climatic conditions' (Stoddart, 1969).

Pedogenic criteria used to decipher past climates, include the analysis of palaeosols and fossils of plants and animals therein. The alluvial soils buried in older flood plains give indication of moist climate and the dominance of fluvial processes. The older loessic soils indicate dry climate and the dominance of aeolian process. Even the materials involved in the deposition of loess also give clues for climatic conditions. The extensive loess deposits of China having areal coverage of 774,000 km^2 and thickness of 90m to 300m consist of materials blown from deserts of central Asia while the European loess is example of glacial loess. The American loess, extensively found in Illinois, Iowa, Nebraska etc. is partly glacial and partly desert loess. The datermination of palaeoclimates on the basis of fossil remains in soils is accomplished through radiocarbon dating and isotope analysis.

Cryogenic Indicators

Cryogenic indicators are related to the proxy data from the evidences of glaciation, glaciers and icesheets. The science dealing with glaciation and glaciers is known as *glaciology.* The processes of glaciation and deglaciation provide significant proxy data for climatic changes and *fluctuation.* The period of widespread glaciation of larger areas of the globe is called *great ice age* which comprises several glacial and interglacial periods. The glacial period denotes onset of cold climate and advance of ice sheets while interglacial periods indicate relatively warmer periods when ice sheets retreat. The geologists, glaciologists and climatologists have identified a few great ice ages (e.g. pre-Cambrian ice age, Carboniferous ice age, and Pleistocene ice age) when major portions of the globe were glaciated and covered with thick ice sheets which are definite indicators of major climatic changes.

Credit goes to European School of Geomorphologists and Glaciologists for indentification and recognition of ice ages. Louis Agassiz (1807-1873 AD) is given credit for an early start in this precarious field and for the recognition and identification of the presence of ice age during Pleistocene period as he presented his ideas in the year 1840. A host of geoscientists, namely Jean de Charpentier, John Playfair (1815), Venetz of Switzerland (1821, 1829), Esmark of Norway (1824), Bemhardi of Germany (1832) etc., studied different aspects of Pleistocene ice age. The Scottish geologist James Geikie postulated the concept of 'great ice age' in 1984 which is comprised of several glacial periods which are separated by warm interglacial periods. A. Penck and Bruckner identified four glacial periods during Pleistocene ice age, e.g. Gunz, Mindel, Riss, and Wurm which were separated by three warm interglacial periods. Similarly, four glacial periods (e.g. Nebraskan, Kansan, Illinoin, and Wisconsin) and three interglacial periods (e.g. Aftonian, Yarmouth, and Sangman) were identified during Pleistocene glaciation of North America. Most of Gondwanaland was glaciated during Carboniferous ice age. The shapes of glaciated valleys, glacial drifts, erratics (glacial boulders), striations etc. provide important clues for climatic changes.

The glaciers and ice layers provide evidences for temperature and precipitation conditions during different phases of ice ages. Advancing glaciers and ice sheets indicate cold phase and lowering of temperature much below freezing point, while retreat of ice sheets heralds increase in temperature and subsequent ablation of glaciers demonstrates onset of warmer climate. The glaciologists have demonstrated the retreat of Alpine glaciers and resultant shrinking in their length and width about 3000 B.C. (i.e. about 5000 ybp, years before present), followed by re-advance of glaciers about 500 B.C. (i.e. 2500 ybp) and again retreat of glaciers. The Alpine and Scandinavian glaciers registered resurgence in the 17th and 19th centuries while they again started to retreat in the 20th century. All these indicate climatic fluctuations characterzed by cooling (advance of glaciers) and warming (retreat of glaciers) periods during the past 5000 years in European continent.

Ice sheets and ice cores are most significant cryogenic indicators of palaeo climates. It may be remembered that the ice sheets are formed by the deposition of several layers of ice. The accumulated snow from the annual snowfall is compressed and is changed to ice wherein the air bubbles and atmospheric dusts are trapped. Thus, every year a layer of ice is deposited upon underlying ice layer. It is evident that the great ice caps of Greenland and Antarctica are comprised of numerous ice layers wherein the then climatic conditions have been preserved. Thus, the study of each ice layer provides ample evidence of the weather and climatic conditions at the time of its formation. The drilling into ice caps is undertaken to obtain ice cores for analysis. The following properties of ice layers derived from the ice cores provide proxy data for temperature precipitation, composition of atmosphere, storminess, volcanic events, atmospheric pollution etc.

(i) thickness of annual ice layers - pecipitation
(ii) chemical analysis of annual ice layers - temperature
(iii) air bubbles trapped in the annual ice layers - atmospheric composition
(iv) dust contents trapped in the annual ice layer - atmospheric circulation and storminess
(v) acid content in the annual ice layer - volcanic events

A few studies of ice cores from Greenland ice sheets (namely, Greenland Ice Sheet Project 1 and 2, GISP-1 and 2) and Antarctica ice sheets have revealed sequences of climatic variations for the last 110,000 years and 160,000 years respectively. The analysis of ice cores to obtain climatic records is based on isotopes of oxygen (two very common isotopes are 160 and 180) wherein the ratio of 160 to 180 provides information about the existing environmental temperatures at the time of the formation of annual ice layers. The isotope analysis of ice cores also reveals relationship between global warming and concentration of greenhouse gases (e.g. carbon dioxide, methane etc.), and level of atmospheric pollution caused by anthropogenic sources since the period of industrial revolution. Besides, the concentration of radioactive elements in the atmosphere and their fallout emitted during nuclear plant disasters (for example, Chernobyl nuclear disaster) and testings of atomic devices, can be measured through the analysis of annual ice layers. The ice core analysis also reveals information about the levels of sulphate content in the air emitted from the volcanic eruption. With the improvements in the technologies of obtaining ice cores, studies are carried out to reconstruct the atmospheric conditions mainly temperature variations and climatic fluctuations for the past 200 years or so on the basis of the analysis of ice cores derived from mountain ice sheets in the tropical and subtropical areas.

The *evidences of periglacial features* and deposits found in such areas which at present do not have periglacial environment, indicate earlier periglacial climate of those areas. The term periglacial literally means around the ice or peripheral to the margins of the glaciers but now this term is used for both *'periglacial landscape'* and *'periglacial climate'*. Periglacial areas are those which are in permanently frozen condition but without permanent ice cover on the ground surface. The periglacial climate is characterized by mean annual temperature ranging between -1°C and -15°C and mean annual precipitation of 120 mm to 1400 mm (mostly in solid form). Many parts of Europe and South-West USA presently do not have periglacial climate but have relict periglacial features which demonstrate that such areas were under periglacial climate in the past.

Tectonic Indicators

The tectonic movements involving plate movements, pole wandering, continental drifts, orogenesis, palaeomagnetism, topographic features etc. and seafloor spreading and sea level changes are significant indicators of palaeoclimatic changes. A. Wegener, a German meteorologist, was primarily concerned with the problem of past climatic changes. It may be pointed out that there are ample evidences, as discussed above, which indicate widespread climatic changes throughout the past history of the earth. In fact, the continental drift theory of Wegener 'grew out of the need of explaining the major variations of climate in the past'. The climatic changes which have taken place on the globe may be explained in two ways as follows :

(1) If the continents remained at their places throughout geological history of the earth, the climatic zones might have shifted from one region to another region and thus a particular region might have experienced varying climatic conditions from time to time.
(2) If the climatic zones remained stationary, the landmasses might have been displaced and drifted.

Wagener opted for the second alternative and postulated the concept of 'continental drift' or 'displacement hypothesis' to explain the global climatic changes on the basis of a number of geographical, geological, palaeontological etc. evidences. The plate tectonic theory as an outcome of post-Second World War advancement (mainly in the 1960s) in geotectonics, the evidences of palaeomagnetism and seafloor spreading have validated the concept of polewandering (shifting of the positions of the poles) which help in the reconstruction of climochronology mainly global climatic changes during Carboniferous and Pleistocene Ice Ages and widespread glaciation.

It has been demonstrated that plate movements during Cenozoic era displaced major continental blocks in the middle and high latitudes in the northern hemisphere and thus such northward displaced continental masses were subjected to the dominance of cold climate resulting into widespread glaciation (such as major parts of North America and Eurasia) during Pleistocene period (known as Pleistocene Ice Age and Pleistocene glaciation). On the other hand, mountain ranges of Tertiary period, formed due to collision of convergent plates and consequent subduction of heavier plate and lateral compression, and high plateaus, became effective barriers in controlling the global atmospheric circulation and generating a few new climatic types such as monsoon climate of south Asia (due to relief barriers of the Himalayas and Tibetan plateau). The investigations regarding the impacts of tectonic movements on climatic conditions have revealed strong relationship between horizontal plate movements leading to seafloor spreading along divergent plate boundaries and orogenesis along the convergent plate boundaries, vertical movement (tectonism) leading to upliftment and subsidence. The changes in global pattern in plate motions caused changes in climates at global scale.

Morphological Indicators

The geomorphological processes and landforms resulting therefrom have been directly related to particular climatic type on the concept that 'each climatic type produces its own characteristic assemblage of landforms and set of geomorphological processes which shape them on the basis of the following themes':

(1) Landforms differ significantly in different climatic regions.
(2) Spatial variations of landforms in different climatic regions are because of spatial variations in climatic parameters (e.g. temperature, humidity, precipitation etc.) and their influences on weathering processes, erosion dynamics and surface runoff.
(3) Quaternary climatic changes could not obscure relationships between landforms and climates.

In other words, there are certain diagnostic landforms which clearly demonstrate climate-landforms relationships.

Thus, on the basis of diagnostic landforms, the inference of climatic conditions at the time of the formation of such landform is drawn which helps in the reconstruction of palaeo climates.

The climatogenetic or climatically controlled landforms are identified and differentiated in two ways, e.g. (i) general observation and acquaintance of whole landscape of each climatic region, and (ii) identification of typical or distinctive landforms which represent the controls of particular climate. The typical landforms are, in fact, main tools of climatic geomorphologists which help them in determining climate-landforms relationships in different climatic regions. Such distinctive landforms are designated as diagnostic landforms which include duricrusts, inselbergs, pediments, tors, glaciated valleys, glacial boulders or erratics, sand dunes etc.

Duricrusts are indurated hardened surfaces of different kinds such as laterites, silcretes, calcretes, alcretes, ferricretes etc., and are normally supposed to have been formed in hot and humid climate of tropical and subtropical areas and hence they are indicative of warm and moist climate. The presence of lateritic crusts in certain parts of Europe (e.g. U.K., Germany etc.) clearly demonstrates the fact that these are not the result of present climate. 'Such crusts are often interpreted as of Tertiary age, or as having been under continuous formation since the end of the Mesozoic. Exposures of silcretes and calcretes are often related to past rather than present climatic conditions' (Stoddart, 1969).

Inselbergs representing steep sided residual hills are considered to be the representative landform of hot and arid and semi-arid climates. It is argued that inselbergs are structurally controlled rather than climatically controlled and most of the present inselbergs were formed before Quaternary period, 'hence present climates are not necessarily those in which inselbergs were formed' (Stoddart, 1969). It may be possible that inselbergs might have been formed when the climate was warm and arid or semi-arid which might have changed after their formation.

Pediments, characterized by low-angle rock-cut surfaces surrounding mountains, are also considered to be the representative landforms of warm arid and semi-arid climates. Pediments are also found in a variety of climatic conditions, e.g. tropical wet and dry climate, subtropical and temperate climates. A few geomorphologists argue that pediments are structurally and tectonically controlled rather than climatically controlled. L.C. King has opined that the process of pediplanation and pedimentation is universal and occurs in all environmental conditions. In fact, 'many arid zone pediments are clearly polycyclic, developed during the complex sequence of Pleistocene pluvials (period of prolonged rainfall) and interpluvials: Many appear to be being destroyed under present climatic conditions, rather than being formed' (Stoddart, 1969).

Tors, 'one of the most controvercial landforms, are piles of broken and exposed masses of hard rocks particularly granites having a crown of rock-blocks particularly granites of different sizes on the tops and clitters (trains of blocks) on the sides' (Savindra Singh, 1977). Tors have been considered of periglacial origin by J. Palmer and R.A. Neilson (1962), of fluvial origin by D. L. Linton (1955), whereas L.C. King has opined that tors are the result of universal processes of pediplanation in different climatic conditions. In fact, the presence of tors right from the Dartmoor of England through Nicaragua to India has complicated the problem of the origin of tors rather than solving it.

It may be concluded that the aforesaid diagnostic landforms are older than Pleistocene climatic changes, so they are definitely not related to present climates where they are found. It may be pointed out that climatic relation of landforms at least in glacial, periglacial and warm desert climates is undoubtedly confirmed. So, such landforms may provide definite clues to reconstruct past climates.

It is an established fact that different geomorphological processes work in different climatic regions because the geomorphic processes originate from the atmosphere and their nature (type, e.g. whether fluvial, or glacial, or periglacial or aeolian etc.) depends on the combinations of temperature and precipitation. Thus, the nature of geomorphological processes in the past geological history of the earth helps in deciphering the climatic conditions in which a particular process was predominant.

The presence of glacial boulders or erratics in a region presently having other than glacial climate, may it be warm tropical or temperate climate, reminds us the fact that the region was glaciated at the time of formation and deposition of these erratics. Similarly, the presence of frost-riven cliffs, tors, attiplanation terraces patterned grounds (having stone rings, polygons, garlands, stripes etc.), thermokarst lakes, nivation hollows, in volutions, pingos, congelifluctate landforms, blockfields (blockmeers), boulderfields etc. tells us the dominance of periglacial processes under periglacial climate at the time of the development of such landforms. The presence of glacial boulders and erratics in the coal seams of Talcher of Orissa (India) Gondwana period clearly indicates an earlier cold and glacial climate in India having presently warm monsoon climate.

Sea Level Fluctuations

Fluctuations in sea levels are considered significant indicators of past climatic changes. Sea level charges are of two types, namely positive change marked by rise in sea level above datum line, and negative change denoting fall in sea level below datum line. Changes in sea level are effected by (i) tectonic movements, and (ii) climatic changes. The tectonic movements cause rise in sea level when there is upward movement of sea floor (ocean bottoms) or down warping of coastal area whereas fall in sea level is caused by down warping leading to subsidence of sea floor or upwarping of coastal lands. It may be mentioned that tectonically induced changes in sea levels are very slow. The climatically induced changes in sea level are rapid and are effected by glacial age and advancement of glaciers and ice sheets (fall in sea levels) and deglaciation leading to retreat of glaciers and ice sheets (rise in sea levels). The sequences of rise and fall in sea levels during Pleistocene Ice Age have revealed several phases of glacial periods (Gunz, Mindel, Riss and Wurm in Europe, and Nebraskan, Kansan, Illinoin and Wisconsin glacial periods in North America) and interglacial periods (relatively warm periods) separating two glacial (cold) periods.

Dendrochronology

Dendrochronology is the science that deals with 'the study of the annual rings of trees in determining the dates and chronological orders of past climatic events'. The ring growth of the trunk and stem of a tree provides significant clues for detecting seasonal rhythms of climate during the life time of the concerned tree. The thickness and spacing, and colour of growth rings indicate the climatic conditions and rate of annual growth of trees. The widely spaced thick growth rings denote warm and moist climate with sufficient rainfall to support luxurious and quick growth of trees, while narrow and closely spaced tree growth rings denote arid climate and poor growth of trees. The colour of growth rings also denote seasonal weather conditions. 'The abrupt change from light to dark-coloured rings (growth rings) delineates the annual increments of growth' (Oliver and Hidore, 2003). It may be mentioned that the study and interpretation of different aspects of tree rings such as, the size (thick or narrow), number, colour shades, symmetry or asymmetry etc. provide information about climatic and environmental variations sustained by the concerned tree during its lifetime and hence the climatic events which happened in the recent past, say 3000-4000 years before present may be reconstructed on the basis of dendrochronological basis.

The study of tree ring growth known as dendrochronology was initiated by A.E. Douglas and his team at the University of Arizona, USA. After the analysis of growth rings of trees in the south-western USA he found close relationship between annual amount of rainfall and growth rings in the adverse climatic conditions of the S. W. United States. The scientists have become successful in reconstructing the climatic conditions and environmental changes for the last 3000 years on the basis 0' analysis of growth rings of living trees, and for the last 5000 years on the basis of fossils of dead trees. The scientists at Laboratory of Tree Ring Research, University) of Arizona, USA, have also derived logical and fruitful inferences about the relationship between weather elements (e.g. temperature, air pressure, and atmospheric circulation patterns) and widths and spacing of tree rings. Such relationships have also been used to demonstrate climatic fluctuations which tool place during the lifetime of trees.

Dendrochronology also helps in reconstructing the advances and retreats of glaciers in the recent past on the basis of shapes of tree growth rings like concentric, (symmetrical) growth rings or asymmetrical rings. It may be remembered the growth rings become concentric and symmetrical as long as the trees remain perfectly erect (perpendicular to the ground surface) but become asymmetrical when the trees are tilted. It is argued that trees are erect when the glacial ice is away from the position of trees but they are titled when ice comes close to the trees. Thus, tilted trees and resultant asymmetrical growth rings may not be always due to glacial advancement, rather it may also be due to landslides, gusty winds, strong storms etc.

REFERENCES

Bloom, A.L., 2002 : *Geomorphology,* Prentice Hall of India Private Ltd., New Delhi, pp. 49-63.

Critchfield, H.J., 2002: *General Climatology,* Prentice Hall of India, Private Ltd., New Delhi, pp. 227-246.

Hobbs, J.E., 1980: *Applied Climatology,* Butterworths, London, Boston, pp. 218.

Miller, *et al.,* 1991 : Unlocking the ice house: Oligocene-Misocene oxygen isotopes, eustasy, and margin erosion, *Journ. Geophys. Res.* Vol. 96, no. B4, pp. 6829-6848.

Morgan *et al.,* 1994: Carbon isotope evidence for the emergence of C4 plants in the Neogene from Pakistan and Kenya, *Nature,* Vol. 367, pp. 162-165.

Oliver, J.E. and Hidore, J.J., 2003: *Climatology, Pearson Education* (Singapore) Pt. Ltd. Indian Branch, Delhi, pp. 261-279.

Singh, Savindra, 1977 : Evolution of granitogneissic tors and cut-off spur mounds of Ranchi plateau, *National Geographer,* Vol. 10, No.1, pp. 93-97.

Stoddart, D.R., 1969 : Climatic geomorphology, review and reassessment, *Progress in Physical Geography,* Vol. 1, pp. 159-222.

16

Greenhouse Effect, Global Warming, Climate Change

T.I. KHAN

Introduction

As the weeks of 1988s summer drought and stifling heat dragged by, a momentous shift took place in the public's attitude towards the global environment. Suddenly, it seemed, everyone knew and cared about a scientific principle long of deep concern to scientists studying the earth system. This principle, known as to scientists studying the earth system. This principle, known as the "greenhouse effect", explains why gases produced by human activity will probably cause the earth's average temperature to increase within the lifetimes of most people living today.

The 1980s were the warmest decade recorded on a global basis, but scientists are still uncertain, and will be for years, whether the warm spell was a normal climatic fluctuation or a response to the billions of tons of carbon injected into the atmosphere each year by human activities. Scientists working in climatology and related fields say that the insulating effects of the greenhouse gases should be clear to all of us within a few decades, and possibly by the end of the 1990s.

One cannot infer from a specific summer that global warming has begun, though a warmer climate would change the probabilities for heat waves and possibly for strong hurricanes. The weather events of 1988 did, however, convey an idea of the kinds of environmental and commercial effects we could expect if the current predictions about global warming come to pass. The north American corn crop was stunted by drought in the grain belt, and productivity fell below consumption (probably for the first time in the U.S. history), so that no grain was added to the nation's reserves. Water levels in thoroughfares like the Mississippi River dropped so low that barges and their cargoes were stranded for weeks. Forest fires burned uncontrollably in America's great natural parks, a super hurricane threatened America's Gulf Coast, and floods in Bangladesh led to the deaths of 2000 people and drove millions of others from their homes. These and other weather events over the course of a single summer highlighted for billions of people that human society is highly vulnerable to extremes in the weather.

Greehouse Gases

Although there may be questions about the causes of a specific drought or flood, there is no controversy about some basic facts about our atmosphere. Trace gases such as water vapour, carbon dioxide, methane, chlorofluorocarbons, tropospheric ozone, and nitrous oxide create a greenhouse effect by trapping heat near the earth's surface, and the concentrations of many of these gases are increasing in the atmosphere. Because of these increase, the gases are expected to trap more energy at the earth's surface and in the lower atmosphere, in turn causing increase in temperature, changes in precipitation patterns, and other as yet unpredictable changes in the global climate.

The principle of the greenhouse effect explains the cold climate of Mars (where water vapour, a highly efficient greenhouse gas, is virtually absent), the hot climate of Venus (where the atmosphere is thick with the carbon dioxide and conditions are so hot that life as we know it could not survive), and the moderate climate here on the earth. Scientists have known for decades that a buildup of carbon dioxide in the atmosphere could warm the earth's climate. They have also known that atmospheric concentrations of carbon dioxide alone have increased by about 26 per cent since coal, oil and gas became the primary sources of energy to fuel the Industrial Revolution. Carbon dioxide concentration are currently increasing by about 0.5 per cent each year.

After water vapour, carbon dioxide is the most plentiful and effective greenhouse gas. It occurs naturally but is also produced in great quantity during the combustion of fossil fuels, particularly coal. When the fuel is burned, its carbon is oxidized to carbon dioxide and released. Carbon dioxide also is released as forests are cleared and the organic matter is burned or allowed to decay. These human activities are injecting almost 6 billion tons of carbon into the atmosphere each year. By comparing this figure with the actual increases in concentrations of carbon dioxide (about 3 billion tons annually), scientists presume that about half of the carbon injected into the atmosphere is being absorbed by oceans and plant life and about half remains in the atmosphere.

Only in the last decade have scientists become aware that other, trace greenhouse gases can also be important contributors to global warming. Concentrations of many of these trace gases are known to vary naturally, but there is widespread agreement that human activities are contributing to the current increases.

Molecule for molecule, the following trace gases absorb infrared radiation much more effectively than carbon dioxide does. Because their concentrations in the atmosphere are much lower than that of carbon dioxide, their individual effect is much smaller. Their combined effect, however, is likely to be equal to or greater than that of carbon dioxide alone.

Methane (CH_4). Methane, also known as natural gas, is produced through bacterial activity in bogs and rice paddies, and in the digestive tracts of ruminative animals and insects such as termites. Most atmospheric methane comes from biological sources. It is present today at roughly 1.7 parts per million and is increasing at a rate of about 1.1 per cent each year. Analysis of gas bubbles trapped in glacial ice shows that the rise in methane levels parallels the growth of the human population. Per molecule it is 25 times as effective as carbon dioxide at trapping heat.

Chlorofluorocarbons (CFCs). Chlorofluorocarbons are a group of synthetic compounds used in refrigeration, insulation, foams, and other industrial purposes. Apart from their role as greenhouse gases, when CFCs rise to the upper atmosphere, or stratosphere, they release free chlorine, which then catalyzes the breakdown of ozone, the protective layer that shields the earth from ultraviolet radiation. The two most prevalent CFCs are CFC-12, which per molecule has 20,000 times the capacity of carbon dioxide, to trap heat, and CFC-11, which ahs 17,500 times the capacity. The Montreal Protocol, an international agreement adopted in 1987 to limit the production of CFCs, will slow but not eliminate the rate of increase.

Nitrous oxide (N_2O). Nitrous oxide is produced naturally, through microbial action in the soil, and in response to the spread of agriculture, the burning of timber, the decay of crop residues, and the combustion of fossil fuels. Agricultural use of mineral fertilizers containing nitrogen presumably accelerates its rate of release. Atmospheric concentrations of nitrous oxide are increasing by about 0.25 per cent per year. It has a long residence time in the atmosphere, and so concentrations would increase for more than 200 years even if emission rates were to freeze at current levels. Scientists believe nitrous oxide levels in the year 2030 will be about 34 per cent more than preindustrial levels. Per molecule, this trace gas has 250 times the capacity of carbon dioxide to trap heat.

Tropospheric ozone (O_3). In the stratosphere, ozone shields the planet from ultral-violet radiation; nearer the ground in the troposphere, the moisture-rich atmospheric layer below the stratosphere, it is an effective greenhouse gas. It is produced through reactions involving hydrocarbons and nitrogen oxides, all released through the combustion of fossil fuels used by motor vehicles and in industry. Concentrations of tropospheric ozone appear to be increasing at many locations in the Northern Hemisphere. Results from studies of the Amazon River basin indicate that tropical forests act as a sink for ozone; if so, their continued destruction could have a significant effect on regional ozone balances.

Although scientists have considerable confidence in the principle of the greenhouse effect and the measurements of increasing greenhouse gases in the atmosphere, two key questions remain surrounded by uncertainties: How quickly will the climate change, and by how much?

Climate Change

Using three-dimensional mathematical models of the climate system, scientists draw a number of inferences about what conditions might be like in the future. The most likely conditions include significant cooling of the stratosphere, warmer surface temperature (which would be felt disproportionately at high latitudes), and changes such as rising sea level, reductions in sea ice, and increases in total precipitation (which again would be non-uniformly distributed around the globe). They also speculate that summers in the mid-continents would be much dryer than they are today.

These response to increased greenhouse gas concentration, as well as their scientific uncertainties, were described by Jerry Mahlman director of the Geophysical Fluid Dynamics Laboratory of the National Oceanic and Atmospheric Administration (NOAA) in Princeton, New Jersey, at the 1989 Forum on Global Change and Our Common Future. Mahlman outlined a list of responses, an earlier version of which appeared in the 1987 National Research Council report, *Current Issues in Atmospheric Change*. The estimates reflect assumptions about future concentrations of trace gases on current trends.

The results of literally millions of measurements and calculations over the past century indicate that the earth is likely to experience significant climate change during the next few decades. The model is predict that because of carbon dioxide and other gases that have built up in the atmosphere since 1860, the earth is probably already committed to a 0.5° to 1.5°C increase in average global temperature. If current emissions trends continue, the combined greenhouse effect of all trace gases would commit us to an "effective carbon dioxide doubling"— the point where carbon dioxide and other trace greenhouse gases combined trap the same amount of energy that carbon dioxide would trap alone if its concentration doubled from the preindustrial level—possibly as early as 2030.

Although the climate models are intricate and require massive amounts of computer time, they are stark, simplistic representations of he complex realties of the real climate system. It is difficult, for instance, to include cloud cover in the models, even though clouds may amplify or moderate the greenhouse effect. Most of the models do not adequately include the dynamics of ocean circulation, an essential determinant of models incorporate the entire range of uncertainties about potential responses of the earth system—the possibility, for instance, that an increase in the temperature could alter cloud cover or increase the rate at which soil bacteria break down dead organic matter and consequently accelerate the biological contribution of carbon dioxide to the atmosphere, or the possibility that climate change could trigger a dramatic shift in ocean circulation that would completely alter temperature and precipitation patterns.

In spite of these uncertainties imposed by both the practical computational limits of the models and the incomplete understanding of the system, scientists cautiously predict how much global average temperatures would rise with an effective carbon dioxide doubling. Many assessments indicate a range of estimates between 1° and 5°C.

Likely Effects of Climate Change

General circulation models predict several important climate changes due to increased greenhouse gases. Jerry Mahlman interprets confidence levels for current models according to the following guidelines: "Virtually certain" means there is nearly unanimous agreement within the scientific community that a given climate effect will occur. "Very probable" means greater than about a 90 per cent (9 out of 10) chances, and "probable" implies that more than about 67 per cent (2 out of 3) chances. "Uncertain" refers to an effect that has been hypothesised but for which there is a lack of appropriate modelling or observational evidences.

Large Stratospheric Cooling (Virtually Certain)

A reduction in upper stratospheric ozone by chlorine compounds will mean less absorption of solar radiation and thus less heating. In addition, increased concentrations of trace gases will increase the loss of infrared radiative heat from the stratosphere. The combination of decreased heating and increased cooling will markedly lower upper stratospheric temperatures, perhaps by 10° to 20°C.

Global Mean Surface Warming (Very Probable)

For a doubling of atmospheric carbon dioxide (or its radiative equivalent from all of the greenhouse gases), the long-term global mean surface warming is expected to be in the range of 1.5°C to 5.0°C. The most significant source of uncertainty arises from the difficulties in modelling the feedback effects of clouds. The rate of warming over the next century will be governed by the rate at which concentration of greenhouse gases increase, natural fluctuations in the climate system, and the intricate responses of the slowly responding parts of the climate system, such as oceans and the glacial ice.

Global Mean Precipitation Increase (Very Probable)

Increased heating of the earth's surface will lead to increased evaporation and therefore to greater global mean precipitation. Nonetheless, precipitation might will decrease in many individual regions.

Northern Polar Winter Surface Warming (Very Probable)

As the sea ice boundary shifts towards the poles, the models indicate that winter surface temperatures in polar regions would be much warmer than they are now. Scientists calculate that the greater fraction of open water and thinner sea-ice in winter will cause the northern polar surface air to be more than 10ºC warmer than the current climate, or 3 times the global mean warming.

Reduction of Sea Ice (Very Probable)

As the climate warms, there will be less total sea ice because of warming at high latitudes of the Northern Hemisphere. However, the NOAA Geophysical Fuild Dynamics Laboratory reports preliminary findings that new model results incorporating a fully interactive ocean indicate little warming at high Southern Hemisphere latitudes over the next century. This would mean little change in sea ice cover there.

Northern High-Latitude Precipitation Increase (Probable)

As the climate warms, the increased poleward penetration of warm, moist air may increase the annual average precipitation and river runoff in high latitudes.

Summer Continental Dryness? Warming (Probable)

Several model studies have indicated that soil moisture in the mid-latitude continental interiors may decrease during summer. This drying is caused mainly by an earlier termination of snowmelt and rainy periods and thus and earlier onset of the normal spring to summer reduction of soil moisture.

Rise in Global Mean Sea Level (Probable)

It is likely that sea level will rise as sea water expands in response to the warmer future climate. Far less certain is how much sea level will be affected by possible melting of glaciers and calving of land ice.

Regional Vegetation Changes (Uncertain)

Climatic changes in temperature and precipitation of the kinds indicated above must inevitably lead to long-term changes in surface vegetation. The exact nature of such changes and how they might affect the climate in turn remain uncertain.

Tropical Storm Increase (Uncertain)

A number of scientist have suggested that a warmer, wetter atmosphere could lead to more frequent and more intense tropical storms, such as hurricanes. However, tropical storms also are governed by other factors such as local wind structure. So far, this effect has not been satisfactorily addressed in the coarse-resolution climate models because tropical disturbances are relatively small.

CO_2 Doubling and Likely Effects

Many assessments indicate a range of estimates between 1° and 5°C. To appreciate what the projected warming really means, consider the numbers involved. When scientists say that on average the global temperature could increase by a few degrees centigrade, they are talking about a very large increase and a tremendous amount of heat. The current average global temperature is about 14°C (57°F). A 3°C rise would create conditions that some organisms have not to contend within the last 100,000 years. If the temperature rises 4°C, the earth would be warmer than anytime since the Ecocene period, 40 million years ago. In the midst of the last glaciation, when much North America was covered by ice, the average temperature of the earth was only about 5°C colder than it is now. Thus, what seems to be a very small average temperature change can have a very dramatic effect. Moreover, the projected rate of warming is 15 to 40 times than the natural warmings after the major ice ages and much faster than what most species living on the earth today have had to face.

The climate model results are based mainly on hypothetical and mostly instantaneous and large changes in concentrations of greenhouse gases. In fact, concentrations of the gases are increasing gradually. Initially, most of the excess heat is absorbed into the oceans, but understanding of the complex interactions between the atmosphere and oceans in incomplete. We can expect that natural, decadal-scale climatic fluctuations due to interactions between the atmosphere and oceans will continue to occur. Mahlman points out the midwestern drought in the 1930s and the high water levels of the Great Lakes in the 1980s are the good examples of the results of such fluctuations. Until such fluctuations can be understood and predicated, it will be difficult to discern the specific signals of more long-lasting climate change as they evolve. Detecting the signals of climate change becomes even more difficult when smaller regions and/or shorter periods of time are considered.

The enormous consequences of the various effects of global warming and the rising clamour for clarification continue to spur the scientific community to refine their mathematical models. Despite scientific uncertainties, these computer models are the only tools available to researchers as they struggle to estimate to what extent economic and social actions to reduce future emissions of greenhouse gases can limit the predicted changes in climate. Stephen H. Schneider, a climatologist at the National Centre for Atmospheric Research in Boulder, Colorado, and Norman Rosenberg, director of the Climate Resources Programme for Resources for the Future, note that another decade or so of observations will enable scientists to assess how well present estimates predicated the sensitivity of climate to increasing trace gases. But, they add, "While scientists debate, the real climate system continues to perform the experiment for us".

All of the predictions about the climate change are based only on five models (although there are many attempts to model portions of the earth system on more limited scales of times and spaces). The five models are the NASA/Goddard Institute for Space Studies (GISS) model, the National Centre for Atmospheric Research (NCAR) model, the NOAA Geophysical Fluid Dynamics Laboratory (GFDL) model, the model developed at the Oregon State University (OSU), and the model developed by the United Kingdom Meteorological Office (UKMO).

These are general circulation models (GCMs) that predict the ways in which temperature, humidity, wind speed and directions, soil moisture, sea ice, and other climate variables evolve through three dimensions and over time. They use mathematical equations to express the basic physical, chemical, and biological processes that govern the global climate system.

The general circulation models agree that change is in the works and that weather systems worldwide are sensitive to increases in greenhouse gases. Their circulations reveal that disruption is all but inevitable and that a wide range of consequences is possible. There are differences between them, however, some versions of GISS and GFDL models now include scenarios of gradual addition of greenhouse gases into the atmosphere, whereas the others assume a massive, one-time doubling of the gases. The GFDL and OSU models attempt to include ocean processes. The GFDL model indicates that some remarkable effects can occur when an active circulation is included. For example, the presence of upwelling circulation in the circum-Atlantic Ocean acts to delay surface warming there for extended periods, perhaps centuries.

REFERENCES

Hidore, J.J.1996. *Global Environmental Changes*. Prentice Hall. New Jersy.

Miller, A. Thompson, J.C. and Peterson, R.E. and Haragan, D.R. 1983. *Elements of Meteorology.* Merrill, Columbus, OH.

Morgan, J.M. and Morgan, M.D.1997. *The Atmosphere and the Science of Weather.* Prentice Hall. N.J.

Oliver, J.E. and Hidore, J.J.2003. *Climatology: An atmospheric Science.* Pearson Education Pvt. Ltd., New Delhi.

World Meteorological Organisation. 2003. *Climate into the 21st Century.* Cambridge, U.K.

17

Potential Threats of Acid Rain on Plants

AZRA AKHTAR AND KALYAN HIRONI

Acid rain is one of atmospheric event due to increased pollution load of gaseous and particulate pollutants. Great attention has been paid particularly in last decades when it was realised that anthropogenic activities such as industrialization, urbanization, automobiles revolution and explosives affects substantial parts of global atmosphere. Acid rain was first identified as a problem in Manchester, England (Baldwin, 1986). Acid rain was first reported in India at Chembur where average pH of rain was 4.8 in 1976 (Khemani, 1993). The main component of acid rain are sulphuric acid and nitric acid and are responsible for acidity of the precipitation of less than pH 5.6.

In India the rapidity of economic development combined with the lack of emission controls, make more serious air pollution problems than similar cities in industrialized nations. NEERI (1980-94) has studied the urban air quality of ten major cities of India to assess the deterioration of air quality due to rapid increase in industrialization and urbanization.

Plants play an important role in monitoring the ecological balance by actively participating in the cycling of nutrients and gases like CO_2, O_2 and air pollutants. Sensitivity of plants to air pollutant is variable. Some are more sensitive and act as biological indicator to air pollution, other, which is tolerant, may act as to air as sink to air pollutants (Bobrav, 1995; Choudhary and Gajghate, 2000).

Sources

Acid deposition is caused primarily by sulphur and nitrogen oxide emissions that are converted to sulphuric and nitric acid in the atmosphere and in the soil. Emission of SO_2 is directly related to the combustion of fossil fuel and the smelting of non-ferrous metals. Motor vehicles and power plants are the most significant source of No_x. The acid rain phenomenon begins with long range transport of these pollutants. The oxides of sulphur and nitrogeon can remain in the atmosphere for several days and can be carried thousands of miles before they are returned to the earth as sulphuric and nitric acid in precipitation or as dry particles.

Effects of Acid Rain

Acid rain causes various type of injuries to the plants depending on the type and concentration of pollutants. The leaf is the first target of pollution on the plants. Pollutants enter in the leaf through stomata during gaseous exchange and damage the leaf morphologically, structurally and psysiologically as given below (Choudhary and Gajghate, 2000) :

1. Nacrosis : Killing or collapse of tissues.
2. Chlorosis : Loss or reduction of chlorophyll resulting in yellow pattern.
3. Abscission: Dropping of leaves.
4. Epinasty : Downward curvature of the leaf due to higher rate of growth on the upper surface.

Acid rain has severe ecological impacts on forest, crops and other vegetation.

Impact on Forest

Forest damages may vary from minor foliar injury to large scale destruction and death of forests stands. In "crown die back" injury leaves needles at the tree top turn yellow, then brown and ultimately drop off (Milson and Duinker, 1987). Vogelmann (1982) reported that acidic deposition has caused the decline of red spouce. Utrich *et al.,* (1980) showed that acid rain leaches important plant nutrients such as Ca, Mg and K from the soil, making them unable to plants. In addition, acid rain mobilizes, the aluminium in forest soil. The fine root of trees are damaged by the toxicity of aluminium. In deciduous trees, leaf discolouration, early leaf fall, death of tree tops, damage to bark as well as lack of natural rejuvenation are observed.

Abrahamsen *et al.,* (1977) reported that decomposition rate of lodge pole pine needles increased at pH 3.0 as compared with pH 6.5. Acid rain reduce organic matter decomposition in forest soil where increase in soil acidity were observed (Abrahamsen *et al.*, 1977, Francis *et al.*, 1980). The air pollutants O_3, No_x and SO_2 and their combinations are considered as the main causative an important predisposing factors responsible for the forest decline in Central Europe (*Guderian*, 1998).

Impact on Crops and other Vegetation

Tamm and Cowling (1976) suggested two pathways direct and indirect, by which acid rain may affect plant growth. Direct pathway through which true growth may be affected offer the greatest possibility for immediate tree growth responses. These could occur due to the formation of cuticular lesions, damage to epidermal appandages, leaching of cations from the leaf and possible alternation of the action of stomatal guard cells. Damage to internal structure, gall formation and alternation of physiological processes such as photosynthesis are other possible direct response mechanisms. Alteration of such physiological processes as photosynthesis, respiration, growth and solute transport could decrease tree vigour, increasing susceptibility to other stresses.

The indirect pathways are by far the most complex, in that the nature of forest soil may modified. This could occur through acidification, increased sulphur and nitrogen deposition, chemical withering, leaching of cations and interferences with nitrogen fixation and litter decomposition.

Based upon the visible effect on foliage it has been suggested that plants sank in sensitivity to simulated acid rain from high to low in the following order : herbaceous dicot, woody dicot, monocots and conifers (Evans, 1980).

Harcourt and Ferrar (1980) have shown simulated acid rain to reduce the growth of phaseolus vulgaries L. Evans and Lewin (1981) observed that simulated acid rain of pH 3.1 and below decrease dry mass of seeds, leaves and stem of pinto beans. Chappelka (1985) found that simulated acid rain at pH 5-6, 4.3 and 3.0 had adverse effect on seedling growth of yellow poplar. Prasad and Rao (1980) carried out that plants exposed to SO_2 + NO_x mixture get adversely affected, leading to decrease growth in terms of various morphological parameters, biomars accumulation and productivity level.

Malviya (1985) conducted experiments with Leonotis nepetaefolia and Ocimum sanctum revealed that acid rain with pH 4.5 and pH 3.0 may adversely affect metabolic processes, growth and medical properties of plants. Ravi Prakash and Naik (1988) analysed ecological effects of acid precipitation on biological materials at Talcher area. They analysed chlorophyll content, root and shoot length from certain crop plants and found adverse effects on these parameters. Nath and Nath (1990) found that level of SO_2 is increasing at alarming rate. SO_2 is known to affect vegetation adversely and villagers have taken alarm at the declining level of crop productivity. Garg *et al.* (2000) studied the effect of the Bansal Amraj Mill dust on vegetation at Jaitwara (Satna) and found that in the trees located near to mill site the chlorophyll content was lesser than the trees locating non-polluted area.

Laboratory experiments on the effect of simulated acid rain on different growth parameters on some leguminous crop plants were observed by performing pot experiments. Simulated acid rain was found to induce visible injury on shoots, nacrosis on the leaves. Plant growth, chlorophyll content (Chl. a & Chl. b) and seed output were also adversely affected with the lowering in pH. Seedling of Pea (*Pisum sativum var. bonnivillae*) and Gram

(*Cicer arietinum vae* C-235) tolerated lowering of pH up to 4.3 without significant damage. The spotting and nacrosis on the leaves were observed at pH 2.8. Dry weight of roots and shoots in both the plants species were also negatively affected with decreased in pH from 7.0 to 2.8. At pH 1.3 the mortality of seedlings were observed. The plants which when exposed to control (pH 6.7) have better growth than plants treated with pH solutions. The lowering in pH affect negatively plant growth and productivity. It is obvious that nutrient balances and gas exchange characteristic may differ significantly among unexposed plants and exposed to pH solution (Akhtar, 1992).

Although direct studies of impact of acid rain on plants and crop vegetation have been carried out in Indian laboratories, however, the role of acidic water and acid rain precursors in physiological processes of plants and their assessment in computing and modelling works are still lacking. Results of various impact assessment studies can be further utilized for more recent research area of impact modelling.

REFERENCES

Abrahamsen, G.R. Horntvedt and B. Tveite. 1977. Impact of acid precipitation on coniferous forest ecosystems. *Water, air and soil pollut* 8:57-73.

Akhtar, A. 1992. *Effect of simulated acid rain on some leguminious crop plants,* Ph.D. Thesis U.O.R. Jaipur.

Baldwin, J.H. 1986, Acid rain : A Global perspective. *Int. J. Ecol. Environ. Sci.* 12:35-57

Bobrav, R.A., 1995. The leaf structures of Poa annua with observations on its smog sensitivity in Los Angeles country. *Am. J. Bot.* 4:467-474.

Chappelka, A.H., B.I. Chevone and T.E. Burk 1985. Growth response of yellow poplar seedlings to Ozone, sulphur dioxide and simulated acid precipitation, alone or in combination. *Environ Exp. Bot.* 25:233

Choudhary, P.R. and D.G. Gajghate. 2000. Assessment of Air Pollution effect on Plants. *A. Review Indian J. Environ. Protection* 20(12) : 925-933.

Evans, L.S. 1980. Foliar responses that may determine plant injury by simulated acid rain In T.Y. Torbara, M.W. Miller and P.E. Morrow (eds.) *Polluted Rain.* Twelfth Annual Rochester International Conference on Environmental Toxicity 239-257. New York : Plenum Publishing Co.

Evans, L.S. and K.F. Lewin 1981. Growth, development and yield response of pintobeans and soybeans to hydrogen ion concentrations of simulated acid rain. *Environ. Exp. Bot.* 21 : 102-113.

Francis, A.J., D. Oslon and R. Bernatsky. 1980. Effects of acidity inmicrobial processes in a forest soil. In D. Drablos and A. Tollan (eds.) *Proceedings of the International Conference on the Ecological of Acid Precipitation* 166-167 Sandefjord, Norway.

Grag, S.S., N. Kumar and G.Das, 2000. Effect of the Bansal Ramraj Mill dust on vegetation at Jaitwara, Distt. Satna. *Indian J. Environ. Protection* 20(5): 326-328.

Harcourt, S.A. and J.F. Farrar. 1980. Some aspects of simulated acid rain on the growth of Barley and Radish. *Environ. Pollut.* (series A) 22:69-73.

Khemani L.T. 1993. Air pollution and acid rain problem in the Indian region. *Indian J. Radio and space physics* 22:207-214.

Malviya, P. 1985. Responses of Plants to simulated acid rain pollution. Ph.D. Thesis, Banaras Hindu University, Varanasi India.

Nath, P. and S. Nath 1990. Air Pollution pp. 36 - 52. *In Environmental Pollution : conservation and planning,* Vol. 2 : 209 Chug Publication, Allahabad, India.

Nilson, A. and P. Duinker. 1987 The extent of forest decline in Europe Environment 29(9); 14-9 & 30-31.

Prasad B.J. and D.N. Rao 1982 Relative sensitive of a leguminous and a cereal crop to SO_2 pollution. *Environ. Pollut.* (Series A) 29 : 57-70.

Ravi Prakash and B.N Naik 1988. Effects of emerging precipitation on biological material at Talcher, Orissa A case study for developing country. *Poll Res.* 7(12) 29-36

Tamm, C.O. and E.B. Cowling 1976. Acid Precipitation and forest vegation. p. 845-855 In : Dochinger, L.S. and T.A. Seliga (eds.) *Proc. Ist Int. Sympo. on Acid Precipitation and the forest ecosystem.* USDA for selv. gen. tech. rep. NE-23 Upper Darby. Pa.

Ulrich, B., R. Mayer and P.K. Khanna 1980. Chemical changes due to acid precipitation in a less derived soil in central Europe. *Soil Science* 130:19-39.

Vogelmann, H.W. 1982, "Catastrophe on Camel's Hump". *Natural History.*

SECTION-IV

ENVIRONMENTAL IMPACT AND SUSTAINABILITY

18

Environmental Impact Assessment as a Tool for Environmental Management

SANTOSH SHUKLA

The main purpose of Environmental Impact Assessment (EIA) is to ensure that the developmental proposals, activities and programmes are environmentally sound and sustainable. EIA in fact is a planning process used to predict, analyse and interpret the significant environmental effects of a proposal/activities and to provide information that can be used in decision making. EIA has to be effectively to minimise and/or prevent adverse effects, at the same time attempts are made to help countries to use the real potential of their resources and maximising the benefits of the proposed development activities. EIA is a process which helps us to modify and improve the design of the proposed developmental activities; to ensure that resources are used properly and efficiently ; to enhance the social aspects related to the proposal; to identify and select the measures for monitoring and managing impacts; and facilitate in decision-making.

On the other hand EIA has an ongoing role during the implementation of developmental activities to further minimise the environmental consequences by initiating the appropriate mitigation and monitoring measures. As a result such a role for environmental impact assessment was formally reconised at the Earth Summit (UN Conference on Environment and Development) held at Rio in 1992. The principle 17 of the Rio Summit states, " Environmental impact Assessment, as a national tool/instrument, shall be undertaken for proposed activities that are subject to a decision of a competent national authority.

EIA has to be applied to all developmental activities likely to cause potentially significant adverse impact or add to actual or potentially foreseeable cumulative effects; as a primary instrument for environmental management to ensure that impact of development are minimised, avoided or rehabilitated; so that the scope of review is consistent with the nature of the proposed project or activity and commensurate with the likely issues and impacts; and on the basis of well defined roles, rules and responsibilities for the key actors.

EIA has to be undertaken throughout the project cycle, beginning as early as possible in the concept design phase; with clear reference to the requirements of project authorisation and follow-up actions, including the impact management, consistent with the application of best practicable science and mitigation technology. It has to be implemented in accordance with established procedures and project-specific term of reference including agreed time frame. It has to provide a meaningful public consultation with communities, groups and parties directly affected by, or with an interest in the project and/or to known its environmental impacts.

EIA has to attempt wherever and whenever necessary to all related and relevant factors, including social and health risks and impacts; cumulative and long-term, large-scale effects. It has to suggest design, locational and technological alternatives to be made for sustainable considerations including resource productivity, assimilative capacity and biological diversity.

EIA process helps in providing accurate and appropriate information as to the nature, likely magnitude and significance of potential effects, risks and consequences of a proposed undertaking and alternative to it. The preparations of an impact statement or report that presents this information in a clear, understandable and relevant form for decision making too becomes possible with EIA. This also provides the references about the qualifications, confidence limits in the prediction made and ongoing problem solving and conflict resolution to the extent possible during the application of the process.

EIA should help in providing the basis for environmentally sound decision making in which terms and conditions are clearly specified and enforced. It also helps to design, planning and construction of a acceptable development projects that meet environmental standards and resource management objectives. It provides an appropriate follow-up process with requirements for monitoring, managemnt, audit and evaluation that are based on the significance of potential effects, the uncertainly associated with predication and mitigation, and the opportunity for making future improvements in project design or in the process of application.

The composition of components, and the responsibilities for an EIA process mainly depends upon the requirements of the respective region, state and country, considering the nature and level of developmental activities. However, most of the EIA processes normally have a similar structure. The main stages of EIA including screening, scoping, assessing, mitigating, reporting, reviewing, decision-making, monitoring and managing and public involvement.

Progress in the adoption of EIA as a tool for environmental management has been extremely slow specially in developing countries. Only nine out of 121 countries having established framework for its implementation. Even in these selective countries after more than a decade of EIA implementation, the performance of EIA has been dismally poor in spite of considerable effects being made by the Asian and Pacific region. These countries are devoted to the development of cost effective EI methods and techniques. Such a poor performance is mainly attributed due to serious gaps in the legislative, administrative, institutional and procedural framework. To over come these shortcoming efforts have to be made to ensure the effective use of a legislative rather than an administrative options to establish a authority for EIA; to create an independent environmental agency with considerable political influences over different sectoral agencies through placement in the office of the President for Prime Minister or a high profile ministry with a responsibility for national economic planning and budgetary control; with the decentralisation of EIA activities among the ministries of government or sectoral agencies; to ensure the mandatory involvement of the local people, the public and concerned agencies involved in scooping, reviewing, monitoring and auditing; a procedural arrangement characterised by two stages assessment, mandatory scoping and preparation of terms of reference, involving the registered consultants for EIA studies, with a provision for public hearing and fixed time frame for review and the conduct of EIA as a fully internalised element of the planning process rather than a separate exercise divorced from the technical and economic aspects of project planning and design. Attempts have to be made to incorporate all these conditions in the EIA guidelines. Adequate provisions for judicial involvement in reviewing the action of environmental agencies and project proponents have to be made.

General Principles

General principles involved in the functioning of EIA mainly includes—EIA as a tool helps in achieving the sustainable development. EIA is also a management tool to be linked closely to the project lifecycle and its decision points to ensure the availability of appropriate environmental information at the appropriate time. Efforts to be made to have the constant interaction and feedback between the EIA team and Project designers. Similarly attempts have to be made to ensure that design and/or locational changes can be implemented to avoid or minimise adverse impacts to the maximum possible extent. EIA procedures have to be integrated, closely, into exiting development planning and in the authorisaton procedures so that minimum disruption is caused to the existing institutional arrangements and maximum effectiveness for EIA is assured by identifying the appropriate time and/or locations which are linked with the existing procedures. EIA must include an analysis of a number of identified alternatives. EIA must be implemented in a multi and interdisciplinary manner. It has to be characterized by integration of social, economic and biophysical environmental impacts to the maximum (possible) extent.

Aims of EIA

The main aim of EIA is to gauge the environmental impacts of the proposed developmental activity/activities. In many cases the environmental impacts of a project may influence the effective and efficient operation of an installation. For example construction of a dam and/or reservoir may involve resettlement of the villagers and farmers. It has been observed and experienced that in the absence of the proper planning, people displaced often return to a area near to their original homes. In addition with the implementation of large scale projects attract large number of workers along with their families during the construction activities. As and when the work ends, some individuals decide to stay in the locality and settle around the constructed reservoir. As a chain effects of such decision gradually agricultural activities increase on the margin of the reservoir. Removal of the existing natural vegetation and replacement with the crops may increase erosion and, hence sediments into the reservoir. If the land bordering the reservoir is sloping than the effects are further enhanced. Sedimentation in to the dam/reservoir may shorten the operational life-span of a reservoir (like Bhakara Dam), ultimately reducing, significantly, its ability to achieves its socio-economic objectives. In most cases there will be few, if any, impacts of this type. It is wise to consider the likelihood that a project might be adversely affected through such feedbacks.

The most important aim of EIA is to introduce effectively a systematic consideration of environmental issues into all important decision-making stages on specific proposed developmental activities. Practically all proposed developmental activities are subjected to an authorisation process whereby a formal decision and/or a series of decisions, are made by an official body on the future of the proposal(s) for the proposed activity/activities. Here the role of EIA is to ensure that the environment is one of the factors which has to be considered in decision making.

It is very important that environmental issues are considered fully and appropriately along with the financial, economic and technical matters by the project initiator from the initial stages. it will be unwise to undertake any proposal/study, neglecting the environmental issues, and to plan, subsequently for preparing a EIA report. At this stage a site with a project design may become time-consuming and expensive to implement any alternative or even to abandon a project. EIA report suggest that significant adverse impacts are likely to occur and cannot be mitigated to make them acceptable. This is how the whole exercise in fact will be a waste of time as well as money for everybody involved in the process.

EIA, therefore, is a process which has a direct influence at various stages and over a considerable period of time. EIA simply is not an activity which is aimed at producing a set of results to be used at a specific stage of decision-making. However, there is no doubt that the role of the result suggested through the process at the approving stage a very important because at this point of time EIA often enters formal, statutory decision-making systems.

In fact, the overall effectiveness of EIA and other related studies is further enhanced if they incorporate a systematic analysis of reasonal alternatives. Basically, the objectives of the developmental activities can be achieved, often, in diversified ways. Specifically with reference to EIAs, one of the alternatives should be the "no-action" option, that is no development. This will provide an objective baseline against which the other alternatives can be measured.

Scoping and Preparation of Terms of Reference

When implementing an EIA for a proposal it is important to ensure that it focuses only on significant issues for the different stages of development. Normally in all EIAs it is very useful to consider both the construction and operational periods. At times it becomes essential to consider the abandonment and /or even decommissioning phases specifically in the cases related to mining and other related activities, i.e. the restoration, reclamation and the blue print for further activities.

It is wise to ensure the cost effective use of resources to study all possible impacts and the wide range of alternatives. Therefore, the most important outcome of scoping is an agreement among the main agencies including the pubic, on the range of alternative likely to be gauged. The most likely significant impacts have to be assessed, predicted and evaluated. Scoping has to be undertaken by the concerned agency who is likely to pay the costs.

The range of individuals, agencies and organisations likely to be involved include ministries of central government likely to have the areas of responsibilities affected by the proposed project under consideration, i.e. agriculture, natural resources—forest, mining, transport, industries, health, social welfare and human resource development); local self-government bodies in whose territory a project is proposed or a part of whose territory is likely to be effected by the project; traditional decision-making bodies, i.e. councils etc.; private sector organisation such as trade associations and chambers of commerce and industries; non-government organisations; and representative of the public likely to be affected by the proposed project.

All concerned must be provided with information of the proposed audits alternatives, to enable them to indicate the issues which concern them. Various methods have to be identified for obtaining response from those consulted during the scoping. During the scoping process other mechanisms and various combinations have to be identified. The results of scoping have to be analysed and evaluated and a fresh draft of ToR and EIA has to be prepared. ToR is a very important document as it determines not only the relevance and utility of the EIA work for project design and management but the usefulness of EIA results for decision making. Normally in a ToR two key components are included (i) the project proposal and its reasonable alternatives (Including the "no action" options) and (ii) the likely significant impacts to be investigated. Considering the progress of the EIA work it becomes necessary to change the orientation of the proposed work.

EIA Work

Basically, EIA work is concentrated on a systematically prediction and evaluation, to suggest alternative for the impact identified in the ToR. All these impacts have to be compared with each other and with the "non-action" option. Efforts have to be made for establishing the close collaboration between those responsible for the EIA and the persons working on technical design and economic as well as financial issues. Most of the staff and consultants not only are qualified but are aware of the most relevant methods and techniques but are trained to use them properly.

EIA study also includes a prediction and evaluation of socio-cultural, economic and /or health impact too are causing secondary environment impacts. In case this is not done than some potentially harmful environmental impacts might be omitted in advertently from the EIA.

For example a water resource development in a area has caused changes in the hydrological regime of the river down streams of the project. The changes in the quality of water and flow, reduced, significantly an area of reeds. The local villages were making the baskets and other articles from the available reeds. These products were the important source of income. With the removal of the reeds the locally available resources villagers start facing the problem in finding the alternative source of income. Some of them start exploiting trees which will be processed into charcoal considering the availability of market. This is how by exploiting these resources local residents are further accelerating the serious problem of the deforestation as well as the soil depletion and erosion. The chain of such incidents have to be identified if the socio-economic importance of downstream natural resources had to be investigated and likely impacts are timely predicted. In such cases efforts have to be made either to protect the resources through controlled discharges of water or to ensure the availability of alternative economic resources which can be used without enhancing the existing environmental degradation.

Goals of EIA

Most of the definitions clearly portray EIA as a tool meaningfully predicting, assessing, estimating and communicating the environmental effects of proposed development policies, plans and projects (Munn, 1979). That is why efforts have been made in developing countries to stimulate and focus on EIA as a tool to analyse the environmental impacts. EIA is merely a tool and procedure designed to ensure that adequate environment considerations are enters into the decision making process. EIA, therefore, is a tool for shaping policies, programmes and project decisions. Effectiveness of EIA, in this reference, mainly depends on robustness of the methodologies and techniques of environmental assessment, along with the institutional and procedural network, which has been established for its implementation.

Procedural Frameworks

EIA procedures refer to the sequencing of tasks among participants in the EIA process. Procedural issues focus on who sets the rules, who does the screening and scoping, who, when and how is the public participation. EIA procedures, therefore, define the formal structure for carrying out the assessment of the environmental implications of projects and its integration in the project cycle. Wide variations have been noticed in the implementation of the EIA procedure. Some of the countries through initially adopted in the foreign models of EIA and found that these are not suitable due to the non-availability of the required infrastructure. As a result considering the local conditions most of the countries have modified, even have altered the EIA producers, considering the socio-cultural, political and administrative environment and availability of the indigenous resources. Thailand has the most elaborate EIA procedure and is unique in having two separate procedural systems, one for the private and the other for the public projects/activities as compared to the single framework in all other developed and developing countries.

To meet the goals and objectives of the EIA, the procedure must content the following basic elements:

1. To save the time and cost assessment be made in two stages. Irrespective of type, scale, threshold and/or environment and/or screening using the simple method of checklist or Leopold matrix to ascertain whether or not the proposed developmental activities are likely to create environmental impacts. At the second stage activities with potentially serious environmental impacts, detailed assessment have to be attempted for preparing an environmental statement.

2. Scoping has to be mandatory, and the public, and communities likely to be affected along with the concerned agencies have to be involved in the process. Scoping, virtually is an activity designed to identify the key, priority and contentious issues related to the proposed developmental activity and/or the course of action must be the part of assessment process and helps in saving time and cost. Finally, it provides a much needed opportunity for the consideration of local resources, social values, preferences and perceptions, political thinking and socio-cultural conflicts in the utilization and development of the sources. Such EIAs which neglect or evade the socio-cultural issues normally make inaccurate predictions. In order to identify the major issues to be focused during the environmental assessment attempt have to be made to hold the explanatory meetings with the local people, representatives of concerned groups and the concerned government agencies.
3. Terms of reference is a useful document for all concerns as it specifies the EIAs requirement and has to be prepared by the project proponent consulting the EIS review committee.
4. The environmental assessment have to be conducted in house by the project proponent and/or by the consultants, as the project proponent is the most knowledgeable about the proposed projects/activities. That is why they are better placed to utilise effectively the available environmental information to make the needful improvement in the project design as a part of project planning. Involvement of the registered consultant not only ensures high quality environmental research and the systematic reporting. During the review and public hearing easily biases and distortion in the data collection, analysis and presentation are detected.
5. The circulation of the environmental impact statement to all concern government agencies ensures the needful corrections and modification of relevant environmental issues raised in the statement. Such a circulation to all concerned agencies ensures the cooperation essential for the success of the EIA. In this process attempt has also to made to circulate the document to the public so that the socio-cultural issues too are adequately covered at the stage of scoping and/or explanatory meeting.
6. The provision of the public hearing provides a form for a thorough analysis of the conclusions and proposals. This also ensures the direct adequate public inputs and positively helps in taking the decision. Provisions have to be made to have the opinion of the people who actually are directly and/or indirectly experiencing or will experience the effects. The participation with objectivity will depend mainly on the socio-political environment and the level of environmental awareness.

7. Framing of the time limit avoids the undue delay in the project implementation due to the EIA requirements.
8. To ensure the timely execution of the development proposal with the measures suggested in the statement, all concerned agencies must be involved in monitoring the compliance and post-project auditing. Representatives of the environmental agency, concerned agencies, project owner and the local people likely to be affected by the proposed project/activity must be involved in the monitoring and auditing process.

Among the developing countries except China the non-involvement of the public and the concerned agencies in scoping is considered as the most serious weakness of the EIA procedures. Similarly most of the countries, hardly have the provisions for public hearing and public scrutiny and their involvement. During the EIA process efforts have to be made adequately addressed the emerging shortcomings.

Mostly EIA is conducted as a separate activity. EIA becomes a postscript to planning and hardly has any role to influences the decisions. Even if such attempts are made at the initial stage of the project process, no doubt seriously considers the suggested environmental consequences.

Though the progress of EIA as a tool for environmental management has been extremely slow in the developing countries, gradually the awareness among the developing countries about the potential benefits of the EIA system is increasing.

Wide variations in the environmental problems from the direct effects of the pollution of waterways and the air, to the indirect effects such as health problems caused by changes in lifestyle or earthquake resulting from the construction of large (Koyn) dams. Some of the environmental problems though initially may appear to be insignificant but might create adverse consequences if the effects of all such proposed activities are added together as cumulative effects. For example, pollution due to automobile use and deforestation resulting from the gathering of firewood for domestic purposes too has such cumulative effects.

To manage the large scale environmental problems it is necessary to have an understanding of environmental impacts and the different ways in which the impacts can vary. Impact can vary in type and nature, magnitude, extent, timing, duration, uncertainty, reversibility and significance.

In the initial years of EIA, environmental impacts were considered to be impacts on the natural, biophysical environment, i.e. effecting the quality of air, water, flora and fauna, climate and hydrological systems and also noise pollution. However, the focus on biophysical impacts alone did not last long. Gradually to cover many aspects related to social, health and economic sector the definition of the term environment has been broadened.

In the post-World War II period the combination of exponential population growth with the development in medical sciences, urbanisation, industrialisation and increasing tendency of exploiting the available resources rather than using the resources on large scale has caused and is causing pressures on the environment.

By the 1950s and 1960s most of the developmental activities in industrialised countries were diversified on large scale and number. However, most of these countries were unable to assess and/or evaluate the potential impacts with the available techniques. US emerged as the first country, where considering the increasing pressures from the public in 1970 National Environmental Policy Act (NEPA) was proclaimed, establishing an environmental policy. The process adopted under NEPA was known as environmental impact assessment.

Since the mid-1970s the use if EIA processes as the main process for analysing wide variety of impacts, to achieve integrated assessment for decision making.

EIA has undergone a remarkable expansion. At present EIA is widely used as an approach for development planning and management. EIA has continued to evolve through innovations in the law procedure and methodology. The evolution of EIA can be classified into four main phases :

(a) Initial development (early 1970s)—Initially efforts were made to introduce the basic, principles, institutional arrangements and to adopt the analytical techniques for conducting EIA;
(b) Expanding scope (later 70s and early 80s)—gradually attempts were made to include social, risk, health and other related factors. Similarly provisions were made to ensure the public participation and to have a grater emphasis on impact management;
(c) Process redirection (mid-late 80s)—increasing efforts were made to attend the cumulative effects, to integrate the project level considering the EIA policy, planning and regulatory frameworks, and finally, to identify and to establish monitoring, audit and other follow-up procedures; and
(d) Towards a sustainability paradigm (to date)—development of strategic environmental assessment, new requirements of EIA established by international conventions (e.g. biodiversity), introduction of consensus based dispute settlement procedures.

Table 18.1 clearly illustrates the trends in the development of project assessment and also indicates that how EIA has evolved in the recent years.

Table 18.1 : Development of Environmental Assessment

Approximate year	*Innovations in EIA Techniques and Procedures*
Pre-1970	Analytical techniques : largely confined to economic and engineering feasibility studies; narrow emphasis on efficiency criteria and safety of life and properly ; no real opportunity for public review.
1970	Multiple objectives benefit-cost analysis emphasis on systematic accounting of gains and losses and their distribution, reinforced through planning, programming and budgeting review : environmental and social consequences not incorporated.
1970-75	EIA primarily focused on description and predication of ecological/ land use changes; formal opportunities for public scrutiny and review established; emphasis on accountability and control of project design and mitigation.
1975-80	Multi-dimensional EIA, incorporating social impact assessment of changes in community infrastructure, services and lifestyle; public participation became the integral part of project planning, increasing emphasis on project justification in review process; risk analysis of hazardous facilities and to improve technology in core areas.
1980-92	Attempts where made to establish better linkages between impact assessment and policy planning and systematic implementation in phases; research focus on effects monitoring. Post project audit and process evaluation, search begins for more orderly scoping and focusing the procedures and less protected forms of consultation based on negotiation and mediation.
1992 onwards	Role of EIA in achieving objectives of sustainable development. Applying EIA to policies and land use plans. Role of environmental strategies and policies.

Since 1991, the significant international developments made in EIA include :

(i) major reforms were made or/and are pending to well established EIA systems specially in Australia, New Zealand and Canada;

(ii) the European Community Directive on EIA (1985) was widely implemented and has attracted domestic reforms in UK and other number state;

(iii) the World Bank and other agencies assisting in developmental activities not only have framed the EIA policies and procedures but have initiated to introduce the EIA process in many developing countries;

(iv) the Espoo convention identified principles and provisions for EIA of projects and abilities with trans-boundary effects; and

(v) additional requirements were placed on EIA by the UN conventions on climate change and Biological Diversity (1992).

This is how EIA has emerged as a response to the concerns expressed by environmental movements of the 1960s voicing about the serious effects of growing human activities, especially the pollution being created by industries.

While the prevailing regulations of planning and the measure of pollution control were not adequately equipped to monitor and control the emerging changes. Since its introduction in USA from 1969, 1970, with the passage of National Environmental Protection Act which requires the preparation of Environmental Impact Statement (EIS) for "major federal actions significantly affecting the quality of the human environment". In most developed countries, now EIA has become a major tool for environmentally sound management practices and for achieving the sustainable development goals.

However, the progress in adopting EIA as a tool for environmental management has been extremely slow among the developing countries. Although these countries have always had general environment-related laws and regulations which have been exploited for EIA purposes. However, these have had very limited impacts as a environmental protection measures due to serious technical, administrative and political constraints. In fact only a few developing countries so far have used EIA. Number of developing countries though have EIA legislation, but only 19 out of the 121 sovereign third world countries are known to have in place the required necessary administrative, institutional and procedural framework for the smooth implementation of EIA as has been shown in Table 18.2.

Table 18.2 : EIA status in Developing Countries

Region	*Total No. of Countries*	*Countries with EIA systems*
Africa	52	-
Asia and Pacific	24	12
Caribbean	11	-
Latin America	20	7
Middle East	14	-
Total	121	19

Over the period of time number of researchers have commented on the slow pace with which EIA has been adopted in the developing countries. Most of these comments have focused primarily on the technical barriers in the implementations of EIA. On the other hand lack of familiarity with the EIA concept itself and its role in the planning process, lack of knowledge about the EIA methodologies and techniques, shortage of indigenous staff with experience in the science of environmental assessment and availability of adequate baseline data and the appropriate technology for EIA. Even in few of developing countries with more than a decade of EIA implementation and experimentation the gap between the intent and the performance of EIA still remain wide. Such failures remain the reality of the day in spite of considerable attention being paid by these countries for the development of simple cost effective EIA methodologies and techniques.

To ensure the effectiveness of the EIA in developing countries as an environmental protection strategy, legislative approach rather than administrative measure in the process of planning have to be introduced considering the guidelines and procedures.

Since the Stockholm Conference (1972) on the Human Environment, developing countries have created number of administrative bodies and institutional arrangements to promote the objectives of the environmental management. Though over the period of time wide variations in the administrative and institutional measure from one country to the other, have been noticed in many countries.

Most of the environmental agencies in many developing countries either are subsidiary part of one ministry or another as has been shown in Table 18.3. Most of these are circumscribed because of their low status in the bureaucracy, in adequate funding and lack of skilled staff and adequate facilities.

The cost and time involved in EIA start decreasing once experience is gained with the process and the availability of adequate baseline data is assured. Implementation of EIA directly helps in improving the design or setting of the proposed project/activity. On the other hand it also helps in raising the environmental awareness of the personnel involved in the process.

Hence environmental impact assessment as a systematic process helps us in decision-making to consider the potential impacts of human activities on the environment. That is how the process helps in understanding the environmental effects of the proposed project/activity. EIA also provides a forum to discuss the large and growing public concern about environment and health.

Since EIA is a planning and decision-making tool has been helping in determining the potential environmental, social and health effects of the various development activities. It also attempts to assess the physical, biological and socio-economic effects in taking a logical and rational decisions. The suggessed possible alternative for site and/or processes in the EIA report helps in mitigating and/or monitoring the potential adverse impacts.

Considering the significance and role of EIA in the larger framework of policy-making, EIA has been recognised as an important tool with more integrated approaches to environmental management and development planning at the Rio Summit. In fact EIA is a mechanism which adds the efficient use of natural and human resources and has provided to be valuable both to those promoting developmental activities and those responsible for their authorisation. That is why EIA has to be used not only to investigate and avoid harmful impacts, but also increase likely benefits. The incorporation of EIA in the decision-making process creates number of benefits.

Increasing in both developed and developing countries environmental assessment is considered as an ongoing alternative process, which can help us

Table 18.3 : Institutional arrangements for EIA in Developing Countries

Country	*Legislation*	*Environmental Agency*
Argentina	Under preparation	National Directorate of Environment Planning, 1975
Bangladesh	—	Environmental Pollution Control Board
Brazil	National Environmental Policy Law, 1982	The special Environmental Agency, 1982
China	Environmental Protection Law, 1979	Environment Protection Bureau (1979-87) Environmental Protection Authority, 1987
Columbia	National Code of Renewable Natural Resources and Protection of the Environment, 1974	National Institute of Renewable Resources and the Environment, 1968
India	Environmental (Protection) Act. No. 29, 1986	Department of Environment
Indonesia	Environmental Management Law, 1982, Regulation P.No. 29, 1986	Ministry of Population and the Environment
Republic of Korea	Environmental Conservation Law, 1977, 1979,1981,1982,1983,1986	Office of the Environment
Malaysia	Environmental quality Act, 1974, 1985	Department of Environment, 1974
Mexico	Law of Public Works, 1980, Federal Law of Environmental Protection, 1982	Ministry of Urban Development and Ecology
Nepal	—	Environmental Impact Study Project
Pakistan	Environmental Protection Ordinance No. XXXVII, 1983	Environmental Protection and Improvement Agency
Papua New Guinea	Environmental Planning Act, 1978	Department of Improvement and Conservation,1978
Peru	—	National Office for Natural Resources Evaluation, 1962
Philippines	Presidential Decrees, 1211 and 1151 (1977), 1586 (1978) Proclamation No. 1179 (1981)	National Environmental Protection Council, 1977
Sri Lanka	National Environmental Acts, 1980,1984,1986	National Environmental Authority
Thailand	Improvement and Conservation of National Environmental Quality Act, 1975, 1978	National Environmental Board
Uruguay	—	National Institute for Environmental Preservation, 1971
Venezuela	Organic Law of the Environment, 1976	Ministry of Environment and Renewable Natural Resources, 1976

in the formulation of sound environmental management and planning strategies to achieve the sustainable development.

It is essential to understand that EIA should not be regarded just as a procedure or a tool of environmental management only but have considered effectively at the stage of decision-making. EIA has to be regarded as an adaptive process which continues even after the decision are taken. It would ensure that the project and/or developmental activities conforms to the standards referred in the EIA report of study. It would also provide a database for any subsequent impacts to be studied as well as allowing the monitoring and control programmes to be adapted with changing circumstances and/or enhanced knowledge.

REFERENCES

Bisset, R. (1996), EIA : *Issues, Trends and Practice,* The Environment and Economics Unit UNEP, Nairobi.

Biswas, A.K. and Qu Geping (1987), Environmental Impact Assessment for Developing Countries, *Typcooly International,* London.

Boyle, J. and Mubvami, T. (1995) Training Manual for Environmental Impact Assessment in Zimbabwe, *Depart of Natural Resources, Ministry of Environment and Tourism, Zimbabwe.*

Coldwell, L.K. (1978), The Environmental Impact Statement: A misused tool, In *Environmental Impact Assessment - Emerging Issues in Planning,* Eds. Jain, R.K. and Hutchings, B.L., pp. 11-25. Urbana University of Illinois Press.

Lawrence, D.P. (1994), Designing and Adapting the EIA Planning Process, *The Environmental Professional,* Vol. 16, pp. 2-21.

Munn, R.E. (Ed.) (1979), *Environmental Impact Assessment : Principle and Procedures.* J. Wiley & Sons. Chichester.

Rees, W.E. (1998), A role for Environmental Assessment in Achieving Sustainable Development, *Environmental Impact Assessment Review,* Vol. 8, pp. 273-291.

Sadler, B. (1995), Environment Assessment : Towards Improved Effectiveness of Environmental Assessment, *Interim Report and Discussion Paper.* Canada.

Shukla, S.K. (2000), : Environmental Impact Assessment : Needs and Processes, In *Geography : Development and Change,* Eds. Singh, D.K., Dube R.S. and Shrivastava, V.K., Association of Marketing Geographers of India, Gorakhpur, pp. 111-121.

UNEP (1998), *Environmental Impact Assessment : Basic Produçers for Developing Countries,* UNEP Regional Office for Asia and the Pacific.

19

Impact of Urbanization on the Environs of Ajmer City

D.N. SATHI

City of Ajmer lies on the eastern foothills of Taragarh range between 25°-27'.N Latitude and 27°-34' E Longitude. At an altitude of 480 m from the mean sea level. It is situated in a synclinal structure covered with sand, conglomerate, boulders and pebbles. The valley is surrounded by Madar hills in the east, Nag Pahar in the west and Mahabir hills in the north in a semi-circular formation. The eastern boundary is made by the Nasirabad water divide and in the western part by the Puskar water divide.

Physiographically, the landscape comprises of a graded sub-stratum made by eroded materials which rests upon a hard platform of the basement rocks. Slope of the surrounding hills are steep and topography is undulating.

Climatically, the area has semi-arid climatic conditions. Rainfall is scanty, erratic and uncertain. The average ranges between 300 to 500 mm. Summers are hot and dry. Diurnal and annual range of temperature are high. Drought spells are quite frequent, climatic researches have revealed that there has been increase in the temperature. This has been due to expansion in the built up density, increase in the domestic consumption of energy and increase in the volume of vehicular emission.

Health and morbidly of urban environments to a large extent depends upon the area of space it occupies. Because expanding urbanization exerts heavy pressure on its functional structure and its resource base.

The first city wall was constructed by Akbar in the year 1571 with a total circumference of 4045 yards, walled city contained palace, audience hall, Daulat Khana and Dargah Bazar and residence of the nobles. The population was small and the total landscape was rural.

Ajmer came under British rule on 9 July, 1818. Location of the city in the midst of Rajpoot starts and its central position between Bombay and Delhi made it favourable place for trade and commerce. Britishers did not interfere with the old city structure and its regid social order of medieval times. They set up a separate township at the north-eastern corner of the city. Ajmer became an administrative headquarters of the Britishers.

Advent of Railways in 1875 further increased its dimension to accommodate its workshop and the larger number of workers. The urban population at this time was 35,111. It doubled to 68,849 by 1881. New settlement sprang up along the main traffic routes in different directions. The population structure comprised of whites, Anglo Indians and wealthy upper class Indians and lower class Indians. Town morphology was divided into civil lines, walled city, offices, service institutions and Railway settlements. Thus the older township started spreading towards Beawar road the never towards Jaipur road and Madar hills whereas the railway settlements expanded towards Nasirabad axis.

Sites of old gardens of Hazari Bagh, Kaisar Bagh, Kala Bagh and Daulat Bagh, Palvisla all came under the wave of newer expansion. Chief impact of population growth resulted into more demand for residential land causing humanization of agricultural land like vegetable and fruit growing area of Hathi Bhata, Palvisla, Jonesganj, Ramganj, Ashaganj Adarsh Nagar and Gulabbari.

After 1947, during the period of post independence, urbanization became unprecedented. This phase of urbanization was characterized by wealth, political interference, market oriented land use and wave of population immigration. Better inhabitable areas were already owned by the influential people, Fuedal jagirdars of Deshi Rajawaras, Sardars and wealthy merchants. Therefore, the bulk of subsequent urbanization took place either on unsuitable lands or got concentrated at few selected places preferably around religious sites. This caused chaotic scatting, congestion, insanitary conditions.

Old city area became badly congested with mixture of new and old dilapidated structures. Inter spaces narrowed down to became blind alleys filled with domestic garbage.

Monetary motives gained upper hand over all considerations of environmental aesthetics. Thus the population of 24000 of 1818 touched the height of six million approach by the end of the 20th century. The total build up area increased five times over the period of past 10 years. Residential areas extended to Lohagal in the north, Makarwati NW, Bhutiya Dungar in the east Urban expansion at present is overtaking those areas which are ecologically must valuable.

Tremendous increase has been noticed in the area of urban occupance from 550 acre in the 1818 to approximately 150 sq. miles by 2001. There were only 4 municipal wards in 1886, which rose to 20 by 1939, 45 in 1970.

Plight if Environmental Degradation

It is true that up to a point bigger also looks better but beyond that bigger starts becoming worst. Leggett has also remarked that during the past few decades urban places have become points of conflicts and disturbed natural conditions.

Intensive misuse of land in view of its economic gains has assaulted its environment and altered its many processes, forms and materials.

Environments of Ajmer city is also suffering from the hazards of excessive soil removal, vegetation clearance, sedimentation and climate fluctuation. Here the drainage basins have been converted into human occupance zones. Whereas green areas, parks, open spaced, recreational areas have shrunk in size. Slopes stability has been affected.

The principal impacts are visible on the forms, patterns and process of natural components, traceable through the effects of deforestation, desertification, floods, reactivation, salinization, destabilization, sinking water table, droughts and through biotic depletion and climatic change.

Deforestation

Ajmer environs are virtually devoid of vegetation cover. Soil cover are thin and coarse, lack of fertility and moisture content. Scanty plants cover is unable to absorb atmospheric pollutants. Aboriginal variety of plants have mostly disappeared. Due to large scale clearance by rural migrants to satisfy their fuel and forage needs.

Climate

During the last 25 years, climate condition in and around Ajmer city has remained unstable. Rate of evaporation has increased. Shifts have also been noticed in the amount of maximum and minimum temperature and in the quantity of annual rainfall.

Soil erosion

Developmental activities like road construction, stone quarrying, laying of transmission lines, newer colonization, loose soil affected the ground stability condition. Loose soils are easily removed by wind and water action.

Drainage

Natural drainage channels have been obstructed at many places. Natural water outlets have either been completely erased or artificially blocks or lined and made longer and circuitous.

Sedimentation

Intimately connected with erosion is the sedimentation problem. In rainy season during peak run off period, huge quantity of loose debris is flushed into city area and spread on roads, sewerage drains or deposited in low lying areas or lake bottoms.

Water Scarcity

Water problem in Ajmer city has been precarious since long. Local water sources have fallen much short. Water level in the wells has sunk to the deeper levels. Surface water reservoirs of Ana Sagar and Foy Sagar are dried.

Deterioration

In this way it is evident that at the present juncture of time grunary is continuously vanishing, water supply is falling shorter, forested and open area are shrinking in size, drier spells are becoming longer, rainfall is erratic and uncertain and soil and moisture are not sufficient to sustain plant life.

Similarly compactness in built-up density, larger per capita availability of land, water, open space and urban comfertability all have alarmingly deteriorated the environmental conditions in the Ajmer city.

Excessive human interference and his mismanagement unchecked wave of urbanization are the twin factors primarily responsible for all these happenings. As a result Ajmer city now is in an urgent need of thoughtful rehabilitation to restore its lost environmental sustainability.

Environmental Aspects of Mining in Meghalaya

R.K. Rai

Environment is a basic attribute and pivot on which life and economic system evolves. In last three decades the vitally important environment has become a prominent global issue. The environmental crisis is multidimensional running from local to global and on a time scale from present to the indeterminable future point of time. If environmental deterioration continues the whole system including all life will be in danger. The developmental activities of the day with more modern advanced technology without any regard to man and his environment are proving ecologically destructive. The over exploitation of resources causes a wide range of environmental degradation. Mining is second only to agriculture in the world in general and in the state of Meghalaya in particular. It is the oldest and most important activity. In the present paper an attempt has been made to evaluate the mining activities and their impact on environment.

The state of Meghalaya is one of the seven states of north-eastern India bordering Bangladesh in the south and Assam plains to its north and east (Fig. 20.1). The state is known for its mineral wealth and its mining for a long time. The state is rich in numerous multi-belts of limestone, sillimanite, coal and atomic minerals.

Geologically, the Meghalaya plateau is the north-eastern extension of the Indian Peninsular Shield. It is separated from the main shield by the Malda Gap. The mineral composition of the rocks and structural settings of the Meghalaya Plateau resembles with those of the Chotanagpur Plateau of Bihar and Bengal (Fig. 20.2). The Plateau comprises of (a) Pre-Cambrian Archaean Gneisses and Granites, (b) Shillong group of rocks, (c) Lower Gondwana rocks, (d) Sylhet traps, (e) Cretaceous-Tertiary sedimentary rocks, and (f) Alluvium deposits. Marine transgression affected the southern parts of Meghalaya Plateau during the Cretaceous to Early Tertiary period and deposited thick sedimentary sequences in association with pulses of fluvial phases. The sedimentary pile contains limestone and coal in the state.

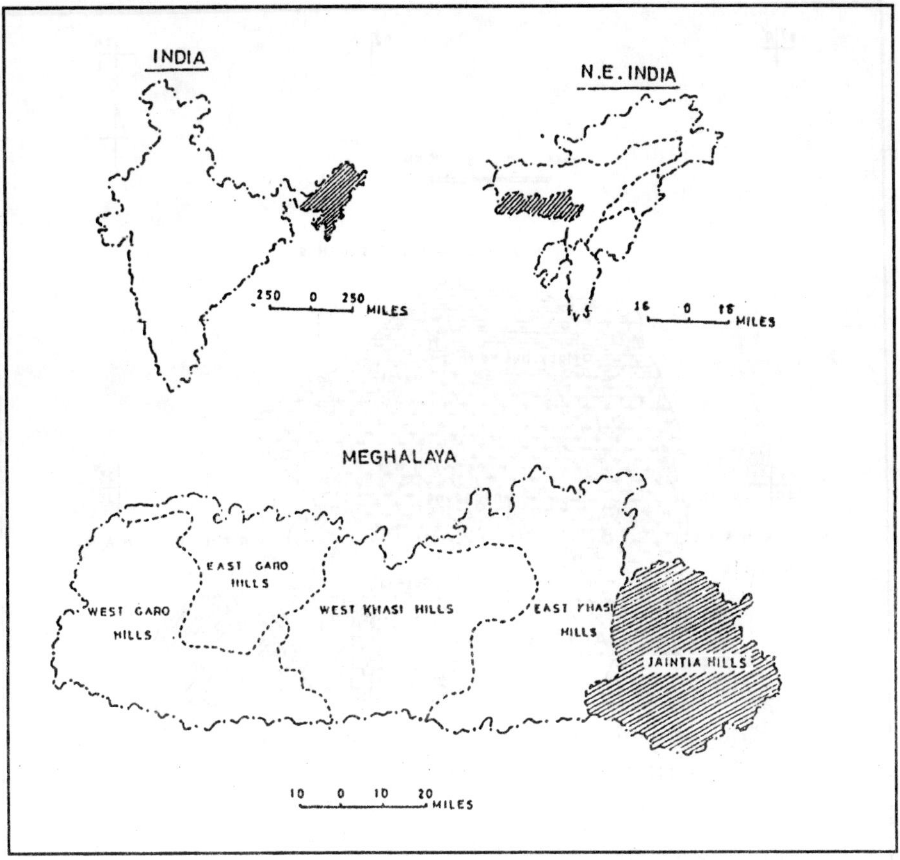

Fig. 20.1 : Jaintia Hills Location

By and large, the exploitation of the minerals are privately controlled by small-scale ventures and exploited with primitive skills and techniques. The existing mining regulations of the Government of India remain inoperative in the state due to the prevailing land tenure/ownership system. Consequently the mining operations are controlled by individuals and scattered throughout the state as per land holding size and ownership rights.

Mineral Exploitation

(a) Coal : The state is endowed with coal bearing sedimentary formations along its southern fringe. It is confined in the upper part of the sedimentary pile, conspicuously restricted to the Eocene segment. A minor patch of Gondwana rocks occur in Singrimari area of Garo Hills located at the western tip of the state. In the state the coal reserves are about 600 m tonnes mostly in the Jaintia

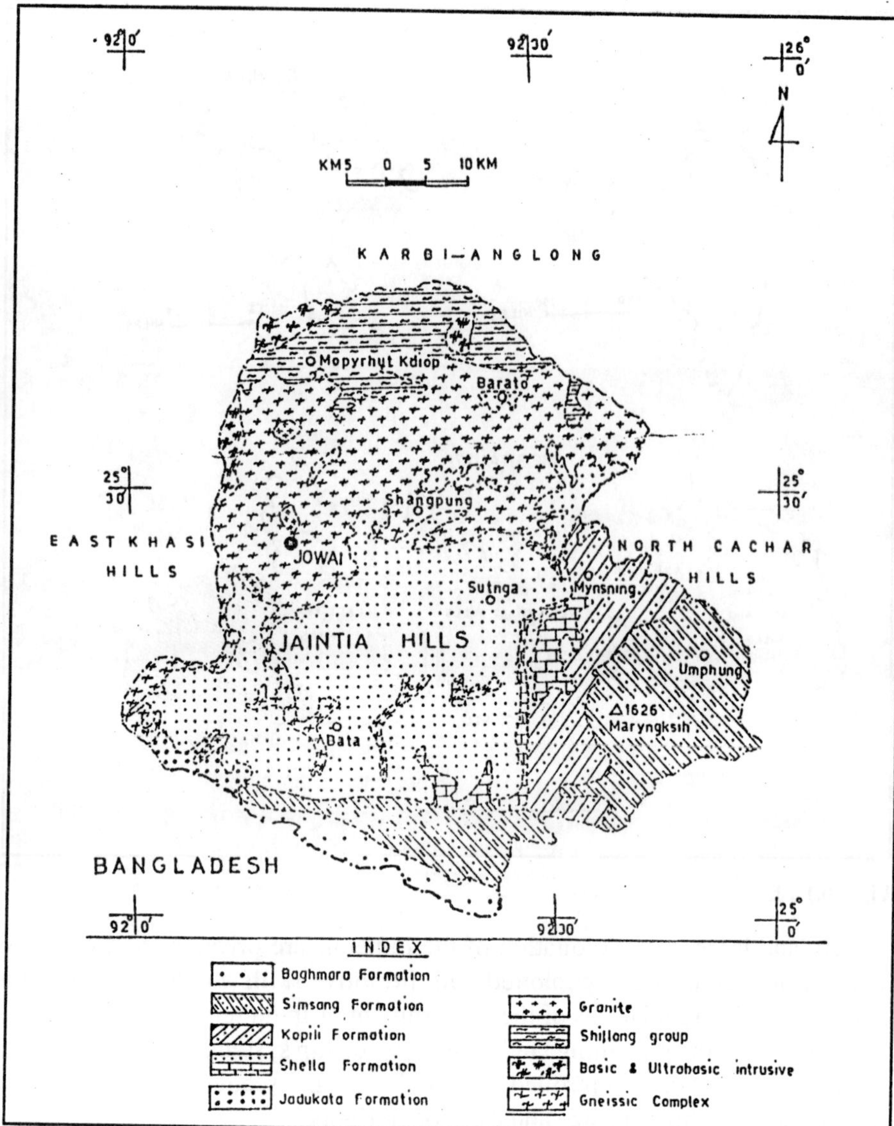

Fig. 20.2 : Jaintia Hills District Geology

and Garo Hills. The coal of Cherrapunji area is almost exhausted leaving the area completely deforested and degraded. The present scenario is that of a barren topography. Invariably, in all the coafields, the seam thickness varies from 0.1 to about 3.5 m. The coal is lignite to bituminous in nature. It is hard, lumpy with a metallic black lustre. The moisture content in general is low (not more than 3%), the ash content is less than 10 per cent, it varies from 30 to 40% and the sulphur content is higii (upto 7%) (Fig. 20.3A).

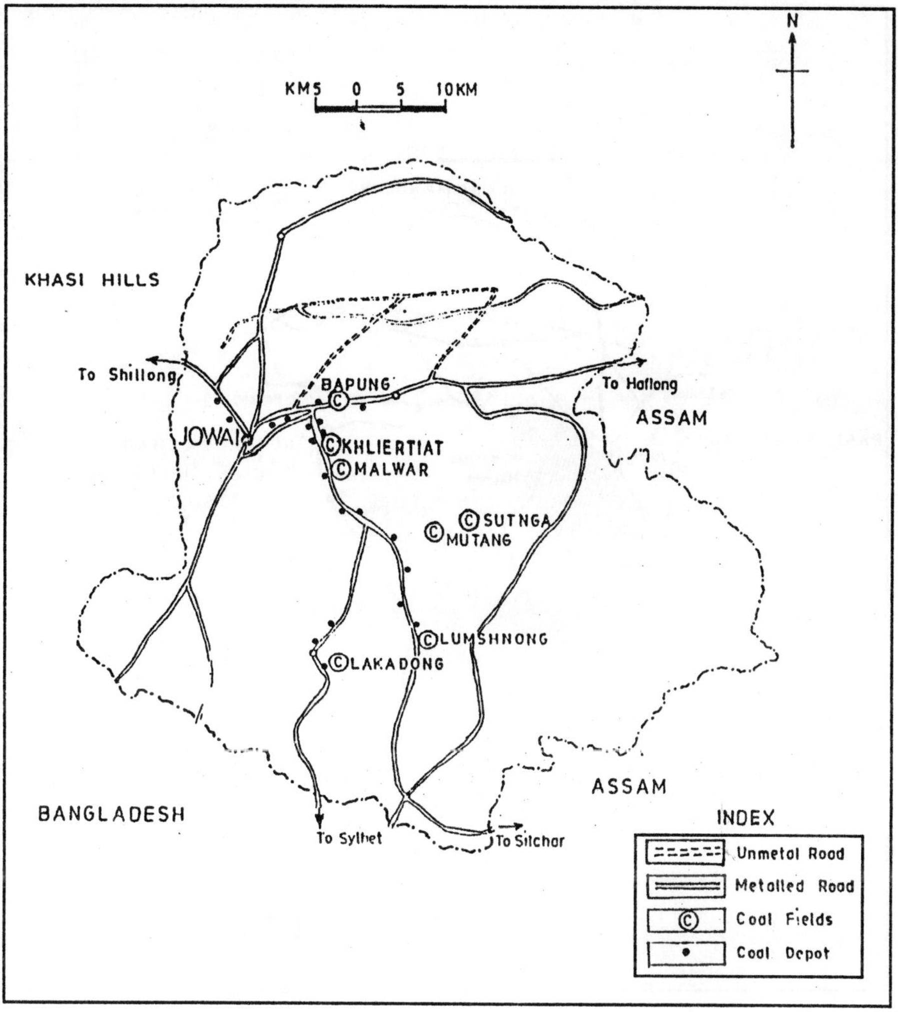

Fig. 20.3A : Jaintia Hills District Coal Fields

The coal is being exploited by local landholders (private hands) operating on small scale with primitive skills and techniques. Locally these ventures are called "rat hole" mining because literally the hole has 1m opening along which the miner crawls and excavate coal (Fig. 20.3B). The coal miner goes inside upto 50-100m in length from the opening. The coal is brought outside in small wooden barrows and finally, to roadside by head load shifting. At present most of the coal is extracted by first digging a well of about 20m diameter. The depth of well varies from one to another. As soon as the layer of coal seem is visible the workers start extracting coal through the rat hole mining.

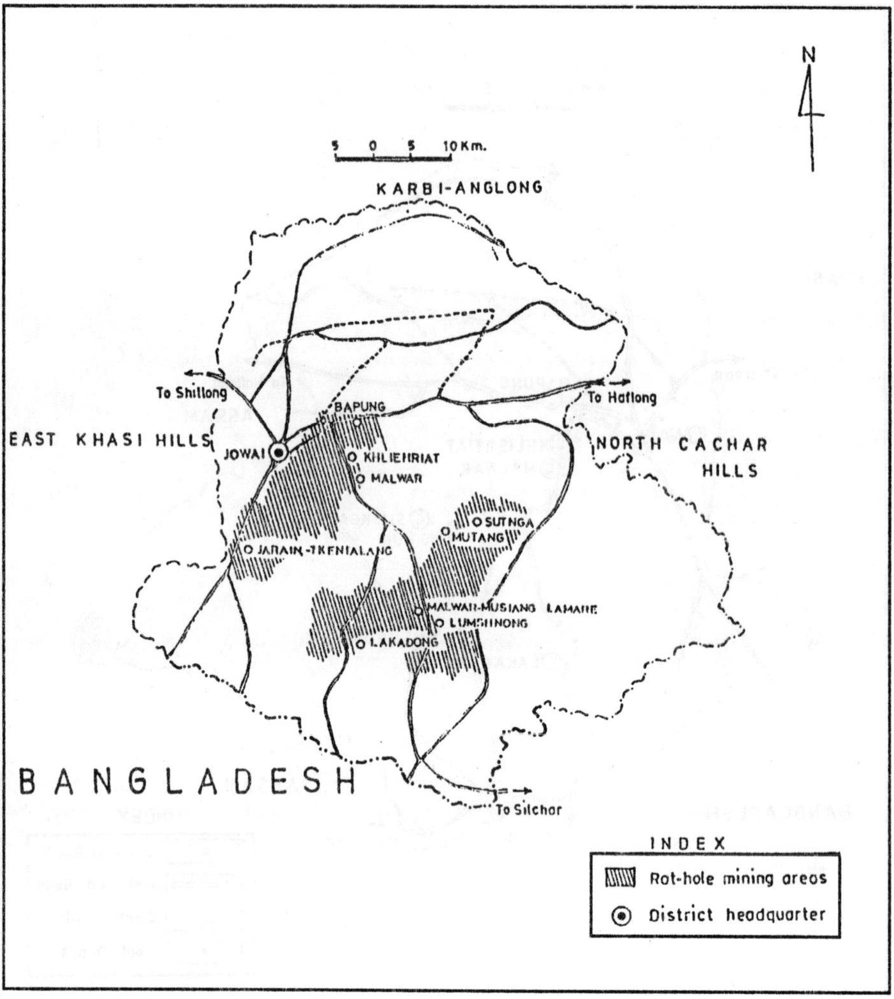

Fig. 20.3B : Jaintia Hills District Areas under Rat-Hole Mining Operations

(b) Limestone : Huge limestone deposits occur in the southern part of Meghalaya. There are three limestone horizons alternating with three sandstone horizones as given in Table 20.1.

Table 20.1 : Limestone and Sandstone Horizon

	Formation	*Thickness (m)*
(i)	Sylhet limestone/Prang Limestone/Siju Limestone	60 to 150
(ii)	Nurpuh Sandstone	15 to 26
(iii)	Umlatdoh Limestone	20 to 110
(iv)	Lakadong Sandstone	35 to 250
(v)	Lakadong Limestone	25 to 60
(vi)	Therria Sandstone	20 to 80

The exploitation of limestone is distributed along three topographic-cum-stratigraphic levels. The highest level corresponds to Cherrapunji, the lowest occurs along the lower southern slopes bordering Bangladesh. Since the deposition of the limestone is horizontal to sub-horizontal on a regional scale, these level difference are maintained all along. Consequently, much of the limestone from the lower levels is sold to Bangladesh along innumerable waterways. This export is facilitated due to easy accessibility. The limestone is being exploited mainly along roadsides by opening 50 to 150m long quarries depending upon individual land holding size.

The quarries have little regards to bench heights and safety. The quarries are left abandoned, once working face recedes away from the road. Large quantities of undersized productions (4/5 cm size) is left behind as waste and large areas are deforested around quarries. According to Anon (1974), most of these limestone deposits though of superior grade is exported mainly to Bangladesh through private entrepreneur. This is mostly due to lack of industrial base in the state. At places near the quarries, highgrade limestone is being utilized for lime making by traditional methods for local consumption.

(c) Sillimanite and Corundum : The world famous Sonapahar sillimanite corundum deposit is located near Nongstoin (West Khasi Hills District). The sillimanite cordundum occurs in Archaean-Proterozoic Migmatite-Geneissic complex as small lenses. Much of the sillimanite-corundum occurs as "floats" confined in about 24 deposits. It is under exploitation since 1922. Out of the estimated reserves of 0.5 m tonnes, only about 0.05 m tonnes of sillimanite-corundum is remaining in the area. These deposits are at the verge of exhaustion. The long continued mining of sillimanite-corundum in the area has caused wide scale deforestation leading to severe soil erosion. The down slope movement of waste debris has rendered lower level agricultural fields unusable. At present the biggest problem is to reclaim these wastelands for productive use.

(d) Building and Construction Materials : About 95 per cent of the state has hard rocky terrain. As a result, the construction activities depend only upon them. The various building stones available in the state are : sylhet trap, quartzites, slates phyllites, khasi greenstone, laterised and friable sandstone. Coarse-grained granites and gneisses.

The requirement of pebbles and cobbles are met from crushed rocks and the resulting sand size material in place of river sand. Sand is also supplemented from weathered-altered-friable sandstone of the Shillong group.

Earlier, most of the houses were the typical Assam type requiring large quantities of wood, but now the inductance of seismic structures in construction has made the RCC pattern popular. Burnt bricks/hollow bricks and rock materials are used for these structures. The demand for burnt bricks is met from outside the state, while the making of hollow bricks has evolved as a local small-scale industry in the state.

Conclusion and Recommendations

The various mining operations in the state have created a number of environmental problems. Broadly these problems could be grouped as :

(A) Landscape Deterioration

- (a) Primitive, unscientific and unsystematic rat hole mining and quarrying of limestone is destroying aesthetic beauty of landscape.
- (b) Large scale deforestation and associated soil erosion.
- (c) Encroachment of forest and agricultural land.
- (d) No back filling of abandoned mines.
- (e) No land reclamation

(B) Health Hazards and Pollution

- (a) Acid mine discharge into potable water sources.
- (b) Noise and dust pollution.
- (c) Open abandoned mines.
- (d) Roof collapse, sudden failure of open pit sidewalls and major subsidence associated during mining activity.

(C) Mineral Conservation

- (a) Left over undersized minerals.
- (b) Non-appropriate gradewise utilization of minerals.

The above listed problems are interrelated and have been compounded mainly because of unscientific mining activities in the state.

Since minerals are the basic requirements of industrial base, exploitation of minerals would keep on going along with the developmental activities. The damage already done to the environment is irreparable but efforts can be made to mitigate the same.

The following corrective measures should immediately be taken up to check further environmental degradation in the state.

1. Land ownership system should be reviewed.
2. Implementation of mine regulations should be made compulsory.

(3) Mandatory afforestation programme should be taken up mine holders.
(4) Rat-hole mining should be stopped.
(5) Reclamation of wasteland should be taken up.
(6) Acid mine discharge to be diverted away from the potable sources.
(7) Proper health care should be provided to the mineworkers.

REFERENCES

Anon (1974) : Geology and Mineral Resources of North East states of India. *Geol. Surv. Ind. Misc. Pub.* No. 30, Pt. IV, 124 P.

Ghosh, A. (1988) : Environmental Impact of Coal Mining in Eastern India, In : (Joshi, S.C. and Bhattacharya, G., ed.), *Mining and Environment in India,* H.R.G. Nainital, pp. 43-51.

Ghosh, M.K. (1989) : Land Reclamation and Protection of Environment from the effect of Coal Mining Operation, *Minetech.*, V.10 No, pp. 350-359.

Ghosh, S.K. *et al.* (1984) : Quality of Mine Waters in Jharia Coalfield, *IAWPC Tech., Annual* V. XI pp. 25-28.

Khoshoo, T.N. (1986) : Environmental Priorities in India and Sustainable Development, Indian Science Congress Association, New Delhi.

Mazumder, S.K. (1976) : A Summary of the Pre-Cambrian Geology of the Khasi Hills, Meghalaya, *Geol. Surv. Ind. Misc. Pub.* No. 23, p. 311-334.

Raja Rao, C.S. (1981) : Coalfields of India, *Geol. Surv. Ind. Vol. I. Bull.* Ser. A. No. 45 76 p.

21

Environmental Impact Assessment

A Special Reference to Kota Thermal Power Station

B.L. Sharma, L.L. Agarwal and Pooja Paur

Any development process is bound to have its impact on the environment. Environment is a subject of immediate relevance and of immense practical importance to the international community as a whole. The famous hymn in the world's oldest scripture, the Rigveda, portrays the beauty of the morning (Usha) and worship its glory.

Environmental pollution has become an emerging threat and of great concern in today's context pertaining to its effect on the ecosystem.

Pandit Jawaharlal Nehru has quoted in 1957 "We have many large scale river valley projects which are carefully worked out by our engineers. I wonder, however, how much thought is given before the project is launched, to having an ecological survey of the area and to find out what the effect to the drainage system or to the flora and fauna of that area. It would be describable to have such an ecological survey of these areas before the project is launched and thus avoid an imbalance of nature." This statement confirms the relevance of EIA. Thus the objective of environmental impact assessment is to ensure that development is sustained in the minimal environmental degradation.

Environmental impact assessment is a tool which allows us to assess the impact of proposed development on the environment of a place. The impact assessment is the most transport possible method, so that the values arrived at for the impact assessment can be subject to check and re-evaluation.

Thus Environmental Impact Assessment (EIA) is used to identify the environmental and social impacts of a project prior to decision making. It aims to predict environmental impacts at an early stage in project planning and design, find ways and means to reduce adverse impacts, shape projects to suit the local environment and present the predictions and options to decision makers.

Objectives

1. To know EIA of KTPS in view of importance of KTPS for Kota city in Rajasthan.
2. Identification of impacts and their effects.

Methodology

1. The primary and secondary information was collected from questionnaire and from different departments respectively.
2. Help have been taken from the Air Pollution Department to study the effect of pollution on the surrounding flora, fauna and human beings.
3. Water samples collected were being tested in PCB's laboratory of Kota.
4. For the demarcation of areas toposheet 54 C was referred.
5. NEERI analytical report was referred for the comparative research study of the area.
6. To know the identification of impacts in the area through questionnaire socio-economic survey was conducted.

Key Map of Kota District

1. Location of Kota City: 24°32′ and 25°51′ North Latitude and 75°37′ and 76°34′ East Longitude.
2. Location Site of KTPS : West of Kota Barrage on Chambal River.

Table 21.1 : Commissioning of Units of Kota Thermal Power Station Stage I, II, III, IV

Stage	*Units*	*Capacity*	*Expenditure (in crores)*	*Date of Synchronisation*
First	Unit-I	110 MW	143	17-01-83
	Unit-II	110 MW		13-07-83
Second	Unit-III	210 MW	480	25-09-88
	Unit-IV	210 MW		01-05-89
Third	Unit-V	210 MW	480	26-03-94
Fourth	Unit-VI	195 MW	635	Aug. 2001

Note : Total planned capacity of KTPS is about 1045 MW.

Record of Excellence

KTPS of RSEB is one of the best, efficient and prestigious power stations of the country which feeds 50 to 95 per cent of the demand of electricity in Rajasthan.

The Kota Thermal Power Station, a subsidiary of the Rajasthan State Power Generation Corporation, has set a record in power generation since 2000.

Scope of Eia

It is envisaged that KTPS a power generation project of such a magnitude and based on coal would lead to a number of impacts on environment, namely, atmospheric pollution by emissions, solid waste generation by way of ash dumping, water pollution by way of waste effluents as well ash in the nearby water, impacts on agriculture due to ash disposal on ground as well as seepage of waste effluent and health impairments due to atmospheric pollution. Besides this, the socio-economic environment in the area would also be affected by way of land acquisition for the project rehabilitation, influx of population, strain on existing infrastructure, increase in business and industrial output, impact on aesthetic and cultural environment and above all health impairments.

Baseline Environmental Status and Identification of Impacts

1. Air Environment
   - (a) Winter Season
   - (b) Summer Season
   - (c) Post-Monsoon Season
2. Noise Environment
3. Hazardous Waste
4. Surface Water (River and Dams)
5. Zoo Planktons
6. Phyplanktons

Air Environment

Fifteen sampling locations were selected by the air pollution department, Kota, during the reconnaissance survey depending upon the importance of site and/or sensitive receptors around KTPS, viz. Residential areas and also based on wind profiles available for the region from previous meteorological observations around the power plant. The samples were collected round the clock during study period in seasons covering winter, summer and post-monsoon.

Observations

Three major air pollutants, namely Suspended Particulate Matter (SPM), sulphur dioxide (SO_2) and nitrogen dioxide (NO_2).

Winter Season

The variation of SPM recorded at different stations is primarily governed by the local phenomena such as wind blown dust storms, traffic, local and agricultural activities leading to lifting up of settled dust.

Windrose : Summer

Summer Season

Dust storms are frequent phenomenon during summer months. Concentration of SO_2 and NO_2 for all the sites were found to be well below the limits prescribed by PCB (80 ug/m^3) for residential/rural area.

Post-Monsoon Season

In post-monsoon pollutants in air do not get lifted up because of rains, high average SPM concentration of 493 ug/m^3 were observed.

Effects

The most effected areas are those of Kota town within 7 km radius including Tipta, Sabji Mandi, Kishorpura, Dadabari, Keshavpura and Mahaveer Nagar. Increase of SO_2 in air has resulted in instantaneous irritation of nose and throat. The study shows that about 34 persons out of every 100 persons in these areas are suffering from the infectious diseases like bronchitis, asthma and lung infections. Release of Nitric Oxide in high concentration has resulted in various respiratory problems, gum inflammation, internal bleeding, oxygen deficiency, Pneumona and lung cancer.

Noise Environment

The noise from KTPS is very complex in nature as it is generated from multiple sources. The generated noise is distributed over the entire frequency spectrum. The important sources of noise in KTPS complex are : pumps, boilers, turbines, compressors, air cooling tanks etc.

In Plant Noise Levels

About 21 major noise sources were identified within KTPS complex. It was observed that the noise level from all important process sources varied between the range of 68-103 dBA. The noise levels measured at other locatiosn in KTPS, viz. Administrative office building, control rooms of the plants, plant floors, fixing floors, main gate, coal mill area ranged between 49-86 dBA.

Effects

The nearest residential area, i.e. the KTPS colony is the worst noise effected area which is about 1.5 km away from KTPS complex and the Kota urban centre which is about 3 km from the KTPS. According to the survey conducted in the area about 40 per cent of people in this area are suffering from speech interference, annoyance, 70 per cent from sleep interference.

About 30 per cent of teenagers are suffering from permanent hearing loss, muscular tensions, nervous irritability, strain and neurotic mental disorder.

Hazardous Waste

Ash generated during coal combusion constitute the chief solid waste at KTPS. About 90 per cent of the ash generated in the process is captured as flyash which is subsequently sluiced with waste and sent for dumping in marked dyke. A thorough analysis of ash samples from the abandoned dyke as well as the dyke as well as the dyke currently in operation was undertaken.

Effects

The localities situated close to the old ash dyke in the event of seepage through the subsurface stratum are likely to receive lower dosages of heavy metals in the ground water.The predominance of heavy metals in the waters of Munji Kund on Nanta road near the currently operating dyke, is supportive of the above observation. However, the shallow bore well water samples at Ganeshpal indicated the highest incidence of heavy metals among all the samples analysed for the purpose.

It is noted that the toxicity characteristics leaching procedure (TCLP) provides leachate emanating from the ash and rock contain metals at concentration levels of less than 100 times those of drinking water standard.

Surface Water (River and Dams)

River Chambal has been dammed to form three dams—Gandhi Sagar, Rana Pratap Sagar and Jawahar Sagar. In 1980, the river again impounded to form reservoir by building a barrage across, i.e. Kota Barrage to provide water for the then proposed Kota Thermal Power Station (KTPS).

Except for the water consumed within the plant, water supplied to KTPS colonies and water used in preparation of ash slurry, all other effuents from KTPS are discharged back into the reservoir.

Most of the sewage/sludge from old city and new city, i.e. Dadabari, Kishorpura, Basant Vihar and Shivpura is discharged into the reservoir down stream and upstream, on the right bank of the river. Left bank of the river is sparsely populated and discharge of treated sewage from KTPS colony and untreated sewage from Kunhari village are the only major areas.

Sewage from the newly developed areas, of Dadabari, Vigyan Nagar etc. gets collected in Kansua and Sazi Nalla, which ultimately join river chambal.

The KTPS draws water from the barrage for all its requirements. Cooling water from the major part (more than 95%) of water drawn from the barrage, most of which is returned back into the barrage without any appreciable change in its physico chemical properties except in temperature.

The other discharge, i.e. the back wash from DM regeneration, boiler bleed, flow down and floor washing etc. are collected through a network of open drains and common drain discharge the waste into cooling water return channel.

Zoo-Planktons

In order to evaluate biological characteristics of surface waters surrounding 30 km area of thermal power stations, samples were collected during pre-monsoon from nine sampling stations for the enumeration of zoo planktons. Among them, seven groups were represented in surface water, where in copepoda was dominating in all surface water. In sampling station at Talera river, protozoa was dominating group where as in station Allaniya river, caldocera was a dominating group during premonsoon.

It was noted that copepoda was dominating in all surface water where as at Talera river, protozoa and at Alania river, Caldocera was the dominating group, thus indicating the unaesthetic conditions of water bodies.

Phytoplanktons

Water samples collected from surface water sources, Kota barrage, up stream and down strea of Akelgarh water works, sarovar lake, left main canal at Girdhar Pur, ashpond effluent talab and Alania river reveals that during winter season there is higher algal went at Kota barrage compared to other surface water sources indicating nutrient availability through organic source of pollution compared to influents. But as we go from Kota barrage to Akelgarh on the right bank of the river there are several patches of obnoxious weed, Eichhornia Crassipes, which is considered a nuisance for it chokes water bodies effecting the aquatic fauna and flora of river Chambal.

Socio-Economic Survey

Socio-economic survey is conducted with the help of a questionnaire and sampling is done on random proportionate and stratified basis considering the adult males and females. The survey is conducted at four levels :

(a) Residents of the Plant Township.
(b) Residents (within 3 km radius) of village Nanta and may be termed as project affected population (PAP).
(c) Residents (within 7 km radius) of Kota city and project site may also be termed as project affected population (PAP).
(d) Villages within 10-25 km radius of project site.

Residents of the Plant Township

Resident engaged in different professional skills like managerial, technical, administrative, skill were interviewed. It was observed.

Almost all the respondents are satisfied with the infrastructure facilities provided by the management except some dissatisfaction shown by type D residents regarding housing.

As regards environment pollution there are no specific complaints this is possibly because of the fact that residential complex is close to the plant and does not get affected by emission as the chimneys are very high (180 m high).

The suggestions from the people involve strengthening of medical facilities, trees plantation an cultural activities.

Expansion activities are welcome because of modern technology leading to pollution mitigation.

Nanta Village within 3 km Radius

This village being close to the plant and more over to the dumping site, the villager's complaints need careful reviewing. The complaints registered were :

Ash from the dumping sites during rainy season flows into the agriculture field reducing the agricultural yield.

This ash also pollute the water source and cause many skin disease like itching, rashes on body etc.

There is pollution in the air particularly during summer and this leads to breathing discomforts.

Nearly forty per cent respondents were farmers, about the existing project nearly 75 per cent respondents gave 'good' opinion but there is no awareness about the expansion activities.

Kota Town Within 7 km Radius

The observation noted were –

- Tipta : The village is very close to the plant site and there are complaints about ash and coal particulars in the atmosphere leading to discomforts in breathing as well as dirt accumulation in houses, clothes etc. However 80-90 per cent people favour the expansion expecting job opportunities and increase in business.
- Sabji Mandi : There are some complaints regarding air pollution but people on the whole have no knowledge about expansion.
- Kishorpura : Is very close to the plant site. Due to ash dust and coal contents in the air, many people in this area have breathing ailments and skin diseases; thus most of the people do not have a favourable opinion about the plant.
- Dadabari : Complaints regarding air pollution, coal dust and ash in the air making the atmosphere dirty and causing health discomforts are registered. However increase in business and income make the people favourable about the project as well as expansion.
- Keshavpura and Mahaveer Nagar : Air pollution is experienced by the residents.

Villages within 10-25 km Radius

The observations noted were—

- As from the personal inventory queries it appears that farming is the main occupation followed by labour and a good number of respondents.
- In almost all the villages there is favourable opinion about the project, however, expansion activities are not known to the villagers.
- As regards specific aspirations from the project it is seen that except for Nanta village which is close to the site and where 75 per cent of the respondents opined that project would create job opportunities, in all other villages it is felt that there would be no jobs made available due to the project expansion.
- Significantly in almost all the villages people in general feel that the project causes environment pollution. The villages which are comparatively closer, namely within 3-11 km almost 80-100 per cent respondents have opined that there is environmental pollution due to the thermal power plant. This needs proper consideration.
- In the village Seenta, land was acquired by the management for railway track affecting 10-12 families. It was stated by the villagers that due to

the Railway line being so close houses in the villages get vibrations and cracks have been developed on the walls. The land compensations issue is also not settled satisfactorily.

Environmental Management Plan

The proposed expansion facilities are not going to significantly change the trends as well as levels of existing air pollution in view of the fact that :

- A high efficiency electrostatic precipitator (ESP) with a collector efficiency greater than 99 per cent will be installed in the flue gas circuit ahead of the chimney to restrict suspended particulate matter in the flue gas to emission levels lower than those specified in the submission regulation (1984) of the central board for the prevention and control of air pollution.
- A chimney height of 180 m. Has been established in accordance with the aforementioned regulations to mitigate the effects of sulphur dioxide discharge by effective disposal resulting in acceptable ground level concentrations.
- Coal to be utilized in the power stations has a low sulphur percentage.
- Furnaces and boilers should operate with minimum excess air so that fuel consumption is reduced and No_x emission is minimized.
- Water spray systems to be provided for crushed coal stock yards, discharge ends of conveyors and wagon tippler coppers (water and chemical sprays) for suppression of fugitive dust emission.
- Implementations of recommendations on green belt within KTPS boundary specifically around the newly installed expansion unit and outside area convening the boundary wall shall attempt thc pollutants from KTPS.

Noise Environment

To control noise pollution following points should be taken into consideration :

- All the machineries should be maintained regularly for keeping the sound power levels at design level.
- The silences and mufflers of the individual machineries should be regularly checked.
- Provisions of ear plus or ear muffs should be made for in plant workers near the identified high noise generating equipment to avoid exposure to high levels.

Solid Waste

The solid waste generated from the KTPS colony is being used for filling in the low lying areas. The present method of disposal of domestic solid waste is in sanitary and may have adverse impact on the environment. It is hence necessary to adopt sanitary land filling method for disposal of household solid water or it should be recycled by using aerobic composing.

- Performance evaluation of proposed effluent treatment units as well as sewage treatment plant should be under taken at regular intervals.

Conclusion

In view of the widespread effect of environmental pollution by KTPS, it may be predicted that in the near future, alarming adverse effects could paralyze the healthy life of human beings in due course of time. The study reveals the fact that in areas within 7 km radius including Tipta, Sabji Mandi, Kishorpura, Dadabari, Keshavpura and Mahaveer Nagar almost one out of every ten persons is suffering from bronchitis. The result of study shows that about 34 persons out of every 100 persons of the city are the victims of air pollution as they are suffering from diseases like rhinitis, allergy, bronchitis and asthama. Hence the study confirms a close relation between SPM and respiratory diseases as seen in other parts of the world.

Environmental education is essential for generating widespread awareness on environmental problems among the villagers. To promote environmental education participation of each and everyone, namely self-help groups, local level institution, NGO's very essential.

Environmental management policy of KTPS should be able to define how much environmental protections can be insisted upon without grossly compromising the social aspiration of important quality of life while enhancing development.

Thus, it can be concluded that the whole KTPS management and community needs in integrated approach to sustain and save our ecosystem from different levels of environmental pollution. There is, therefore, a powerful need to implement the four R's Action Plan : Refuse, Reduce, Reuse and Recycle.

REFERENCES

Chaphebar, S.B. 1981; Significance of growth changes inplants is pollution indication, WHO workshop on biological indicators and indices of environmental pollution, cont. bd. Prev. cont. water poll, Osmania University, Hyderabad, 19-25.

Environment Pollution and Urban Administration; (Dec., 1977); Indian Institute of Public Administration, New Delhi.

Mandan, S. and Saxena, M. (1998) Boc. Regi. Workshop-cum-sympo, on flyash disposal and utilisation at KTPS.

Papers of Refresher Courses on Water Pollution Control; (Nov. 5-12, 1976); *NEERI*, Nagpur.

Pollution Control Law Series – PCLS/4/2001, Environmental Standards for Ambient Air, Automobiles Fuels, Industry & Noise.

Prospects of Thermal Power Plant in Rajasthan, (1975) : Kota.

22

Geoenvironment and its Influence on Social Factors

A Case Study of a Rural and an Urban Area on West Bengal

JAYATI DAS AND LAKSHMI SIVARAMAKRISHNAN

Introduction

The people of Kolkata Metropolitan Area face the challenge of accelerating environmental change linked to rapid unplanned urbanization, high potential for industrialization and continued growth of metropolis. Poor hygienic sense, indifferent attitude among the population exacerbates and complicates the impact on health due to multiple environmental risks. The result is likely a heavy burden of ill health.

Indiscriminate and unregulated activities in industries including those of innumerable small industries have increased the concentration of pollution in the air. Uncontrolled vehicular densities on badly maintained road space (hardly 5%), lack of adequate parking facilities have profound effect on traffic and air pollution.

The rural scenario has been discussed with respect to Sunderbans which is the largest inter-tidal area in the world. The Indian Sunderbans at the apex of the Bay of Bengal lies between 21°31′ and 22°40′ north latitude and 88°10′ and 89° east longitude is located at the southern fringe of the West Bengal state covering major portion of Southern 24 Parganas and parts of north 24 Parganas (Fig. 22.1).The region is bounded by Bangladesh in the east Hugali River in the west, the Dampier Hodges line in the north and Bay of Bengal in the south. (Fig. 22.1). The Sunderbans now designated as the Sunderbans Biosphere Reserve covers an area of 9630 sq. km (Mitra, 2001). This mangrove ecosystem comprising of 102 islands of which 56 are inhabited supporting 3.5 million people mostly farmers, fishermen, wood cutters, honey collectors, boatsmen and others (Bandopadhayay, 2000). Extensive clearance of the forested area to support the excessive pressure of population has brought into light how this fragile environment is being threatened by the activities pursued by man in this region. The reclamation of land which started in early 1830s continued and today literally most of this mangrove forest has been cleared and the very

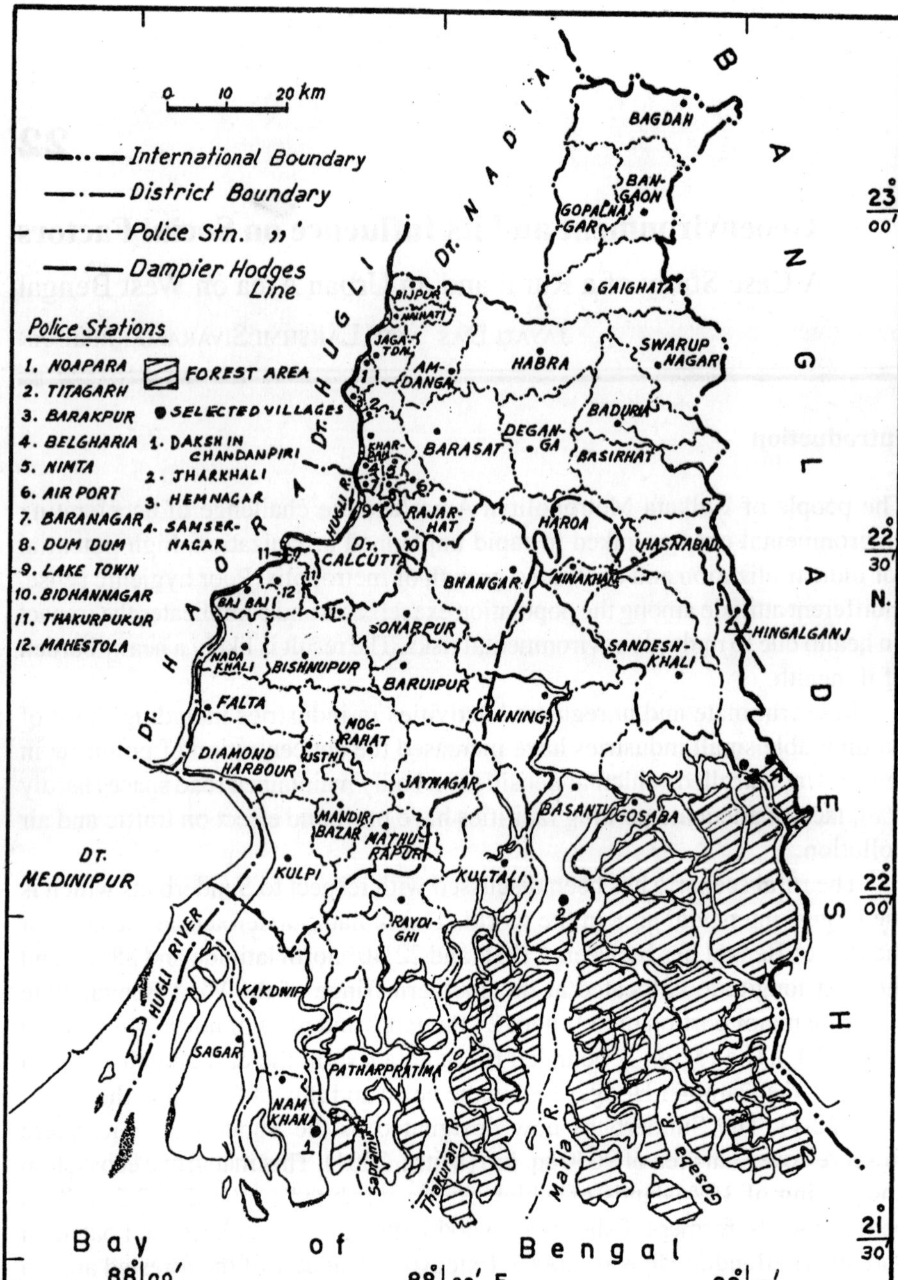

Fig. 22.1 : Sundarban Biosphere Reserve

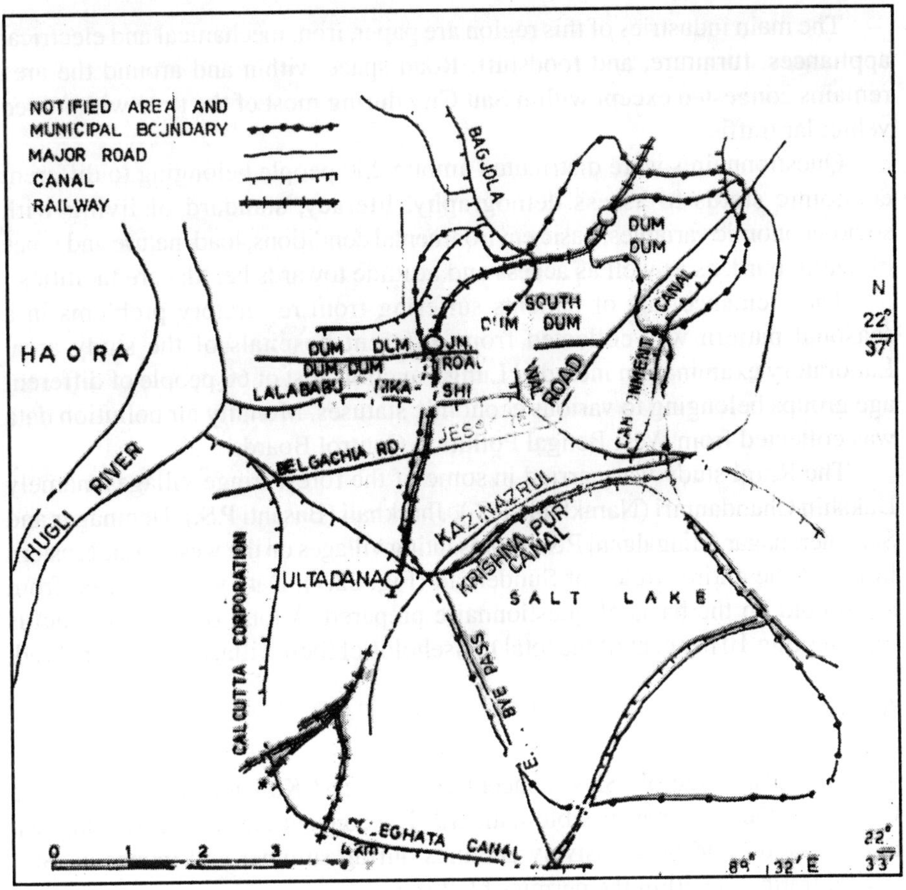

Fig. 22.2 : Location Map

existence of the remaining 4263 sq. km is being threatened by human activities. New occupations being by done people has raised some major environmental problems particularly in the recently reclaimed marginal areas which lie in between the uninhabited forest blocks and the old reclaimed mature land. These changes in the occupational pattern not only have environmental implications but have also brought into light many socio-economic and socio-cultural problems.

Methodology

The urban study was conducted among the people of all age groups belonging to high, middle and low socio-economic status with an area of 43.56 sq. km, with a population of around 4.5 lakhs (2001) of Dum Dum, Municipalities and Salt Lake City (Fig. 22.2).

The main industries of this region are paper, iron, mechanical and electrical appliances, furniture, and foodstuff. Road space within and around the area remains congested except within Salt City during most of the time with mixed vehicular traffic.

Questionnaires were distributed among 250 people belonging to different economic status to assess demography, literacy, standard of living with socio-economic variables, basic environmental conditions, load, nature and types of health burdens as well as access and attitude towards health care facilities.

The incidence rate of patients suffering from respiratory problems in a seasonal pattern was collected from different hospitals of the study area. Laboratory examination included Lung Function Test of 60 people of different age groups belonging to various economic statuses. Monthly air pollution data was collected from West Bengal Pollution Control Board.

The Rural study was carried in some of the forest fringe villages, namely Dakshin Chandanpiri (Namkhana P.S.), Jharkhali (Basanti P.S.), Hemnagar and Samsheranagar (Hingalganj P.S), representing villages on the western and eastern limits of the entire stretch of Sunderban (Fig. 22.1). Data was collected from household on the basis of questionnaire prepared. A total of 350 households representing 10 per cent of the total households of these villages were surveyed.

Results

The annual average of SPM concentration in east Kolkata except Salt Lake City is far above the permissible standard. The concentration of SPM is highest during winter and progressively declines during summer and monsoon. SO_2 concentration is within the permissible limit.

The NO × CO, O_3 and Pb are reported to be within the standards. (WBPCB, 2001).

The scenario is a little different in Salt Lake City, which is much cleaner and greener area. The mean annual SPM and RPM are higher than the NAAQs for residential area, where as Pb are below.

The daily mean concentration exceeds the ambient standard of SPM, RPM, NO_x and Pb on 24 days (23%), 25 days (24 %), 15 days (14 %), and 2 days (8 %), respectively while SO_2 concentration was always below.

Table 22.1 : Annual average AIR Quality Data of the Study Area

(In ug/m³)

	SPM	*RPM*	*NO_x*	*SO_2*	*CO*	*PB*	*O_3*
Overall	300.03	203.39	62.45	30.51	4639	2.03	125.26
Salt Lake City	141.70	72.55	49.52	7.35	3395	0.05	58.98

Source : WBPCB, Kolkata, 2001.

The annual Air Pollution Index (API) value indicates medium air pollution at Salt Lake City while API value of other areas of east Kolkata are quite high indicating high air pollution.

Table 22.2 : Air Pollution Index At East Kolkata

	Mean API	*Maximum*	*Minimum*	*Summer*	*Monsoon*	*Winter*
Overall	15.92	185.02	66.59	59.35	38.42	113.75
Salt Lake City	15.92	137.02	53.62	48.68	30.32	95.17

Source : WBPCB, Kolkata, 2001.

Effects of pollution on health studied worldwide have established a clear relationship between air pollution and mortality (Lipfert, 1994; Dockery,1999). Eight thousand people die everyday from air pollution in the world. Air pollution is associated with increased risk of acute respiratory infections, the principal cause of infant and child mortality in the developing countries (Bendahmane, 1997; Romieu and Borja- Aburto, 1997, Zhang *et al.,* 1999). Short-term exposure to high level of pollution may result in irritation of the eye and respiratory tract.

Seasonal variation of respiratory disease show high incidence rate in winter months and minimum in the rainy season is again consistent with the seasonal averages of SPM and CO concentration in air which show a distinct seasonal variation. The repeat event of acute respiratory illness (ARI, asthma attacks) as well as government hospital visits for the same show also a similar variation in their month-wise distribution (Table 22.3).

Table 22.3 : Seasonal Variation in Incidence Rate of Respirator Disorder

Months	*Salt-lake City*	*Overall*
January	124	250
February	29	75
March	35	78
April	12	39
May	18	45
June	7	27
July	10	32
August	13	24
September	19	32
October	25	51
November	152	258
December	235	390

Sources : Bidan Nagar Government Hospital and Dum Government Hospital.

Lung function test revealed less then 70 per cent predicted values for FVC, FEV and PEFR in 22 per cent, 15 per cent and 14 per cent respectively in the more congested areas of east Kolkata while in Salt Lake City these were 12 per cent, 7 per cent and 6 per cent respectively. Over population with high pollution with cumulative effect of longer stay (4 hour or more) in indoors affects lung conditions very much.

The following Tables 22.4 and 22.5 show the changes in the occupational pattern of the surveyed households both with respect to primary occupation and secondary occupation in Sunderban areas.

Table 22.4 : Primary Occupation

Name of Village *Occupation pattern* *(% of households)*	*Samsher-nagar* *Past* *present*	*Hemnagar* *Past* *present*	*Jharkhali* *Past* *present*	*Dakshin Chandanpiri* *Past* *present*
1. Agriculture and allied activities	35.00 12.62	33.00 11.32	52.30 28.75	50.00 42.97
2. Shrimp fry collection and agriculture	30.00 16.50	10.00 5.66	5.00 1.67	10.00 9.92
3. Shrimp fry collection	25.00 33.04	28.00 63.21	1.00 8.33	50.0 14.05
4. Crab collection	— 5.13		5.00	3.0
5. Labour	2.00 12.32	16.00 4.72	10.75 18.75	15.00 15.70
6. Miscellaneous bee keeping, fry business forestry, fishing etc.	8.00 20.39	13.00 15.09	30.95 37.50	17.00 17.36
Total	100.00 100.00	100.00 100.00	100.00 100.00	100.00 100.00

Source : Field Survey (Past: 1992 and Present: 2002).

Table 22.5 : Secondary Occupation

Name of Village *Occupation pattern*	*Samsher-nagar* *(% of households)*	*Hemnagar* *(% of)* *households)*	*Jharkhali* *(% of* *households)*	*Dakshin Chandanpiri* *(% of* *households)*
1. Shrimp fry collection	6.32	7.84	21.85	7.41
2. Agriculture and fry	45.65	31.37	27.16	12.96
3. Crab collection, honey, forestry, etc.	13.05	5.88	5.96	3.70
4. Labour	28.26	33.33	25.17	51.86
5. Fishing	—	—	11.92	22.22
6. Miscellaneous	6.52	21.58	7.94	1.85
Total	100.00	100.00	100.00	100.00

Source : Field Survey (Past: 1992 and Present: 2002).

It is clear from Table 22.4 that there has been a decline in percentage share of household engaged in agriculture operations and increasing in the percentage of households pursuing other occupations particularly shrimp fry collection

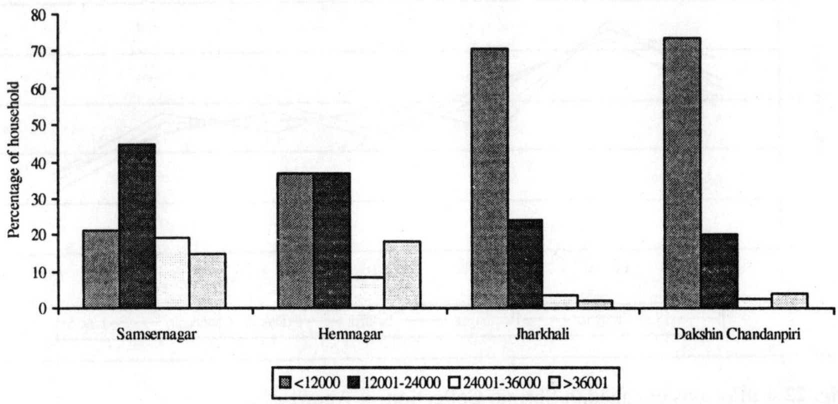

Fig. 22.3 : Income of the Households

Table 22.6 : Income of the Household

Name of Village *Annual income of the househilds*	*Samsher-nagar* *(% of households)*	*Hemnagar* *(% of) households)*	*Jharkhali* *(% of households)*	*Dakshin Chandanpiri* *(% of households)*
<12000	21.30	36.67	70.55	73.75
12001-24000	44.70	36.67	23.93	20.00
24001-36000	19.20	8.33	3.68	2.50
>36001	14.80	18.33	1.84	3.75
Total	100.00	100.00	100.00	100.00

Source : Field Survey, 2002.

and miscellaneous activity that is fry business, forestry etc. Shrimp fry collection has gain importance as secondary activity also as can be seen from Table 22.5.

The surveyed households were divided into four income groups as shown in Table 22.6. It can be seen that in Jharkhali and Dakshin Chandanpiri where agriculture still the most important primary occupation more than 70 per cent of the households are in the first income group that is < 12000 per annum (Fig. 22.3). But on the other hand household which have shifted to fry collection as in Sameshnagar and Hemnagar have higher percentage of households in higher groups that is 12000-24000 per annum.

In order to assess the type of labour that has shifted to these activities and the age of the people engaged in these activities (mainly fish fry and related

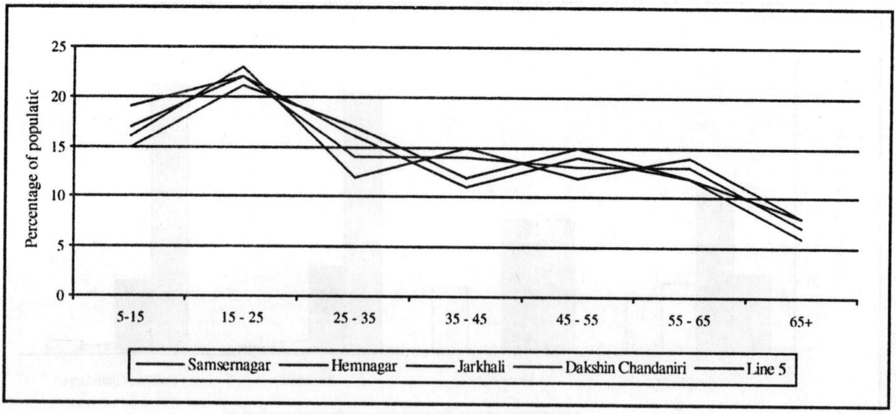

Fig. 22.4 : Persons engaged in Shrimp Collection — Agewise

Table 22.7 : Percentage of Households Engaged in Fish Fry and Related Activity - Age Group Wise

Name of Village *Age group*	*Samsher-nagar (% of households)*	*Hemnagar (% of households)*	*Jharkhali (% of households)*	*Dakshin Chandanpiri (% of households)*
5 - 15	15.00	19.00	17.00	16.00
15 - 25	21.00	22.00	22.00	23.00
25 - 35	17.00	16.00	14.00	12.00
35 - 45	12.00	11.00	14.00	15.00
45 - 55	15.00	14.00	13.00	12.00
55 - 65	12.00	12.00	13.00	14.00
65+	8.00	6.00	7.00	8.00
Total	100.00	100.00	100.00	100.00

Source : Field Survey, 2002.

activities) were taken in to account which is given in Table 22.7. It can be seen that a large percentage of children and youth are engaged in fish fry collection about 30 to 35 per cent in all the villages (Fig. 22.4). This particular economic activity has led serious socio-culture problems in this area and some of them have been highlighted.

Some of the major problems include alcoholism and gambling which seems to be on an increase. Easy money or ready cash, which comes with shrimp fry collection has made people particularly the youth addicted to drugs and alcohol. Gambling has also increased which also leads to family conflicts. From the surveyed result it can be seen that alcoholism has greatly increased in recent years.

More than 73 per cent of the sample population have been addicted to alcohol in the last five years. So probably easy money coming with this occupation is the cause of this problem.

Table 22.8 : Addiction to Alcohol

No. of years	*% of people*
Recently < 1 year	27.00
2 -5 years	46.00
5 - 10 years	20.00
More than 10 years	7.00
Total	100.00

Another serious problem seen was school dropouts. It is already seen that 16.75 per cent and 22 per cent of the people in the age group 5-15 and 15 - 25 are engaged in fry collection indicating that the children and youth either do not go to school regularly during the collection section and completely dropout during the peak collection period that is post monsoon season when the prices of these fingerlings is high. While interacting with the families it was seen that most parents want their children to go to school but want of money forces them to send them for fry collection. These children are exposed to saline water for long periods of time so they prone to many diseases. This may be due to the long hours of catching fry, which upsets the food timings, and exposure to saline water causes skin diseases. The removal of the seed from the fish larvae is also a tedious process and strain on the eyes. Many of the children also complained of eye problems.

Family conflicts due to negligence of family also common. In most cases woman and children are engaged in fry collection throughout the year either a primary or secondary activity. As a result the family suffers as the women and children are not able to give enough time.

The social impact of the occupation carried on in this environment is been keeping and collection of forest products. A large section of the population has to venture into the deep forests for honey collection and often fall prey to tigers. This is also because of tiger straying into villages. The forest area is shrinking so the tiger stray in search of food. Another problem, which these people face, is that due to strict and rigid forest policy their assess to the forests has decreased. So illegal collection of wood from these protected forests goes on and often this is the bone of contention between the villagers and the forest officials.

Conclusion

In the urban infrastructure air pollution is a multidisciplinary subject and the community health aspects should be given priority in future policy making. The study showed that a small rise in the respirable particulate matter inhaled in air creates a corresponding sharp rise in incidence of respiratory problems.

Residents of East Kolkata are presently suffering from respiratory diseases and might suffer in near future due to deteriorating air qualities. Coupled with this are the episodic swings in air pollution that would be related to occurrence of several hundred extra deaths per year.

Factors as yet unidentified may have possible implications in respiratory diseases due to air pollution. For example, the prevalence and pathogenecity of certain infectious disease like tuberculosis and pneumonia could be related to the level of air pollution.

In future policy planning and implementation the point should be recognized that there is no entirely "safe" level of air pollution and control strategies are not stable. As rightly pointed out by the noted epidemiologist Lipfert, "it is not what is a safe level of pollution but rather, what is a tolerable or acceptable of sickness from pollution" (1994).

In rural Bengal deltaic region with mangrove forests is witnessing a change in the occupational pattern. The collection of fish fry is no doubt remunerative and helped to improve the economic conditions of the people but along with as number of environmental and social problems have cropped up. Over exploitation of this resource is causing rapid depletion. Tiger shrimp collection by the villagers by netting out the shrimp seeds leads to over exploitation and damage to other fishes, prawn and crab species. It also causes damage to the riverbed because of repeated walking over by these people. Also associated with it are many socio-culture problems like gambling, alcoholism, school dropout's diseases etc.

At the end it may be said that whether it is a rural or urban environment man's intervention and development activities bring about rapid economic and social changes and these may have far-reaching effects on the society as a whole.

REFERENCES

Bandopadhayay, D. (2000) : *"Can the Sunderbans be saved." Economic and Political Weekly,* Vol. XXXV, No. 45 Nov. 4 -10, pp. 3925-3926.

Mitra, A. and Bhattacharyya, D.P. (2001) : "Biological Diversity of Mangrove Wetlands in Sunderbans and Scope for alternative Sources of Income." *Everyman Science,* Vol. XXXVI, No. 2 July-Sept., pp. 72.

Mukherjee, A., and Mukherje, G. (1999) : "Ventilator function in some school children within Calcutta Metropolitan Area." *Man and Environment Annual,* Vol. 10, pp. 82-84.

Schwartz, J. (1993) "Particulate Air Pollution and Chronic Respiratory Disease." *Environment Resource,* 62 pp. 7-13.

West Bengal Pollution Control Board and Department of Environment and Government of West Bengal (2002) : *Health Effects of Air Pollution - A study of Kolkata.* pp. 47-49.

West Bengal Pollution Control Board (1995-96), *Air Quality Monitoring in and Around Calcutta during the period October 1995 - October 1996,* Project Report by Centre for Man and Environment, pp. 1-11.

23

Natural Resource Management for Sustainable Development

Causes and Concern

S. Padmaja and Kausalya Ramachandran

Introduction

Natural resource management is the heart of the issue of sustainable development. A proper and scientific management of resources is required to achieve sustainability. Unfortunately many of the developing and poor countries are facing the grave problems of land degradation and environmental pollution due to increase in population growth which in turn results in greater pressure on natural resource base. The paper highlights the causes and concerns of natural resource management and indicates a few solutions for achieving sustainable development.

A computer looks ahead and shudders: Study sees disaster by year 2100; scientists warn of global catastrophe. This may prove true if we do not care for our mother earth which has given us plenty to survive. This holds true, keeping in view, that 'humanity must live within the carrying capacity of the Earth. There is no other rational option in the longer term. Unless we use the resources of the earth sustainably and prudently, we deny people their future. We must adopt life styles and development paths that respect and work within nature's limits. We can do this without rejecting the many benefits that modern technology has brought, provided that technology itself works within those limits.'

Because of the way we live today, our civilizations are at risk. Of the 5.3 billion people alive now, especially the 1 billion in the developed countries, are misusing natural resources and seriously overexploiting the earth's ecosystems. World population may double in 60 years, but the earth will be unable to support everyone unless there is less waste and extravagance and a more open and equitable alliance between the rich and poor. Even then, the likelihood of a satisfactory life for all in remote unless present rates of population increase are drastically reduced.

Our new approach must meet two fundamental requirements. One is to secure a widespread and deeply held commitment to new ethic—the ethic for sustainable living, and to translate its principles into practice. The other is to integrate conservation and development. Conservation to keep our actions within the earth's capacity; and, development to enable people everywhere to enjoy, long, healthy and fulfilling lives.

If this seems to be an ideal way to exist, where are we going wrong? Is it in the way we are exponentially increasing our numbers, or is it that we are overusing the resources at our command? Maybe both these aspects are equally responsible and the answer may lie in trying to seek a solution to these.

Concept of Sustainable Development

The word 'sustain' derived from Latin word 'sustainere' (sus meaning—from below, and tenere meaning—to hold) means to keep in existence or maintain; it also implies a long term support or permanence (John Ikerd, 1990).

Human interactions with complex ecosystems/natural environment results in long lasting changes in environment, some of which may have long-term detrimental impact on the productivity potential of natural ecosystem. If nature is considered the storehouse of all resources, human action would facilitate or impair this capacity of the earth which would be remain sustainable. Another aspect to this relationship is the making of correct choice of resource use at a rate and a manner considering the process of natural rejuvenation and carrying capacity of land for sustainable development. For instance, it would be an informed choice to buy a car which would run on CNG and if possible creating a car pool thereby cutting vehicular pollution, saving fuel and also not sacrificing the convenience.

Historical Perspective

Sustainable Development was first developed as a concept in 1987 with the publication of the Brundtland Report. But it was not until 1992 at the Rio Earth Summit that nations around the world came together to push for concerted action to try and reach an agreement on the best way to slow down, halt and reverse environmental deterioration. Out of Rio summit came Agenda 21, the Framework Convention on Climate Change, the convention on Biological Diversity, Principles of Forest Management and norms to reduce Air Pollution.

In 1987 the Brundtland Report, also known as our Common Future, alerted the world to the urgency of making progress towards economic development that could be sustained without depleting natural resources or harming the environment. This report defined *sustainable development* as [Development that] meets the needs of the present without compromising the ability of future generations to meet their own needs. The report primarily focused on the needs and interests of humans and an equitable redistributing resources towards poorer nations to encourage their economic growth and ensuring the availability of resources to meet the demands of future generations. The Report also suggested that social equity, economic growth and environmental maintenance are simultaneously possible and that each nation is capable of achieving its full economic potential whilst at the same time enhancing its resource base. However, it recognized the achieving this equity and sustainable growth would require technological and social change.

The UN General Assembly convened the United Nations Conference of Environment and Development (UNCED) five years after the Burndtland Report was published. The UNCED is popularly referred to as the Rio Earth Summit which was held in Rio de Janeiro in Brazil in 1992. It was the largest environmental conference ever held and its objective was to build upon the hopes and achievements of the Brundtland Report, in order to respond to pressing global environmental problems and to agree upon major treaties on biodiversity, climate change and forest management. And for the first time the a more nature-centred approach towards environmental problems was adopted. However, despite its environmental focus, the biggest debates in the summit centred around finance, consumption rates and population growth. Legislation was passed and many agreements made, committing nations and become more sustainable. These agreements and guidelines are still adhered today and are influencing many political and business decisions.

Agenda 21 is the outcome of the UNCED/Earth Smmit/Rio Summit and is in purport the blueprint for sustainability in the 21st century. It is a commitment to sustainable development by world's governments and is monitored by the International Commission on Sustainable Development (ICSD). It addresses the development of societies and economics by focusing on the conservation and preservation of our environments and natural resources. While the Conventions, Principles and Declarations of the Earth Summit, provide guildelines to deal with the problems of poverty, hunger, resource consumption and the deterioration of ecosystems, Agenda 21 provides a format for this transformation of civil societies, detailing an action plan for sustainable development and establishing targets for actions that ensure economic development with environmental protection.

For instance, prior to Agenda 21, a nation's financial standing was gauged by its money alone. However, the agenda stresses the valuation based on full value of a nation's natural resources. It also encourages nations to consider the costs of environmental degradation. It stresses on the need to eradicate poverty that strikes at the root of process of sustainable development. The agenda urges the member nations to work 'towards international agreements which respect the interests of all and protect the integrity of the global environmental and development system.'

In 1997 at Kyoto the US refused to cut back emission and therefore the movement of sustainable development suffered a setback. The Marrkesh Accord to ratify the Kyoto Protocol, the Doha (Qatar, November 2001) conference for discussing multilateral trading system and the Monterrey (Mexico) conference for financing for developing (2001), all preceded the WSSD at Johannesburg, South Africa.

The WSSD was held in Johannesburg, South Africa in September 2002, as a five-year follow-up. Although the world community professed their undying support to the cause for sustainable development, there was little progress in the direction of an agreed global plan of action to cut emission or save biological biodiversity. Instead it was a show of between the 'haves' and the 'have-nots'. The powerful North insisted on linking foreign aid and policy to governance and reduction of corruption in developing South. This was an effective subterfuge to simple business preposition which unacceptable to the G77 grouping of developing countries. In the final analysis, the WSSD was huge gathering of the environmentalists whose efforts were effectively scuttled by the powerful North, and the issue of Sustainable Development remained a mere ideology. To consider that all is lost, may be wrong, for the chimera of Sustainable Development is now well and truly embedded in the world psyche.

Evidently, the UN body, at present, is ineffective as far as deliverance of goods concerned. The lack of financial resources is at the heart of the problem. There is a need for Institutional reform within UN for the sake of environment and sustainable development. Though the conferences have yielded little but its importance cannot be underestimated as to evoke strong reaction from the various parties concerned and keep the issue alive. Needless to say that sustainable development must not merely be an agenda for conferences but must form an integral part of human survival strategy on a daily basis.

Indicators for Measuring Sustainable Development

Indicators provide a way to measure sustainable development and its progress. Baseline information of a given economy, environment and society, is required to make improvements and monitor progress. Indicators are pieces of information that summarise the characteristics of a system. For example greenhouse gas emissions and global temperature are indicators of climate change. Collecting data from a broad section of sustainability issues, over a period to time, will provide important information that will play significant role in decision and policy making. Some of the factors which provide valuable indicators of sustainable development are—rate of consumption of non-renewable resources, energy consumption and production, ground water exploitation, poverty reduction, cut in greenhouse gases, reduction in emission, the check on soil erosion, level of education, availability of health care facility and housing, rate of crime, air pollution, ozone depletion, wild life, available forest cover etc.

Natural Resources Management of the Issue of Sustainable Development

It was the nature that was existing much before the man came into picture. Slowly, then he started using the resources for his survival and step-by-step technology became a domineering agent in the taming of nature of his endless desires and demands. A chronological glance at the philosophy and purpose of geography brings into light this of superialism either in the name of initially prevailing determinism or a subsequent development in the name of 'possibilism'. As the time progressed better sense prevailed upon and a new concept 'neo determinism' came into reckoning which is more viable, realistic in the sense that Mother Earth cannot be unduly tamed endlessly and there should be a concept of co-existence. Though this was to be taken seriously but the immediate momentary gains overshadowed the long term repercussions. This means a hapless overexploitation of resources and consumption of natural resources, which has become a topic of much concern.

These natural resources fall clearly under two major categories—renewable and non-renewable which are infinite and finite, respectively. Consequently, the depletion of renewable which are infinite and finite, respectively. Consequently, the depletion of non-renewable resources are of much concern and their high levels of consumption would limit growth or development opportunity.

Evidence do suggest that technology development is yielding more efficient means of providing light and other energy related services from renewable resource. Therefore this may be an issue tackled by technologists by providing alternative solutions to the use of non-renewable resources that are existing today in terms of limited our potentials. Renewable resources are also unwisely thought to be indefinitely renewable. Yet some are highly location-specific or their supply depends on the quality of finite resources, viz., land. As natural systems are asked to supply more and more resource and to absorb even more waste and pollution, degradation of their productive capacity and even the possibility of ecosystem collapse cannot be ruled out. In many areas, exploitation of both biological and physical resources already exceeds the regenerative ability of natural system. Thus not only are many renewable resources increasingly scarce, but damage to the underlying systems that sustain or renew them threatens the immediate economic and human development of many nations.

Non-Renewable Resources

For minerals, economists distinguish between resource base and reserves. " A traditional measure of the abundance of non-renewable resources, is the reserve to production ratio or reserve life index, which expresses the number of years of production at current annual rates that proven reserves will sustain. Since new resources are continually discovered and new process technologies increase the portion of those that can be economically recovered, estimates of reserves are continually adjusted. Nonetheless the reserve to 'production ratio provides a snapshot of the perceived abundance of a mineral at a given time.' Changes in the ratio overtime, measures whether the perceived abundance is increasing or decreasing."

A comparison of the world reserve to production ratios for nine major metals and for three major fossil fuels shows that the ratio varies from about 20 years (for zinc, lead and mercury) to well over 100 yrs (iron, aluminium and coal). Petroleum reserves constitute a 40 years supply at current production levels. An additional and critical factor for many non-renewable resources is the prospect of substitution. In long distance communication system, copper is increasingly being replaced by optical fibres made of glass, advanced plastics and composite materials are reducing the amount of iron and steel used in automobiles and the amount used in airplanes. Increasing sophistication in the design of materials at a molecular level implies the ability to develop entirely new methods of meeting human needs and to engineer around any shortages of many minerals. More efficient methods of converting fuel into electricity from gas turbine/combined cycle technologies to fuel cells—promise to reduce the energy needed to meet a given level of demand for power. In 1990s, fossil fuel consumption in the industrialized world was 48 per cent, developing countries 28 per cent and the transition countries consumed 24 per cent. Hence in the next century, alternative resources such as biomass, solar, wind and nuclear may have to supply a significant fraction of energy needs, reducing pressure on fossil fuel resources. Deliberate policies, especially in developed 'North' nations is required to restrict the demand for energy resources to reduce the level of ozone depletion and consequent health hazards.

Renewable Resources

Many of the renewable resources like sunlight, air, and a diversity of plants and animals are traditionally taken for granted as "free goods" of nature. Because of lack of awareness of impending shortages, these resources and their resource base are most endangered as they are being severely degraded and depleted in many regions. For instance, forest cover loss in Asia is severe accounting for

30 per cent while it is 20 per cent for the world as a whole and 18 and 19 per cent in Africa and Latin America, respectively.

Clean Air

In a survey of 20 of the world's largest cities, air pollution, at levels that exceeded WHO guidelines, was widespread (WRI, 1995). All cities surveyed exceed the guidelines for at least one air pollutant and 14 exceeded the guidelines, for two pollutants. Suspended particulate matter (SPM) along with Sulphur-dioxide (SO_2) is considered to be more hazardous. In developing countries, where urban areas are growing at 4 per cent per year or more, the technical infrastructure for emission controls often lags behind and strict enforcement of emission control is lax. Clean air may become an increasingly rare resource for a large fraction of the earth's population. According to WDR report of 1993, over 119 million persons worldwide suffer from acute respiratory infections annually (WRI, 1998).

Clean Water

Water is the most priceless of resources for which there is incessant demand, however, its supply is limited. According to WRI (1998) the world has a water reserve of 9.8 per cent of which 3300 km^3 is withdrawn, while Asia, North America, Europe and Africa with 13, 11, 9 and 3 per cent reserves withdraw 1750, 550, 500 and 150 km^3. In urban areas in the developing world over 170 m persons lack access to clean water for drinking, cooking and washing while in rural areas, over 855 m lack clean water. Agriculture dominates the water use pattern. While agriculture in Asia, Africa, North America and Europe accounts for 85, 87, 40 and 30 per cent of water use respectively, industries and domestic sectors account for the rest.

Water supplies are contaminated with human wastes, toxic chemicals and heavy metals. According to WRI (1998) water pollution is the most serious environmental problem facing developing countries because of its direct effect on human welfare and economic growth. Sewage and water treatment technologies are available but capital intensive. In addition, the capacity of rivers to support aquatic life and of coastal fisheries to maintain their productivity is threatened by pollution, loss of oxygen associated with the decomposition of pollutants. Due to contamination of surface water bodies, many people have turned to ground water resource which are being depleted faster than they can be replenished. Although opportunities exist to use and reuse water more efficiently there is no substitute for water, new sources of supply (such as desalinization) tend to be expensive and energy extensive.

Soil Fertility

The Global assessment of soil degradation study conducted for the UN environmental programme found that in recent decades, nearly 11 per cent of the earths fertile soil has been so eroded, chemically altered, or physically compacted as to damage its original biotic friction (its ability to process nutrients into a form unable by plants) about 3 per cent of soil has been degraded virtually to the point where it can no longer perform that function. The continued loss of soil fertility, combined with rapidly rising populations in most of the developing world poses the threat of insufficient food, fibre, fuelwood supplies in the future. More intensive cultivation higher yielding crop varieties and novel sources of food from biotechnology may ultimately substitute for losses in soil fertility and shortages of arable land. But substitutions could come at a high social cost, reducing agricultural employment and food security for the poorest segments of developing societies.

Degraded soil translates into less food for all. During the period 1945-1990, human induced soil degradation affected 579 m ha due to vegetation removal, 133 m ha due to over-exploitation, 679 m ha due to over grazing, 522 m ha agricultural activities and 23 m ha due to industrial and bio-industrial means. Asia accounted for the maximum loss due to the combined impact of all these causes with vegetative losses degrading 300 m ha land while in case of Africa over-grazing impacted 245 m ha of land.

Biodiversity and Ecosystem Services

Survival of human kind is infinitely dependent on the health of natural ecosystems which is often taken for granted. An attempt to valuate the share of ecosystem services to the process of development indicates that over 65 per cent of goodies come from nature amounting to over 33.3 trillion US $ (Costanza,1997) while human effort accounts for only a third (18 trillion US$) which is also dependent on the quality of environmental factors.

Ecosystem services would broadly include the following—

(a) Soil formation and microbial recycling of soil nutrients,
(b) Flood prevention and erosion control in watersheds,
(c) Maintenance of world's stocks of plants and animals and gene pool, and
(d) Climate regulation and atmospheric gas balance,
(e) Habitat.

A 6-8 per cent extinction from the tropical forest species pool over the next 25 years will cause an accelerated loss of biodiversity. Under half of the world's

250,000 flowering plant species occur only in areas that will be largely deforested or otherwise disturbed over the next three decades. Tropical forest land, a particularly rich habitat was converted to other uses at an estimated 15.4 m hectares (about 0.8%) per year during the 1980s. Wetlands are also under ever growing pressure. Despite stepped up efforts to create wild life and nature preserves, gennplasm banks and genetically managed zoo populations pressure on habitats is a serious threat to the earth's genetic heritage and to the ecosystems in which diversity flourishes—shortages of renewable resources and depletion of the resource base can impede both short-term development and long-term sustainability.

Issues and Concerns

Natural resource consumption and environmental degradation : Natural resource utilization would lead to environmental degradation if undertaken in a rate and manner which is not in tune with its natural rejuvenation process. The consequences of such an act by human being would be felt either locally or globally depending on the magnitude. Mining activities causing mine spill dumps and manufacturing processes leading to accumulation of waste cause environmental degradation. Some of the processes that have a lasting impact on environmental quality are fossil fuel consumption, mining of metals and minerals, forest products and agriculture, intensive fishing and whaling.

Utilization of Fossil Fuels

In the sphere of resource use and its implication, fossil fuel has been in the forefront. A wide range of environmental degradation is associated with fossil fuel production, transport and use. Local effects include:

(a) Land degradation from strip mining of coal and the disposal of deep coal mine tailings and from oil and gas production and transport in fragile environments;
(b) Fresh water pollution from acid mine drainage, oil refinery operations and improper disposal of used petroleum products;
(c) Marine pollution from oil spilled or deliberately flushed out of tankers, and
(d) Air pollution from fossil fuel combustion (SO_2 and particulates), refinery operations (toxic emissions), coal combustion (dust and soot), industrial and automobile fuels (urban smog).

The consequences of these could be on regional scale resulting in acid precipitation from coal and oil combustion with its impact on monuments, buildings, forests and other vegetation, soil and lakes and on global climatic change from CO_2 and methane emission during the production, transport and use of natural gas.

Mining of Metals and Minerals

Extraction, refining, disperse use, disposal of metals and industrial minerals may cause significant local environmental problems like—

(a) Mining can degrade land creating quarries, vast open pits and huge amount of solid waste. In 1991, 1000 million metric tons of copper ore were dug up to obtain 9 million metric tons of metal.
(b) Air pollution is caused due to dust from mining, acidic gases from smelting and refining and CO_2 from cement production. Fine particles of toxic trace metals which accumulate in soil and aquatic ecosystems and in animal and human food chains are often dispersed during mining and refining operations.
(c) Water pollution is often caused through leaching from tailings or abandoned mines and the disposal of chemicals used in refining.
(d) Improper use of minerals like Lead, Chromium based paints, Lead plumbing, Gasoline additives, etc. can harm human health.

Forest Products

Consumption of forest resources can lead to environmental problems as well as loss of critical habitat and species. Fuelwood consumption when exceeds supply can contribute to deforestation and devegetation. The demand for fuelwood, leaves and foliage used as cattle fodder, is estimated to be six times the sustainable yield of forests. Tree plantations do not support as much biodiversity as natural forests, and pulp and paper industry is a significant source of water pollution. Other plantations like coffee and oil palm can also cause environmental degradation as their cultivation is undertaken in steep terrain by means of forest clearing. In South-east Asia shifting cultivation is a menace which is causing climatic change.

Agricultural and Livestock Products

Consumption of farm produce in the form of beef has risen. As a result, the world's livestock population is growing much faster than its human population. A range of environmental problems are associated with livestock husbandry and industry, viz., leather processing. Overgrazing contributes to soil degradation and loss of vegetative cover from agricultural lands fuels the process of desertification in the vulnerable SAT regions of the world. In developing countries, chemicals used to tan hides are dumped into streams causing water pollution. Cotton, a high-valued commercial crop is also the cause of a number of local environmental problems most of which stem from the heavy use of pesticides. In India 56 per cent of all pesticides used in the country, is applied to cotton which ultimately gets into the food-chain.

These may be the general problems linked with resource consumption. One of the great challenges India faces over the next several decades is how to accelerate the pace of economic growth without exhausting the resources of growth.

Environmental Issues of Immediate Concern to India

India with its diverse resource base aims to achieve environmental health, good soil for providing food and fodder and to generate high income from its exportable cash crops. However the obstacles to achieve these are—

(a) Rate of population growth—India with a population of 1 billion persons in 2000 would attain a score of 1.4 billion by 2025 which would make it the most populated country in the world. With only 2.3 per cent of land resources and 1.7 per cent of worlds forest stock, it would be a Herculean task to achieve sustainable development.

(b) Climatic vagaries linked with south-west monsoon are plenty. Over 296 M ha m rainfall or 74 per cent of all precipitation is received in the form of south-west monsoon increasing the dependency for agricultural activities, more so in case of rainfed agriculture.

(c) Ethnic diversity is a challenge which brings in regional variations that could impede the process of sustainable development.

(d) Demographic and social landscape interwoven with economy is another matter of concern where poverty remains a serious problem which is strongly linked with health—Over 40 per cent of India's population is affected by malnutrition. More vulnerable groups are the landless agricultural labourers, urban slum dwellers and remote tribal communities.

(e) Industrial pollution is another issue of concern. The Bhopal Gas tragedy still haunts many. Most industrial air pollution is in the form of suspended particulate matter (SPM). Emissions from major industries are already on the rise especially from textile, chemicals, iron and steel, cement, fertilizer and paper. Industrial waste (inclusive of solid and hazardous waste) from petrochemical, pharmaceutical, pesticides, paint and dye, petroleum, fertilizer, asbestos, caustic soda, inorganic chemical and general engineering industries is highly uncontrolled both on public land or municipal solid waste dumping sites is common. The case of Patancheru in Medak district of AP, India provides a disturbing case study.

(f) Forest management is another issue linked with faulty policy formulation and implementation in India. Encroachment and elimination of tribals from their natural habitat are causing impediments to the cause of sustainable development.

(g) Water supply and sanitation remain major problems.

(h) Salinisation and water-logged of cropland affecting approximately over 5 to 12 m ha in Haryana, Punjab, West Bengal, Andhra Pradesh and Maharashtra is a matter of grave concern.

(i) Loss of soil nutrients is a major problem. Minor nutrients viz., zinc, copper, ferrous, manganese and boron are lacking in soil thus adversely affecting yield.

(j) Ground water depletion and pollution—Over 16 per cent of administration blocks in India have over-utilized ground water resource. Simultaneously, heavy leaching of pollutants (N_2 fertilizers) has been observed which is higher than the national standard.

(k) Indoor air pollution in rural areas is a common problem which is caused due to smoke from wood and other biomass fuels. In the urban areas the level of SPM have increased drastically especially in case of Bombay, Calcutta and Delhi (GEMS, 1993). In Bombay due to higher levels of S_2 there has been an increase in respiratory and cardiac problems.

(l) Vehicular emissions in the form of unburnt HC, is also on the rise.

Conclusion

Sustainable Development must cease to remain a slogan alone; it is essential to realize now that time is running out and that concrete action must be taken to ensure sustained development. In India a significant step in this direction would be to take action in the following direction:

1. *Effective implementation of policies* : viz., clean air, water, conservation of species, forest management etc. Although in India we have policies, at present these policies are not being implemented seriously.
2. *Effective population control* : At present the growth has dipped to 1.9 per cent, however, it is still higher than world mean of 1.4 and the Chinese mean of 0.9 per cent (Census of India, 2001). India is required to take drastic steps if the benefits of development has to reach the poorest of poor. Only then can sustainable development be ensured.

3. *Food policy and access to food:* Food production has reached 198.5 million tonnes and there is a large buffer stock spilling over. However, we still have pockets where the people are too poor to access these food resources. Hence steps must be taken to ensure that every Indian has an access to adequate nutrition and malnutrition can be eradicated. Only then the sustainable development can be ensured.
4. *Education to all :* The law on access to universal education in India is at the moment merely a slogan. The rate of school dropouts is significant and some innovative methods must be adopted to keep the students in school. Also newer avenues of vocation and employment must be opened so that the educated youth are not disillusioned by the process of development. The development process must ensure the participation of the people who have been left out from the main stream.
5. *Optimum utilization of land resource* : Land use must be based on land capability, suitability and its carrying capacity. This would conserve resources and check degradation which is at the heart of the problem of unsustained development.

REFERENCES

Alexandrtos, N. (1995) : *World Agriculture: Towards 2010.* UN-FAD & John Wiley & Sons, NY.

Balmett, V., Payner, R., and R. Steiner (1995): *Agricultural Sustainability: Economic, environmental & statistical considerations.* John Wiley & Sons, UK. 266 p.

Bossels, H. (1998): *Earth at a Crossroads: Paths to a sustainable Future,* Cambridge Univ. Press, UK, 388 pp.

Census of India (2001): *Provisional population totals.* Series I, Paper !-2000. RG&CC of India, New Delhi.

Chopra, K. & B. Goldar (2000): *Sustainable Development Framework for India: The Case of Water Resources* (Mimeo). Institute of Economic Growth, Delhi.

Costanza, R.(1997): The value of the world's ecosystem services and natural capital, *Nature,* Vol. 387, p.256

Crosson, P. & J.R. Anderson (2000): Land Degradation and Food Security : Economic impacts of watershed degradation. In Lal, R. (2000): (ed) *Intergated watershed management in the global ecosystem.* Soil and water conservation society. CRC Press. Washington DC., pp. 291- 303.

Dasgupta, P. & S. Lele (2002): Water Resources, Sustainable Livelihoods and Ecosystem Services, *Economic and Politically Weekly,* 37 (18) 1709-1711.

Doran, J.W., Sarrantonio, M. & M.A. Lieheg (1996): *Soil health and sustainability, Advance in Agronomy,* Vol. 56, pp. 2-55.

FAO (1995a): *Sustainahility issues in agricultural and rural development policies.* Vol. I: Trainee's reader. Training material for agricultural planning 38/1.

Ikerd, J.E. (1996) : *The Health of Agro-ecosystems: A Socio-economic Perspective.* First International Ecosystem Health and Medicine Symposium, Canadian Wildlife Service of Environment, Canada, Ottawa.

Lynam & Herdt, (1989): Sense and Sustainability: Sustainability as an objective in International Agricultural Research, *Agricultural Economics,* Vol. 3: pp. 381-398

WRI (1995): *World Resources—a guide to the Global Environment,* Oxford University Press. p.182;

WRI (1998): *World Resources—a guide to the Global Environment,* Oxford University Press. p. 75; p. 157; p. 186.

Re-evaluating the Sustainability of Irrigated Agriculture in Arid Zones

A Case Study of Indira Gandhi Nahar Pariyojana*

RAJESH BHAKAR

Introduction

Indira Gandhi Nahar Project, a gigantic human endeavour to transform a part of desolate Thar Desert into a land of prosperity and plenty has often been hailed by planners, administrators, and politicians as a "decisive victory over harsh nature". The transformation brought about by the project in poverty alleviation, improving agricultural productivity, providing livelihood, settling people, checking desertification, and providing drinking water, etc. has been remarkable. At a first glance, one is tempted to see in the project achievements a vindication of "Possiblist Tradition" of geographical scholarship. However, a critical examination of project performance dispels such misgiving and reveals that nearly four decades after its initiation, today the project is reeling under its own weight. While the introduction of irrigation in the arid lands of Rajasthan has opened up vast possibilities for development of agriculture, it has simultaneously posed several intriguing planning, management and environmental issues which put a big question mark over the very sustainability of irrigated agriculture in this arid environment.

World Bank is, more often than not, at the receiving end of scathing criticism by environmentalists for its advocacy and funding of multi-purpose river valley projects and for its neglect of adverse impact of these schemes on the environment. However, in the case of IGNP, after associating with the project for nearly 13 years, it became apprehensive about the success of the project and stopped further assistance. Bank Mission observed that developmental planning of IGNP was contrary to the accepted principles of irrigation demanded that further expansion of Project should be stopped. Bank desired a review and re-planning of the project, pending which it stopped further assistance (IGNB,

* This article was presented at the XXIV Indian National Cartographic Association (INCA) International Congress, held at Service City, Kolkata, from 6-9 October, 2004.

2002: 358). Interestingly, the World Bank assistance has not been revived till date.

In view of Bank Mission's observation, it was decided by Planning Commission (Government of India) to initiate about 16 planning studies. These "Planning Studies" have been commissioned to many organizations and among them studies by Water and Power Consultancy Services India Ltd. (WAPCOS), Institute of Development Studies (IDS), Operations Research Group (ORG), National Council of Applied Economic Research (NCAER), National Institute of Hydrology (NIH) and Mott MacDonald Limited (MML), are important. Review of these comprehensive studies indicates, that if the project has to survive as a viable prospect on a sustainable basis, its developmental plans have to urgently take account of the following findings:

- As the Project Command lacks drainage, extensive areas are bound to be affected by water logging and salt buildup in the long run.
- Stage-II areas are prone to strong winds: consequently aeolian processes render the water distribution network and agriculture fields vulnerable to sand encroachment.
- The assumed water resource availability, on which Project Planning is based, is not reflected in the observed flows of the Ravi-Beas river system over last 25 years.

This article is an attempt to analyze whether the current planning sincerely addresses the concerns identified above. A conscious attempt has been made to carry out this analysis by restricting, as far as possible, to the data published by project authorities and to the findings by "Planning Studies" commissioned by Indira Gandhi Nahar Board.

Waterlogging and Salinisation

Nearly all major irrigation schemes situated in desert environments are affected by waterlogging and salinisation; for example, Indus Valley Irrigation system in Pakistan; lower Colorado River lands including areas of Mexico; Nile River system in Egypt and many parts of Bhakra and Beas Irrigation System in Punjab, Haryana and Rajasthan. In fact, long-term failure is built-in feature of the irrigation projects in desert regions from the start. The ancient hydraulic civilizations of Fertile Crescent, viz., Sumeria as well as Babylonia, which relied on irrigation of crops in a desert land, withered in the long run due to salinisation, waterlogging and silt accumulation in agricultural lands (Strahler *et al.*, 2002: 593).

The command area of IGNP is much more complex and hazard prone as it lacks surface drainage, i.e., has no natural outlet to sea. Introduction of irrigation in such an area means an annual input of around 6-7 million acre feet (MAF) of water. Not all of this evaporates and the result is that command area has turned into a huge sink. Within a few years of start of canal irrigation the effect on ground water regime became visible in form of rising water table and instances of waterlogging in patches. Systematic ground water level observations were initiated in Stage-I of the command in 1981 and based on their analysis *Review of North West Drainage Study* indicated, that in year 1995-96 around 10 per cent of Culturable Command Area (CCA) of Stage-I became waterlogged. It also reports that potentially sensitive area has increased from 9 per cent in 1981 to 25 per cent in 1990, and expressed the fear that by year 2000 it may go up to about 50 per cent of the command, if no drainage solutions is provided for (Mott MacDonald, 1999 e). The *Ground Water Wing* of Command Area Development Authority (CADA, 2003) has observed following trends in regional ground water table:

- A rise in water table of up to 1.17 m/year was observed during the period 1972-82.
- During the period 1982-92, the net rise in water table has been reduced to 0.76 m/year but a large area came under potentially sensitive and critical categories. The rate has reduced because sizeable areas were waterlogged, or became critical, enabling more water to evaporate from soil profile.
- In the decade of 1992-2002, the trend has been fluctuating, initially water table fell during 1992-94 due to scanty rainfall and meager flow in canals, then it rose during 1995-97, then again there was a decline. But there has been a steady rise throughout the period in the regional water table in areas away from canal.

The latest information available, presented in Table 24.1, indicates that since1997-98 the waterlogged and critically sensitive areas have declined substantially. But there is not much to cheer about it as much larger areas have gone out of cultivation due to scanty rainfall and very poor availability of water in the conveyance system (this is discussed later).

In addition to the water imbalance, there is question of salt imbalance, although the quality of water in Indus River system is first class, there is nonetheless 700-ppm salt count and the annual volumes of 6-7 MAF are huge. It has been estimated that IGNP brings in about 1,500,000 tonnes of salt a year within the irrigation water supply (Mott MacDonald, 1999a: 11-12). The uptake of salts by plants is negligible, i.e. there is no loss of salts by evapotranspiration. The system as it stands, has absolutely no way of rejecting the buildup of salts.

Table 24.1 : Extent of Water Logged Areas in IGNP Command

(Area in Hectares)

Types of Area		*92-93*	*93-94*	*94-95*	*95-96*	*96-97*	*97-98*	*98-99*	*99-00*	*00-01*	*01-02*
Potentially Sensitive Area (Water table within 1.5 to 6.00 mtrs.)	Stage-I	202960	160760	98643	207300	297820	310056	2987600	280023	225153	179170
	Stage-II	—	—	—	—	17303	18067	18067	17500	12184	18304
	Total	202960	160760	98643	207300	315123	328123	316827	297523	237337	197474
Critical area (Water table within 1.00 to 1.50 mtrs.)	Stage-I	22000	17760	18970	20670	24140	28760	27960	26430	13425	11355
	Stage-II	—	—	—	—	3610	3792	3792	3700	2229	1261
	Total	22000	17760	18970	20670	27750	32552	31752	30130	15654	12616
Water logged area (Water table within 0.00 to 1.00 mtrs.)	Stage-I	13750	9680	10192	14750	17220	22008	19492	19051	12672	10098
	Stage-II	—	1000	800	1000	1243	1243	1243	1190	369	78
	Total	13750	10680	10992	15750	18463	23251	20735	19340	13041	10176

Source: CADA, 2002-2003.

However, salinity of soil is not perceived as an immediate problem in IGNP area as capillary rise of water in sandy soils is minimal, except where secondary salinization has developed due to waterlogging problems.

The response of Indira Gandhi Nahar Board (IGNB) to the environmental hazards of waterlogging and salinization is most interesting and appears contradictory. First, taking note of the seriousness of the problem, IGNB have decided to spread water as thinly as possible in the largest possible area. As Rajasthan is "water scarce" and "land surplus" state, the decision to distribute this "scarce" water as extensively as possible, serves the government well, as it

Table 24.2 : Proposed Water Allowance and Irrigation Intensities

Stage	*Proposed Water Allowance*	*Proposed Irrigation Intensity*	*Existing Water Allowance*	*Existing Irrigation Intensity*
Stage-I	3.5 cusec/000 acre	100 %	5.23 cusec/000 acre	110%
Stage-I Lift	3.0 cusec/000 acre	80 %	3.0 cusec/000 acre	100%
Stage-II Flow	3.0 cusec/000 acre	80 %	3.0 cusec/000 acre	80 %
Stage-II Lift	2.0 cusec/000 acre	60%	2.0 cusec/000 acre	60%

Source : GOR. 2003. Indira Gandhi Nahar Project: Annual Report 2001-2002 & 2002-2003.

caters to the ideal of "equity" and "well-being of as large a number of people as possible". To put this into effect they have reduced water allowance and irrigation intensities.

Next, IGNB argues that under currently redesigned "water-allowance" and "irrigation intensities", chances of development of waterlogged conditions are slim. It claims, "considering the large depth of ground water table varying between 30-60 m in greater part of command and depth of annual irrigation supplies of about 510 mm; when compared to average annual potential evapotranspiration of the area between 1800-2500 mm; it is difficult to evaluate and ascertain the need of drainage arrangements that would be needed. Hence it is better to monitor the ground water system effectively, and plan the drainage as and when required" (GOR 2004, p. 9). However, these formulations of IGNB are simplistic, as they have conveniently ignored the hydrogeology of the Project-Command. In Stage-II, where most of the expansion is planned and which will have about 71 per cent CCA as against 29 per cent of CCA in Stage-I, the potential hazard of water logging is even more threatening. Because on one hand, the coarse textured highly permeable soils will lead to excessive deep percolation of water supplied by irrigation, and on the other, large areas are underlain by impermeable layers of quaternary fluvial deposits or "hard-pan" at shallow depths, which will give rise to perched water table. The *Ground Water System Study of Pilot Areas: IGNP Command Stage-II* reported that about 33.4 per cent of flow area and 76.4 per cent of lift areas have hardpan or impermeable layer between 0-10 m from ground level (ORG, 1989: 21). Command expansion work is going on and the project, which originally had no provisions of lift schemes, now has seven lift schemes and has even been extended to parts of Desert National Park, where a perched water table develops even in rain-fed conditions.

Areas with hardpan up to 5 metre below ground level (mbgl) are prone to get waterlogged and hence unsuitable for agriculture (WAPCOS, 1993: XII-4). Initially IGNB decided to exclude from allotment areas with hardpan up to 5 mbgl (Rathore and Swaminathan, 1991: 71). Surprisingly, later on it was decided to allot areas having hardpan between 1-5 mbgl, if their radial hydraulic conductivity is more than 2 mm per day (IGNB, 2002: 201). The State Government in its latest decisions has decided to consider only that area as prone to get waterlogged where the hydraulic conductivity of hardpan between 1-5 mbgl is less than 1 mm per day (IGNB, 2002: 175). There is no scientific basis for such categorization, as the "Soil Irrigability Classification" of IGNP command is based on top 1.5 metres of soil without any consideration of drainage, on the presumption that drainage will be provided (Mott MacDonald, 1999b). It is not clear as to what has prompted these logic-defying decisions, and these need to be investigated.

Sand Encroachment Hazard

The enlarged scope of the project, with the inclusion of 6 lift schemes and increased CCA in flow command of Stage-II, is firmly rooted on the premise that introduction of canal water in this region would result in replications of the successes achieved in the Bhakra and Gang Canal Commands in Sriganganagar and Hanumangarh districts of Thar Desert. A developmental plan based on this premise may in all likelihood go astray, because it fails to take account of the fact that no two parts of a desert are similar. In fact, aridity is the only common denominator; otherwise arid environments are extremely diverse. Various detailed investigations (WAPCOS, 1993; ORG, 1989; Mott MacDonald, 1999) of IGNP command have revealed that even Stage-I and Stage-II are far from similar in terms of landforms, soils, hydrogeology, and water-balance. For example, the soils of Stage-I of IGNP are part of a former flood plain of Indus river system mixed with sandy aeolian deposits. As one proceeds towards South-west into Stage-II, the soils encountered are sandy and coarse textured. Similarly, the landforms of IGNP show a great variation ranging from the young alluvial plains of Ghaggar in upper reaches of canal, giving way to rather flat old alluvial plains in and around Bikaner. These in turn transforms into a complex system of "semi-stable longitudinal, parabolic and transverse dunes over 50 m tall", mixed with "mobile bare drifts and barchanoids" in Stage-II area of Jaisalmer district. Even the depth of hardpan layer is up to 15 mbgl in Stage-I and rises towards south-west and near the tail of Stage-II; it sometimes is almost at the surface. Potential Evapotranspiration values, an important indication of crop-water requirements, are around 1825 mm/year for Sri Ganganagar, rise to 2151 mm/year in Bikaner and then further increase to 2527 mm/year in Jaisalmer. (WAPCOS, 1993: IV-42). The difference between Ganganagar and Jaisalmer is a huge 700 mm/year, with Jaisalmer recording higher values. Ironically, Stage-I areas in Sri Ganganagar have substantialy higher water allowance and irrigation intensity than Stage-II area of Jaisalmer.

However, from the standpoint of irrigated agriculture it is the vigour of aeolian processes in Stage-II, which transforms it into an altogether different world. Mean wind velocities in Jaisalmer and Phalodi in the months of June-July are 24-27 km/hour, while in Sri Ganganagar and Bikaner the range is 9-10 km/hour and 11-13 km/hour respectively (Sanyu and WAPCOS: 2-19). Wind with speed above 15km/hour is a very effective agent of sand movement. Strong winds and undulating topography of Stage-II command expose it to hazard of desertification. Even the stable sand dunes have a high risk of being reactivated under the slightest biotic pressure or mismanagement. Paradoxically; one of the major objectives of IGNP is to check the extension of desertification.

The menace caused by strong winds is greatly augmented by agricultural activity. Soil loss from a field ploughed with a disc plough is three times more than the field tilled by country plough or "Desi" plough and seven-time more than fallow land (Dhir, 1987). Farmers have to level their land as preclude to irrigation by "surface methods". Land levelling in this region means levelling of the low dunes and, thereby, loosening a greater thickness of sandy bed over a wider area. Such activities, without an assured and ample supply of water to the disturbed lands, undoubtedly lead to accelerated aeolian hazard resulting into formation of new barchanoid forms which engulf cultivable flat areas with mobile sand (Mott MacDonald, 1996: 9).

In addition, there are probably insuperable problems with maintaining water distribution channels open in the dunal areas. The canals in the fill section across the winds behave almost like a wall to the flowing winds, resulting in low-pressure area inside the canal where sand may get deposited. The low-pressure area is reduced if the canal is full (of water) and the sand deposition inside is less. In an empty canal it is much more. The only possible remedy is protection by broad wind breaks/shelter belts which reduce wind speed considerably. The Indira Gandhi Main Canal and its major branches and distributaries are less affected, since broad wind breaks/shelter belts protect them. In the case of minors the situation is different, the shelter belts are narrow, sometimes only 2-3 rows wide. These are also subjected to closures for longer duration. The condition of on-farm watercourses is bad, they are usually above the surface level, and have no shelterbelts. The combined affect of empty channel and higher wind velocity chokes these small canals and watercourses leading to subsequent reduction in carrying capacity and blockages. These constraints place some of the Stage-I (Phase II) area and much of the Stage-II area in the non-irrigable class by surface methods.

CCA of Stage-II is undulating and consists of small patches of plain area between dunes. The allottees discover that only about 25-30 per cent area is plain and rest is impossible to irrigate by surface methods. Expansion of Project in such an inhospitable terrain could be done only with the presumption, that farmer will be able to change the irrigation practice from "basin or flood irrigation system" or surface methods (conducive to higher water allowance) to "sprinkle or drip irrigation practices" or low volume measures. However, these new techniques require expensive operating overheads and technical skills. Ironically, majority of allottees is poor landless labourers. Economic analysis has shown that individually farmer have little to gain financially from "low volume measures" as the installation and operating costs are high and offset the gains made by increased intensity of irrigation. Fuel requirements for pump operation are the major hurdle (Mott MacDonald, 1999g). Fuel costs have escalated very sharply, with the international crude oil prices touching an unprecedented high of $53 a barrel during the year 2004, making this proposition even more unlikely. This means much of the Stage-II area is in the non-irrigable class by any method. This argument finds support from the statistics of adoption of sprinkle irrigation system in the command. The total number of allottees is around 175,000 (80,000 in Stage-II), and less than a mere thousand are using sprinkler system, despite the fact—25 per cent of total costs are provided by the government as subsidy (CADA, 2003: 16).

Water Resource Availability

IGNP Command suffers from lack of reliable irrigation supplies. Given the drastically reduced water allowances and irrigation intensities—in a very harsh environment, the margins for farmers are very thin. Erratic water supplies for even one year mean economic devastation for settlers. An impression has been created by Project Authorities that reason for water shortfall is the "tail-end syndrome" along with low rainfall in the catchments of Ravi-Beas Rivers. They blame Government of Punjab for not releasing water in required amount at crucial periods in the agricultural season. In support of this argument they (IGNB) present statistics of water indented and water released. However, it appears that major share of the blame should go to planning flaw on part of IGNB.

Current planning is based on an estimated availability of 7.59 million acre feet (MAF) of water. The water for IGNP is made available from Rajasthan's share in the waters of Ravi-Beas rivers. Rajasthan's share is determined on the basis of 1981 inter-state agreement between states of Punjab, Haryana, and Rajasthan. This agreement, based on the 1921-60 flow series of Ravi-Beas Rivers, estimated that the mean flows, at the rim-stations of—Madhopur on Ravi and Mandi plains on Beas—, to be 20.56 MAF. After deducting the long established pre-partition uses of 3.13 MAF and losses in transit, the net sharable surplus for allocation among states was 17.17MAF. This sharable surplus resource was allocated among party states on *pro rata* basis in the following fashion :

Table 24.3 : Proportional Share of Parties to 1198 Inter-State Agreement

State	*Water in MAF*
Punjab	4.22
Haryana	3.50
Rajasthan	8.60
J&K	0.65
Delhi	0.20

Source : WCCR, 2002 : 198.

These entitlements form the basis for actual allocation of available water in a particular season by the regulator—Bhakra and Beas Management Board (BBMB). The Govrnment of Rajasthan has fixed share of IGNP at 7.59 MAF out of its total entitlement of 8.6 MAF. Since this allocation is based on 'mean' or 'average' flow values at rim stations, it is fraught with problems on a number of accounts.

- Mean or average values are poor indicators of water resources available because river flows fluctuate not only seasonally but also have year-to-year variations. A significant part of mean flow is summer flood (including occasional very large flood) when the discharge is high, which is recorded at rim stations, but can't be stored by reservoir systems in place and flows down to Pakistan. For example, during 1988-89, the inflows in Ravi-Beas Rivers were 25.526 MAF, but less than 18 MAF could be used and the "sharable surplus" was only 14.55 MAF (WCCR, 2002: 115). It is substantially less than the 17.17 MAF assumed as the sharable surplus by inter state agreement of 1981. Similarly, in the year 1978, while the inflows were 27.2 MAF, after allowing for pre-partition uses, only 13.1 MAF was the resource that could be shared even with the best possible operation of the reservoir system, actual utilization was only 11.3 MAF (Mott MacDonald, 1999 a: 7-9).
- Even seasonal fluctuations show considerable variation over years. For example, the middle of May is an important period in the agricultural

rotation, but during May, 11-20, the discharge in Beas River may be more than 18000 cusecs in one year and less than 3500 cusecs in another. Likewise, in the last ten days of September, another crucial period for crops growth, the rate of flow in Beas in any one year may be as high as 68,000 cusecs or as low as 11,000 cusecs (Gulati, 1973: 30).

- Govrnment of Punjab has recently claimed that total availability of surplus water in the Ravi-Beas Rivers has reduced from 17.17 MAF (assumed under the agreement of 1981) to 14.37 MAF as per the flow series of 1981-2002 (GOP, 2004: 9).
- The actual availability of resource of IGNP for the years 2000-01 to 2003-04 has been 4.5 MAF, 5.8 MAF, 4.69 MAF and 5.35 MAF respectively, and current year's availability of water in the reservoirs are feared to be the lowest ever.
- Mott MacDonald (1999d), after a very detailed investigation of 'controlled surplus waters' available to be shared indicate that in the median year some 6.1 MAF is the discharge that could be expected to be the entitlement of the project. This would reduce to 5.6 MAF if one-year shortfall were used as the planning criterion.
- Surprisingly, against normal dependability criteria of 75 per cent, the IGNP system has been designed for less than 50 per cent dependability for irrigation use. This, to put simply, means that at least half the years water availability would be deficit (Kavadia, 1994: 304).

Thus it becomes amply clear that the figure of 7.59 MAF derived from mean flow, is not very useful for planning purpose. The planning criteria have to be realistic and present figure is highly unreasonable and unfair for the settlers of IGNP command. According to latest Project revisions, 2.19 MAF is to be used in Stage-I, 0.87 MAF is reserved for drinking water, and 4.21 MAF will be provided for Stage-II. However, the above analysis of realistically available resource clearly indicates that the actual resource available for Stage-II, is more than often going to be considerably less than the 4.21 MAF, envisioned for Stage-II, if the entire command of Stage-II of 14.10 lakh ha is to be provided with designed supply.

Ground Reality of IGNP Command

Only, a couple of instances of actual developments in IGNP command is sufficient to highlight the mismatch between proposed plans and ground reality. As is evident from information collected from various sources and presented in Table 24.4, in the year 1999-2000, the irrigation potential created through water courses construction was 8.94 lakh ha and utilization was 9.81 lakh ha (due to higher irrigation intensities or double crops in some areas) which, is the highest recorded utilization till date. Water available for IGNP that year was 7.63 MAF, which, incidentally, is the highest ever availability for the project. Three years later, in 2002-03, the irrigation potential created increased to 9.90 lakh ha, but actual utilization was only 4.73 lakh ha. In Stage-II utilization was only 0.85 lakh ha against created potential of 4.7 lakh ha. The reason—water availability reduced to 4.69 MAF (WCCR, 2002: 14, CADA, 2003: 21). During the year 2003-04, water availability improved slightly and irrigated area statistics improved.

Again, for year 2004-05, the story of 2002-03 appears to be repeating. As per data released by BBMB due to low rainfall in the catchment of Ravi-Beas Rivers, the level of water—Pong, Ranjit Sagar, and Bhakra dam—is at a historic low (Sharma, 2004). This year the *Kharif* season irrigation was restricted to only 1.62 lakh ha. Situation in Stage-II is an eye opener; against a created potential of 4.70 lakh ha merely 3589 ha could be utilized (Agriculture Extennsion Wing, CADA: Personal Communication). Availability of water for *Rabi* is estimated to be around 2.04 MAF for entire IGNP command. This has resulted in the situation that, even if the canals are divided in four groups and only one group is supplied water on the basis of either 3.5 day or 7-day *warbandi*, the irrigation water would be available only up to 27 January, 2005, and after that only drinking water could be made available. This means a water course will run only for seven days a month, instead of assumed 16 or 17 days a month for the entire season. This is a big let down. One may say, these are exceptionally dry years, true, but this situation has developed when only about 9.9 lakh ha of irrigation potential has been created through watercourse construction and one must not forget that still around 5 lakh ha irrigation potential is yet to be created, and in future, its requirements would have to be supplied with. Only IGNB can answer how will it cater to those demands.

Interestingly, drinking water is a priority and provision has been made on the basis of "Planning Commission" norms of rate of overall population growth at rate of 2.65 per cent for the period 1981-2031. However, whether the adoption of overall rate of growth is a right index to determine the drinking water requirements is an issue that needs to be looked into. As IGNP is a resettlement project, the rate of population growth in the command would be far higher than the overall population growth (Ramanathan and Rathore, 1991:7).

Time and Cost Overruns

The Project has been revised and expended several times on one pretext or another. This expansion has resulted in unprecedented time and cost overruns. Original scheme of 1957, [based on preliminary surveys, envisaged CCA of

1.3 Million hectares (Mha)] was to be completed within 10 years of initiation, at an estimated cost of Rs. 66.67 crore (Rs. 666.7 Million). In 1963, Survey of India, conducted extensive surveys and project was revised. This comprehensive plan provided for CCA of 1.49, for a total sum of Rs 184 crore. Project has undergone several revisions since then. According to the latest revision (GOR, 2004), the envisaged CCA has been extended to 1.9 Mha (.553 Mha in Stage-I and 1.41Mha in Stage-II, and is expected to be complete only by the year 2008) and estimated costs have escalated to a whopping Rs. 7139 crore (Table 24.5). Up to March 2003, Rs. 3798 crore has been spent on the project. It is evident from Table 24.5, that on Stage-I, which account for around 70 per cent-85 per cent of utilization of created potential on average, only Rs. 666 crore have been spent, while the expenditure on Stage-II is Rs. 3329 crore. Although it is crystal clear that there is not enough water even to irrigate the already opened up CCA, IGNP authorities are continuing with the expansion and have plans to spend Rs.3257 crore more on Stage-II.

Table 24. 4 : Expenditure Incurred on IGNP

	Stage-I				*Stage-II*				*Total*			
	Canal Works	*CAD*	*Total*	*% of 11*	*Canal Works*	*CAD*	*Total*	*% of 11*	*Canal Works*	*CAD*	*Total*	*%*
	(1)	(2)	(3)	(4)	(5)	(6)	(7)	(8)	(9)	(10)	(11)	(12)
Expenditure upto March 2003	388.5	286.2	674.7	17.76	2056.8	1067.0	3123.8	82.24	2445.3	1353	3798.5	100
Estimated Costs (1992 prices)	377	286	663	16.32	2267	1132	3399	83.68	2644	1418	4062	100
Estimated Costs (After price escalation)	473	286	759	10.63	3522	2858	6380	89.37	3995	3144	7139	100

Source : Government of Rajasthan, Project Profile, 2004.

Conclusion

Without adequately addressing the drainage needs and sand encroachment hazard, IGNP planners have resorted to project expansion, with a system of varied water allocation in different reaches of the command; much beyond what could be adequately supplied with available water resource. The vital issue for the economic viability of Stage-II is given a certain rate of water supply to a farm of 6.32ha, can the farmer raise crops with sufficient yield to make an adequate living. The uncertain availability of water for IGNP makes this proposition highly unrealistic.

These facts raise question—is it right to extend the project much beyond what is feasible, raise expectations of the people and alter the existing socio-economic processes and expose them to unforeseen economic and ecological disequilibrium. It appears that there are compelling reasons for some serious rethinking on the sustainability of the agrarian strategy adopted by IGNP authorities. Evaluating the performance of Multipurpose River Valley Projects, D'Souza *et al.* (1998:297), have observed that, "...the complete lack of a comprehensive audit at post-construction stage and monitoring of costs and benefits actually accruing ex-post has allowed the Project authorities to flaunt the ex ante data as an example of successful implementation". Present study reinforces this observation. A detailed investigation of the Project's performance *vis-à-vis* 'victory over nature' claims is urgently needed to ascertain the social, economic and environmental costs and benefits actually accruing.

Gilbert White, the leader of environmental hazards research in Geography, drawing upon his experience in river basin planning, found that planners emphasized 'technological fixes' (dams, for example) and left the possibilities for behavioural changes with respect to hazards, for example, floods, largely unexplored. For White (1957:129), usually such situations are result of separation of knowledge and practice. But the planning process in the case of IGNP itself seems to be separated from knowledge and succumbing to political pressure and vested interest.

REFERENCES

CADA. 2003. *Indira Gandhi Nahar Project: Annual Report of Integrated Development – 2002-2003.* Bikaner : Commissioner Command Area Development.

Dhir, R.P. 1987. "Flux in Indian Desert", In *Desert Ecology,* ed. I. Prakash. Jodhpur: Scientific Publishers.

D'Souza, R. Mukhopadhyay, P. and Kothari, A. 1998. "Re-evaluating Multi-Purpose River Valley Projects: A Case Study of Hirakud, Ukai and IGNP". *Economic and Political Weekly* 33 (6): 297-302.

Government of Punjab.2004. "The Truth behind Punjab's Waters". Advertisement in *The Times of India.* New Delhi: July 17.

GOR. 2003. *Indira Gandhi Nahar Project: Annual Report 2001-2002 and 2002-2003.* Jaipur: Government of Rajasthan.

GOR. 2004. *Indira Gandhi Nahar Project: Project Profile.* Government of Rajasthan.

Gulati, N.D. 1973. *Indus Water Treaty—An Exercise in International Mediation.* Bombay: Allied Publishers.

IGNB. 2002. *History of Indira Gandhi Nahar: A Venture to Turn Desert of Rajasthan into Granary.* Bikaner: Indira Gandhi Nahar Board.

Kavdia, P.S. 1994. "Perspective Plan for Water Management in Indira Gandhi Canal Project". In *Planning for Sustainability in Irrigation,* eds. R. Hooja and P.S. Kavadia, pp. 303-314. New Delhi: Rawat Publications.

Mott MacDonald Limited. 1996. *IGNP Studies for the State of Rajasthan:Interim Report 6- Afforestation for Sand Deposit Control and Biodrainage.* Government of Rajasthan.

Mott MacDonald Limited. 1999a. *IGNP Studies for the State of Rajasthan : Main Report.* Government of Rajasthan.

Mott MacDonald Limited. 1999b. *IGNP Studies for the State of Rajasthan : Annex-A—Sustainable Land Irrigability Classification.* Government of Rajasthan.

Mott MacDonald Limited. 1999c. *IGNP Studies for the State of Rajasthan : Annex C—Drainage Study.* Government of Rajasthan.

Mott MacDonald Limited. 1999d. *IGNP Studies for the State of Rajasthan : Annex F—Review of Operational Technology for Supply of Ravi-Beas Water to IGNP.* Government of Rajasthan.

Mott MacDonald Limited. 1999e. *IGNP Studies for the State of Rajasthan: Annex G—Review of North West Drainage Study.* Government of Rajasthan.

Mott MacDonald Limited. 1999f. *IGNP Studies for the State of Rajasthan: Annex H—Environmental Review.* Government of Rajasthan.

Mott MacDonald Limited. 1999g. *IGNP Studies for the State of Rajasthan : Supporting Report 2—Assessment of the Financial and Economic Returns from Sprinkler Irrigation and Subsurface Drainage in IGNP.* Government of Rajasthan.

ORG. 1989. *Groundwater Systems Study of Pilot Area: IGNP Command-Stage-II.* Baroda: Operations Research Group.

ORG. 1997. *Settlement Motivation Programme in IGNP Command.* Baroda: Operations Research Group.

Ramanathan, S. and Rathore, M.S. 1991. *Equity and Productivity Issues in IGNP—A review of Literature.* Jaipur: Institute of Development Studies.

Sanyu Consultants and WAPCOS. 1993. *Remedial Measures Project (Phase-I) for Indira Gandhi Main Canal.* New Delhi: Sanyu Consultants and WAPCOS.

Sharma, H.B. 2004. *Fansley Ganva Chukey Hain; Aab Beenjon Ki Bari Hain* In *Danik Bhaskar.* Bikaner: 15 September.

Strahler, A. and Strahler, A. 2002. *Physical Geography:Science and Systems of the Human Environment.* Singapore: John Wiley and Sons.

WAPCOS. 1992. *Indira Gandhi Nahar Project, Stage-II : Feasibility Report.* New Delhi: Water and Power Consultancy Services.

WAPCOS. 1993. *Ecological and Environmental Studies for Integrated Development of Indira Gandhi Nahar Project, Stage-II : Final Report.* New Delhi: Water and Power Consultancy Services.

WAPCOS. 1996. *Study of Availability of Water in IGNP.* New Delhi: Water and Power Consultancy Services.

WCCR. 2002. *Important Data and Decisions of Ravi and Sutlej Complex, IV Edition.* Chandigarh: Water Controller Cell for Rajasthan, Government of Rajasthan.

White, G.F. 1957. "The Changing role of Water in Arid Lands". Reprinted in *Geography Resources and Environment Volume I: Selected Writings of Gilbert White,* eds. R.W. Kates and I. Burton, pp.125-142. Chicago: University of Chicago Press.

25

Groundwater Quality of Various Villages Near a Fertilizer Plant in Kota and Baran Districts

MEENAKSHI VIJAY, SURESH C. SHARMA, ASHU RANI AND KALYAN M. HIRONI

Groundwater is one of the major source of potable water in rural areas of Kota district. Ground water quality survey is conducted in villages around one of the largest nitrogenous fertilizer urea manufacturing industry situated approximately 35 kilometres away from Kota. All these villages are situated within the 10 km periphery of the industry in all four directions. Ground water samples are collected from various sources such as handpumps, openwells and tubewells during 2000-2001. Approximately 10 samples from each village are collected and analyzed for various physico-chemical parameters such as pH, Turbidity, Conductivity, T.D.S., Total Hardness, Chlorides, Sulphates, Nitrates, Calcium and Magnesium ions. Forest cover flora and fauna are also studied during the survey.

Villages under study—surveyed villages covered under two districts and three thesils of Rajasthan. Two districts are (1) Kota and (2) Baran. In Kota district 12 villages and in Baran district 4 villages are selected (Fig. 25.1 and Table 25.1).

Geographic, Physiographic and Climatic Features

District Kota and Covered Villages

District Kota is situated in South-eastern part of Rajasthan between 23.56° to 25.51° North latitude and 75.37° to 76.38° East longitude. It has dry climate sub-humid. Temperature ranges from 7.0 to 48°C. Relative humidity recorded for the study year is 39 per cent, normal average rainfall is 80.87cm and real rainfall is 53.65cm. Wind velocity varies from 2-20 km/hr.

Total area of Kota district is 5198.15 square km which includes urban area of 274.64 square km and rural area of 4923.505 square km. Total population is approximately 12.5 lakhs which is 1.5 per cent of the total population of

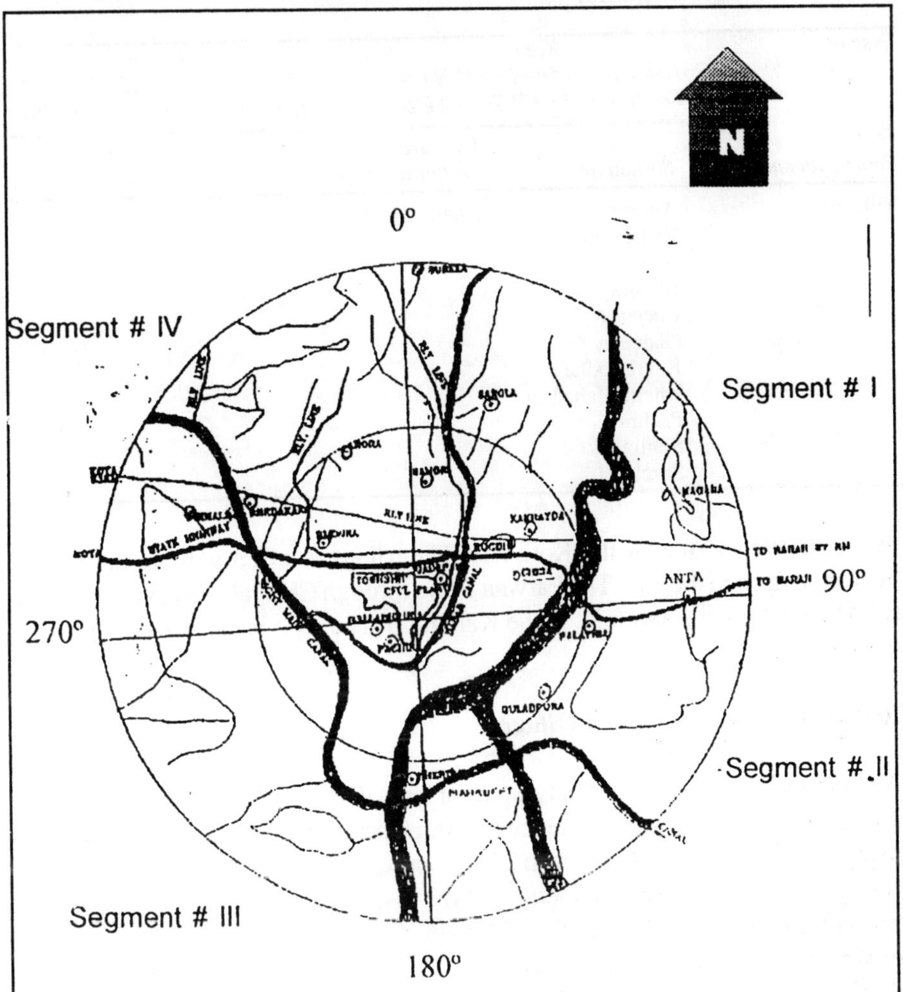

Fig. 25.1 : Fertilizer Plant and Environs

Rajasthan. Population density is found 235 per square km by the office of the District Statistical Department, Kota. Out of total population of Kota district, Tehsil Digod is having 99052 rural population and Tehsil Ladpura is having 85039 persons.

Soil in the region present a great variety, viz. driftsand, sandy loams, silts, sandy clay and clay loams (having *kanker* layers with adequate alkali and lime).

The Chambal is the principle perennial river of the district. Its tributries include the Kalisindh, the Parvati, the Parwan and the Andheri. Chambal enters this district from the west. The Parvati is about 65km in length and enters the

Table 25.1 : Villages under Study

District	Kota Latitude: 23.56°-25.51° North Longitude: 75.37°-76.38° East		Baran Latitude:24.25°-25.25° North Longtitude : 76.12°-77.26° East
Tehsil Panchayat samiti	Digod Sultanpur	Ladpura Ladpura	Mangrol Anta
Villages	Amora Ballabhpura Bamori Bhonra Cheensa Gadepan Kakrawadha Khelri Mahadeet Pachra Simlia Surela	Khera-Kheri	Anta Gulabpura Nagda Palaitha

district in the south near the Kota city. The tributries of Parvati are the Sukri, the Banganga and the Kul. The Parwan flows through Chhipabarod, Atru, Sangod and Mangrol tehsils and falls in the Kalisindh. The Ujar is another tributory of the Kalisindh.

District Baran and Covered Villages

The district Baran was previously a part of Kota district under the Hadoti region of Rajasthan. It came into existence on 10th April, 1991. It is located on south-eastern border of the state between 24°25′ and 25°25′ north latitudes and 76°12′ and 77°26′ east longitude. The length of the district from west to east is 120 km and width from north and south is 110 km. Climate of the district is dry sub-humid and average rainfall is 82.0 cm. The temperature fluctuates between 9.5°C to 25°C in winter and 28°C to 45°C in summer.

The total geographical area of the district is 69965 square km. The total population of the district is 810.32 thousand persons. Rural population is 84.8 per cent while urban population is 15.2 per cent. Population density is 92 and 116 for rural and urban areas respectively.

The soils of the district are mainly kacchari dark black and laterite. The perennial rivers in the region are Kalisindh, Chambal and their tributaries.

Land use pattern of two districts covering areas of villages under investigation shows an increasing trend in forest and cultivated area. Fig 25.2 shows land use patterns of two districts as per the data collected from Rajasthan Government sources.

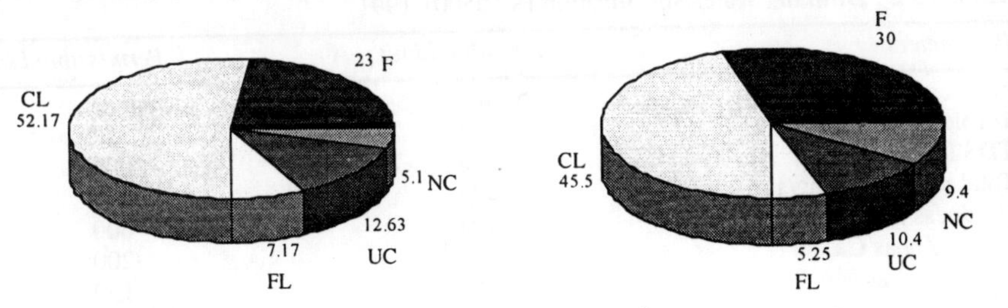

Fig. 25.2 : Land use Pattern (As per 1999 Govt. data)

Groundwater Analysis Results

Samples were collected from handpumps, openwells and tubewells from each village. Samples are analyzed as per the standard APHA methods of water analysis. Water quality parameters such as pH, Turbidity (NTU), Conductivity (umhos/cm), Total dissolved solids (T.D.S.) (mg/1), Total alkalinity (mg/1), Total hardness as $CaCO_3$ (mg/1), Ca, Mg(mg/1) as Cl, SO_4^{2-} (mg/1) as SO_4^{2-}, NO_3 mg/1 as No_3^-, NO_2(mg/1) as NO_2^- are determined for each sample. Results of Table 25.2 gives the average value of these parameters for each village.

On comparing the results of present study with Environmental Impact Assessment conducted earlier, we find the increasing trends in almost all the phsico-chemical parameters. However, the positive increase finds in forest and cultivated area implies a better affect of fertilizer plant on the soil quality.

The ground water which is also used for drinking purposes is not found free of pollution in all the villages. Higher values of anions like Chloride, Sulphates and Nitrates in compare to Indian Specifications (Table 25.2) make the groundwater unhealthy, as they can't be removed by simple filtration or settlement techniques of purification of water which is normally adopted in the study area.

It is also observed that water samples collected from agricultural fields are having higher anions. We can conclude hereby that fertilizer plant may cause more of air pollution through fumes and flyashes than ground water or soil pollution. Water analysis results do not show any abnormal changes in the values as compared to the other areas away from the impact periphery of the fertilizer plant. Secondly, any impact of water quality will be more on surface water due

Table 25.2 : Drinking Water Specification IS 105001 1991

Parameters	*Desirable Limit*	*Permissible Limit*
pH	6.5 - 8.5	none
Turbidity (NTU)	5	10
TDS (mg/1)	500	2000
Total alkalinity (mg/1)	200	600
Hardness as $CaCO_3$	300	600
Ca (mg/1) as Ca	75	200
Mg (mg/1) as Mg	50	150
SO_4 (mg/1)	200	400
NO_3 (mg/1)	45	100
NO_2 (mg/1)	—	1.0

to settlement of polluted particulate load than on ground water. The further investigation on surface water and leaching possibilities of pollutants through soil will give detailed information of water pollution in the area.

REFERENCES

1. Gujral, Seema, Vikas Sharma and Ashu Rani, Assessment of Dispersion of Ambient Suspended Particulate Matter and Meteorological Conditions near Kota Thermal Power Station. *Indian J. of Environ. Protec.*, 21 (3), 2001, pp. 238-249.
2. *Office Records,* Office of Statistical Department, Kota.
3. *Standard Methods for the Examination of Water and Waste-water,* Published by American Public Health Association.

26

Influence of Water Availability on Powerty in India

R.B. SINGH AND S. SEN ROY

Introduction

One of the major determinants of the quality of life of the people in a particular area is the optimum development and efficient and utilization of water resources. Water is required for agricultural production, for drinking purpose by human beings and animals, for municipal and commercial uses. Poverty is a multidimensional social phenomenon. One's status and location affect perceived causes of poverty. Definitions of poverty and its causes vary by gender, age, culture, and other social and economic contexts. Many factors converge to make poverty a complex, multidimensional phenomenon. Poverty is routinely defined as the lack of what is necessary for material well being—especially food but also housing, land, and other aspects. Poverty is the lack of multiple resources leading to physical deprivation. Poor people's definitions reveal important psychological of poverty. Poor people are acutely aware of their lack of voice, power, and independence, which subject them to exploitation. Their poverty also leaves them vulnerable to rudeness, humiliation, and inhumane treatment by both private and public agents of the state from whom they seek help. Their inability to fully participate in community life leads to a breakdown of social relations. The absences of basic infrastructure—particularly, road, transport, water and health facilities have emerged as critical. In India about 525 million people are living below the $1/day per capita income level which is one of the highest in the world (Table 26.1). This has been attributed to various reasons both physical and human. Some of factors are uneven relief, low levels of technology, predominantly agarian society combined with high dependency of agricultural productivity on the varying seasonal rainfall conditions.

Study Area

India is a land of many rivers and mountains. Its geographical area of about 329 m. hec. is criss-crossed by a large number of small and big rivers. The rivers and mountains have a greater significance in the history of Indian cultural

Table 26.1 : The world's poorest nations

Country	*People below $1/day poverty line (millions)*
India	525.0
China	353.0
Bangladesh	36.4
Nigeria	31.8
Brazil	30.9
Indonesia	30.7

Source : Computed from UNDP data in HDR, 1999.

development. The presence of the great mountain mass formed by the Himalayas and its spurs on the north and of the ocean on the south are the two major influences operating on the climate of India. India has a very great diversity and variety of climate and an even greater variety of weather conditions. The climate ranges from continental to occenic, from extremes of heat to extremes of cold, from extreme aridity and negligible rainfall to excessive humidity and torrential rainfall. Rainfall in India is dependent in differing degrees on the south-west and north-east monsoons, on shallow cyclonic depressions and disturbances and on violent local storms which form regions where cool humid winds the sea meet hot dry winds from the land and occasionally reach cyclonic dimension. Evaporation rates closely follow the climatic seasons, and reach their peak in the summer months of April and May and the central area of the country displays evaporation rates during this period. With the onset of monsoon, there is marked fall in the rate of evaporation. Against total precipitation of 400 billion cubic metres per year, India's annual internal renewable water resources are estimated at 1953 billion cubic metres, out of which utilisable water resources are taken as 1086 billion cubic metres per year, comprising surface and ground resources of 690 billion cubic metres and 396 billion cubic metres, respectively (Navalawala, 2000).

Driving Forces of Poverty in India

India was once one of the world's richest countries. That was because it has vast arable lands, sunshine throughout the year, adequate to abundant rainfall over most of the country, hundreds of rivers, and untold mineral wealth. Today it ranks amongst the poorest nations in the world, ranking 132nd among some 150 nations for which measurements are available. In India the poverty stricken regions tend to be in the climatic zones subject to drought and other water related problems as they can least afford alternative sources. In India the poor do not consume as much water as the rest of the population. Water is scarce for drinking purpose as also for agriculture and industries in most parts of India.

Poverty is associated with the specific minimum consumption level and pattern. Whenever this level falls below minimum average then poverty conditions are recognized to exist among that section of the population. Poverty is directly dependent upon the state of the economy. Looking back 45 years or so, progress against poverty in India has been highly uneven over time and space. It looks 20 years for the national poverty rate to fall below—and stay below—its value in the early 1950s. And trend rates of poverty reduction have differed appreciably between states.

Urban poor have been frequently defined as an overflow of rural poor to the urban areas. They are thus composed largely of migrants from the rural areas who have crossed the poverty line (Gopalkrishan, 1996). They move into the urban area with a strong possibility of being absorbed in the expanding service or tertiary and secondary sectors of the urban economy.

Rural poverty in India emanates from historical associations and semi-feudal relations of production in agriculture. Land reform steps were initiated but hardly made any significant change or transformation in the agrarian relations. As a result, all the agricultural labour household and large number of small and marginal farmers as well as non-agricultural landless rural labour continue to live below poverty conditions. Rural poverty has been frequently attributed to worsening of economic conditions through rapid increase of population. This increased the population pressure on land and consequently led to a decline in the real per capita income. This was also followed by decline in farm and industrial productivity and increases in food prices. These are further encouraged by socio-political situations, which prefer short-term resolutions and hence resulting in long-term impacts.

Linking Water and Poverty

The world's water resources are under so much stress and strain that it would result in a world water crisis affecting most severely in the Middle East, parts of Africa and southern parts of Asia (Rodda, 2001). The poverty ratio data distribution for 1999-2000 shows variations that can be attributed to various parameters (Table 26.2). The spatial distribution of poverty in India shows wide disparities in the levels of poverty ratios in different parts of the country. The poorest states are some of the resource-wise richest States of India, which are Orissa (47.15%), Bihar (42.6%), and Madhya Pradesh (37.43%). There are many causes for the comparatively higher levels of poverty in these states with one of the main causes being the availability and utilisation of water. The total subsidies for domestic water supply in India are currently about Rs. 40 billion a year. Of that amount, roughly Rs. 12 billion goes to urban water systems and Rs. 25 to 30 billion to rural schemes. In rural areas, it is generally the better off people who live close to the water sources, while the poor have marginal lands and less reliable sources, and suffer from falling water tables, seasonal variations, etc. In urban areas, most of the subsidies also benefit the non-poor, who typically pay for less than the real cost of water they consume (Singh, 2002; WSP-SA, 1999). Therefore it becomes imperative to draw our attention to the water availability in different parts of India and its effect on the levels of poverty in various parts of the country.

Table 26.2 : India's Poverty Ratios

Sl.No.	*State*	*1999-2000 Poverty (%)*	*Level of Ground Water Development (%)*	*Irrigated Area (%)*
1.	Punjab	6.16	3.0	95.2
2	Himachal Pradesh	7.63	14.0	18.4
3	Haryana	12.72	11.4	15.2
4	Kerala	12.72	11.4	15.2
5	Gujarat	14.07	9.1	32.7
6	Rajasthan	15.28	10.9	32.3
7	Andhra Pradesh	15.70	11.1	40.7
8	Karnataka	20.04	17.9	23.8
9	Tamil Nadu	21.12	18.1	50.8
10	Maharashtra	25.02	20.7	14.7
11	All India	26.10	18.0	38.3
12	West Bengal	27.02	13.5	29.8
13	Uttar Pradesh	31.15	21.9	65.8
14	Assam	36.09	10.7	15.0
15	Madhya Pradesh	37.43	22.0	24.7
16	Bihar	42.60	27.5	45.7
17	Orissa	47.15	22.9	25.8

Source : Government of India.

Relationship Between Poverty and Groundwater Development

The relation between poverty and groundwater development shows a distinct inverse relationship. The states, which are having higher levels of groundwater development also, show lower levels of poverty. Punjab, which is having only 6.16 per cent of its people under the poverty line and it utilizes 98.34 per cent of its groundwater resources. Similarly, in the state of Haryana there is only 8.74 per cent of the people below poverty line and it utilizes about 75.61 per cent of its groundwater resources (Singh, 2000). On the other hand in Orissa and Bihar where the proportion of people below the poverty line is more than 40 per cent here the utilization of groundwater is also very low at about 15.22 per cent and 35.99 per cent.

Role of Surface Water Availability

Access to water is important for bathing and drinking as well as for agricultural production. Population growth and rising living standards are placing increasing demands on the amount of finite freshwater on the earth (Grover, 1998). In India, poverty is directly related to overall agricultural productivity, which in turn is dependent upon water for irrigation. In many villages, poor agricultural output is directly linked to the absence of irrigation facilities. While this affects all farmers within these areas, farmers with particularly small-holdings are most affected. In most of the villages, lack of these facilities is identified as the "root cause" of poverty. Richer communities are identified as those that have access to water.

The surface water availability in India falls into mainly two categories, which are peninsular and extra peninsular river systems. The peninsular rivers are mainly dependent on the monsoon rainfall and are more or less dry in summers. The peninsular rivers can be further subdivided into coastal rivers and inland rivers. The coastal rivers are small streams concentrated along the western coastal plain mainly in the states of Gujarat and Rajasthan having very less potential of water (Table 26.3). The inland rivers consist of river Narmada and Tapi mainly on the west coast forming estuaries while the Kaveri, Krishna, Godavari and Mahandi form fan shaped deltas on the east coast. River Mahanadi is restricted by its course mainly in the north-east in the state of Orissa. The extra peninsular rivers consist of the perennial rivers fed by snowmelt from the glaciers. Most of the rivers traverse the vast Indo-Gangetic Plain before falling into the Bay of Bengal, forming the world's largest delta at its mouth. Though the volume of water fluctuates but it never dries up. The Ganga-Brahmaputra river system has the maximum water potential in the country of 1110.62 cubic km/year (Table 26.3) draining the densely populated Indo-Gangetic Plain. The levels of poverty are to a large extent determined by the surface water potential of the various river basins. Orissa and Chhattisgarh (formerly part of Madhya Pradesh) are the states falling in the low surface water potential areas that also have considerable section of the population under the poverty line. Some of the other states like Tamil Nadu and Gujarat falling in the water scarce regions have been able to lower the proportion of people below poverty line by bringing about economic development through more efficient utilization of water resources. Tamil Nadu is also known for many hydroelectric projects, which have been developed keeping in view the local social and physical conditions. The average runoff in the river system is approximately 1869km^2.

Water is vital for realizing the full potential of agriculture especially for agro-based economics like India and therefore goes a long way in determining the per capita income, levels of consumption and other infrastructural facilities.

Table 26.3 : Surface Water Potential of the Country

Sl. No.	*Name of the River Basin*	*Average annual potential in river (Cubic Km/Year)*
1.	Indus (up to Border)	73.31
2.	(a) Ganga	525.02
	(b) Brahmaputra, Barak & Others	585.60
3.	Godavari	110.54
4.	Krishna	78.12
5.	Kauvery	21.36
6.	Pennar	6.32
7.	East Flowing Rivers Between Mahanadi and Pennar	22.52
8.	East Flowing Rivers Between Pennar and Kanyakumari	16.46
9.	Mahanadi	66.88
10.	Brahmani and Baitarni	28.48
11.	Subernarekha	12.37
12.	Sabarmati	3.81
13.	Mahi	11.02
14.	West Flowing Rivers of Kutch, Sabarmati including Luni	15.10
15.	Narmada	45.64
16.	Tapi	14.88
17.	West Flowing Rivers from Tapi to Tadri	87.41
18.	West Flowing Rivers from Tadri to Kanyakumari	113.53
19.	Area of inland drainage in Rajasthan desert	NEG.
20	Minor River Basins Drainage into Bangladesh and Myanmar	31.00
	Total	1869.35

Source : Central Water Commission, Government of India.

Therefore the optimum development of the country depends upon the efficient utilization of its water resources. The total cultivated area of the country is approximately placed at 193 million hectares of which 161 million hectares is gross sown area with net sown area at about 139 million hectares. The net irrigated area is the maximum under Ganga river basin, which is about 16 million hectares as compared to the other rivers having less than 5 million hectares. Among the peninsular rivers the maximum net area under irrigation is in the case of Krishna river basin with an area of 2.6 million hectares draining the states of mainly Andhra Pradesh. Madhya Pradesh, which is located in the interior of the country and also eastern part of Maharashtra, is not having enough surface water resources. This result in lesser availability of fresh water for various purposes, namely irrigation and other diverse purposes. This further aggravates the levels of poverty in these states. The Western Ghats also has lesser potential groundwater availability, which has a direct impact on the livelihood of the people. Presently, six of India's 20 major river basins fall into water scarce category (Nautiyal, 2002). Some of the river basins that are having acute water scarcity include Kaveri and Mahanadi river basins. Even the glaciers of Himalaya are rapidly melting and are feared to disappear within the next 50 to 80 years. All these will lead to changes in runoff pattern of rivers like Ganga causing floods, loss of life and property and loss of agricultural production. The average per capita water availability in India has decreased over the years from 2177m^2 in 1951 to 1869m^2 in 2000. The relation between net irrigated area and the levels of poverty show a distinct inverse relation. In case of low poverty states of Punjab and Haryana the percentage of net irrigated area is 95.2 per cent and 78.2 per cent as compared to high poverty states of Orissa and Madhya Pradesh where the percentage of net irrigated area is only around 25 per cent. Net irrigated area largely implies, the area, which is not entirely dependent on the natural sources of water like rain, but also they are irrigated by artificial sources like canal tubewell. Therefore to develop artificial means of irrigation in an area necessitates that there should be adequate disposable income of the people to invest on such activities. Poverty affects the entire of institutional and public life right down to individual level.

Water Induced Natural Disasters and Poverty Levels

Natural disasters in India have ranged from rampaging rivers to scorching droughts, which have left long-term impact on the people of India. Floods have occurred regularly in various parts of the country resulting in large scale damages to crops over the years with the worst occurring in 1978 when the total damage costs were about Rs. 41,000 million. The flood-affected areas and the number of people affected by the ravages of floods have increased tremendously since 1950s. The flood affected area increased from 6.48 million hectares in 1950s to more than 9 million hectares in 1980s.

According to the National Commission on Floods, the Ganga basin is the most flood prone followed by Brahmaputra river basin. The most flood prone zone therefore encompasses the states of Uttar Pradesh, Bihar, West Bengal and Assam along with some of the river basins located in Orissa.

Due to the topical and erratic and uneven monsoons the crops have been subject to failure either partly or completely resulting in conditions of scarcity or famine (Singh, 1978). Drought is identified with a situation when the water losses due to evapotranspiration, exceed the effective rainfall. Thus drought results from long continued dry weather and lack or insufficiency of rain which causes depletion of underground water supply and soil moisture resulting in reduction in stream flow. As agriculture provides livelihood to more than 70 per cent of the people in India therefore their dependence is such that when agriculture fails the economy fails which most acutely affects the poor people. The most drought prone areas of India are Gujarat, Rajasthan and Orissa. The acute water shortage has made the life of people more precarious in the state of Orissa. Kannan *et al.* (2000) rightly point out that even some parts of coastal Orrisa, which come under the humid zone, is also facing water problem for drinking and agriculture. The mushroom growth of industries without assessment of demand for water and the depletion of forest cover have deteriorated the situation. Out of these Orissa is prone to both floods and droughts making it highly disaster prone state.

Conclusion

The crisis about water resource management therefore arises because most of the water is not available for use and secondly, it is characterized by highly uneven spatial distribution, which affects the quality of life of the people both directly and indirectly. One of the foremost and crucial steps, which can be taken to control the water scarcity in a region, is water harvesting and watershed management. Technological intervention, probable institutional reforms, community participation, water conservation and re-use, water policy are some of the feasible measure that provide opportunities to evolve a systematic approach to water resources management (Gupta, 2001). India receives 1194mm rainfall, which are 394mm more than the world average of 800mm. Large amount of it is allowed to runoff and not collected properly. Therefore, keeping in mind the prevailing as well as potential after 73rd Constitutional Amendment Watershed Development has been included in the schedule of subjects to be handled by panchayats (village government). This provides opportunities for combining development of grassroots democracy and natural resources in a symbolic manner. Watershed management would ensure supply of water to every field, remove hunger and poverty from poor areas, provide green cover over denuded areas, bring in more rains and improve environment. Watershed management is also described as the programme that holds the key to solve problems of employment, economy, export and equity. As has been rightly pointed out by Water and Sanitation Programme—South Asia initiatives at the community level can also go a long way in reducing water shortage in the affected regions. These include organizing local people into groups, mobilizing funds at the local level and setting up a private water supply system (WSP-SA, 1999). Another effective measure is in the form of capacity building by enabling them to procure materials, services and works by themselves, assisted by support organisations.

REFERENCES

Gopalkrishnan, R. (1996) *Geography of India,* Jawahar Publishers, New Delhi, pp. 33-37.

Grover, B. (1998) *Twenty-five years of International Cooperation in Water-Related Development Assistance, 1972-1997,* Water Policy, Volume 1, Number 1, Eisevier Science Ltd., USA

Gupta, A.D. (2001) Challenges and Opportunities for Water Resources Management in South East Asia, Hydrological Sciences - *Journal - des Sciences Hydrologiques,* 46 (6) December, 2001, International Association of Hydrological Sciences, UK.

Kannan, K. Brahmanand, P.S. Reddy, G.P. (2000) Combating Water Scarcity, *Yojana,* October, 2000, pp. 46-47.

Nautiyal, S. (2002) Water Crisis: Agenda for the Century, *Yojana,* February, 2002, pp. 11-19.

Navalawala, B.N. (2000) Groundwater over Exploitation, *Yojana,* August, 2000, pp. 49-56.

Rodda, J.C. (2001) Water Under Pressure, Hydrological Sciences – *Journal – des Sciences Hydrologiques.* 46 (6) December, 2001, International Association of Hydrological Sciences, UK.

Seth, S.L. (2000) *Watershed Management in India,* Kurukshetra, June, 2000.

Singh, R.B. (2000) *Environmental Consequences of Agricultural Development : A Case Study from the Green Revolution State of Haryana, India,* Agriculture, Ecosystem & Environment. 1637, pp. 97-103.

Singh, R.B. (2002) Sustainable Management of Water Resources in an Agriculturally Developed Region of the Indo-Gangetic Ecosystem, India, FRIEND 2002—*Regional Hydrology : Bridging the Gap Between Research and Practice,* Eds. Henny A.J. Van and Demuth, Siegfried, IAHS Pub. No. 274, pp. 315-22.

United Nations Development Programme (UNDP) (1999) *Small and Private Initiatives in the Water and Sanitation Sector in India, Water and Sanitation Programme – South Asia,* UNDP, New Delhi.

27

Utility of Integrated Approach of Groundwater Management in Depleted Areas for Sustainable Development

D.D. OZHA AND F.M. GOLANI

Introduction

Water is the key input to our agriculture, which is the prime arbiter of living standards in the rural areas. It is also prime component and an integral part of environment. Due to growing population and consequently increasing demands of water, both as supplier for diverse needs and as carrier of various wastes, this environmental component has been under constant strain. It has been assumed that if there is anything that the world will see in the 21st century, it will be decline in proper management of water resources.

It is obvious that ground water resources, their exploitable potential, yield of wells and chemical quality of water mainly depends upon hydrogeological formations. Due to indiscriminate pumping of ground water for various purposes, depletion in their level have been taken place and consequently the potential zone have been converted to "Dark zone" category. The situation in drought or desert areas are much more severe. Therefore, now the time has come to think and act accordingly by way of adopting different methods to stop declining water resources.

Constraints of Ground Water Resources in Desert

Depletion in Ground Water Level

It is obvious that water levels vary with rainfall, physiography, drainage and climatic conditions. In case of Rajasthan the trend of water level during 1984-99 has been analysed on the basis of water level records during pre-monsoon 1984 and pre-monsoon 1999[1]. According to which out of 32 districts, 27 districts show a depletion of water level. However, only in 5 districts a rise in water

Table 27.1 : Decadal Ground Water level Situation January (1990-99) - January 2000

S.No.	District	Stations Analysed	Rise Range (m) Min.	Rise Range (m) Max.	Rise <2m %	Rise 2-4 m %	Rise > 4m %	Decline Range (M) Min.	Decline Range (M) Max.	Decline < 2m %	Decline 2-4 m %	Decline > 4 m %
1.	Ajmer	33	0.25	8.73	15	12	3	0.96	9.50	21	12	36
2.	Alwar	51	0.16	12.54	33	8	6	0.08	12.44	35	12	6
3.	Banswara	26	0.07	6.69	35	12	4	0.27	3.54	31	19	0
4.	Baran	19	0.03	1.27	26	0	0	0.02	4.45	53	16	5
5.	Barmer	79	0.16	11.55	24	6	5	0.04	19.81	41	6	18
6.	Bharatpur	41	0.05	4.35	49	10	10	0.21	7.26	24	2	5
7.	Bhilwara	30	0.05	8.66	7	0	3	0.16	14.94	27	20	43
8.	Bikaner	57	0.02	14.77	33	16	21	0.01	16.30	18	5	7
9.	Bundi	18	0.15	0.89	17	0	0	0.20	18.13	56	6	22
10.	Chittorgarh	23	0.56	0.99	9	0	0	0.13	7.86	43	13	35
11.	Churu	57	0.09	13.37	35	2	11	0.01	7.43	42	5	5
12.	Dausa	25	0.01	2.17	32	4	0	0.28	24.24	36	16	12
13.	Dholpur	20	0.10	4.79	25	10	5	0.36	13.28	35	10	15
14.	Dungarpur	18	0.17	1.72	17	0	0	0.26	4.97	44	22	17
15.	Ganganagar	40	0.33	13.75	20	25	55	—	—	0	0	0
16.	Hanumangarh	33	0.04	19.20	36	24	27	0.04	1.69	12	0	0
17.	Jaipur	64	0.30	1.58	13	0	0	0.12	19.29	38	36	14
18.	Jaisalmer	63	0.02	21.82	33	22	11	0.04	14.16	22	5	6
19.	Jalore	21	—	1.41	5	0	0	2.06	11.49	0	43	52
20.	Jhalawar	25	0.01	3.85	36	16	0	0.02	4.02	32	12	4
21.	Jhunjhunu	34	0.05	2.54	21	6	0	0.01	14.94	26	21	26
22.	Jodhpur	56	0.08	10.99	32	4	5	0.09	15.44	34	13	13
23.	Karauli	22	0.06	5.65	18	9	18	0.02	2.60	41	14	0
24.	Kota	20	0.03	1.24	20	0	0	0.10	6.16	65	5	10
25.	Nagaur	52	0.13	2.40	27	2	0	0.05	15.16	25	19	27
26.	Pali	33	0.23	4.69	27	0	3	0.36	10.88	24	21	24
27.	Rajsamand	27	2.03	5.78	0	7	7	0.20	13.85	22	11	52
28.	Sawai Madhopur	24	0.19	3.78	38	4	0	0.07	2.79	38	21	0
29.	Sikar	32	—	0.56	3	0	0	0.09	7.12	50	34	13
30.	Sirohi	15	—	—	0	0	0	0.74	17.89	27	7	67
31.	Tonk	28	0.32	2.12	32	7	0	0.02	5.76	21	32	7
32.	Udaipur	49	0.17	8.27	8	0	8	0.31	24.75	16	35	33
	Rajasthan	1135	0.01	21.82	24	7	7	0.01	24.75	30	18	16

level has been observed. A decadal picture of ground water level in the state (1990-2000) is shown in Table 27.1.

The prime cause of water table depletion is uncontrolled withdrawal of ground water to meet out ever growing demand for agriculture, domestic and industrial purposes, through huge number of structures in potential areas. The sustainability of vagaries of rainfall has further fallen prey to unplanned development. Thus resultant is the shrinking of ground water resources due to which the wells are getting dry, deterioration in chemical quality and increase in lifting cost.

Challenge of Alarming Increase in Potability Parameters

In Rajasthan, particularly in desert zone, there are three main quality oriented problems in ground water resources, viz. salinity (Barmer, Bikaner, Sri Ganganagar, Nagaur, Churu), high nitrate (Nagaur, Churu, Barmer, Jhunjhunu, Sri Ganganagar) and high fluoride (Nagaur, Sirohi, Pali, Barmer, Jalore, Sikar, Jaipur, Ajmer, Dausa, Sawai Madhopur, Tonk and Dungerpur). Since water and human health are interlinked. Therefore, the excessive concentration of health affecting constituents, e.g. nitrate and fluoride, are causative factors for several diseases, viz. Methemoglobenemia, Cancer, Dental and Skeletal fluorosis. The maximum value of potability parameters in desert zone of Rajasthan are tabulated in Table 27.2. Therefore, for sustainable health, immediate ameliorative measures are needed to combat the problem.

Table 27.2 : Maximum value of potability parameters

District	*Potability parameters (mg/l)*		
	T.D.S	*Nitrate*	*Fluoride*
Jodhpur	13060	2000	22
Pali	16000	1020	14
Jaisalmer	17950	1480	12
Churu	15625	2400	32
Nagaur	25440	4750	34
Barmer	15000	1900	18
Jalore	16800	1400	14
Bikaner	29140	1600	22
Sikar	9050	2155	15
Jhunjhunu	7300	1000	12
Sri Ganganagar	18060	1500	8
Hanumangarh	19080	1700	6

Problem of Irrigation

Ground water of desert region of Rajasthan are enriched with soluble salts and thereby imparting saline character. Irrigation with brackish water causes physical and chemical degradation of soils resulting in the decrease of crop yields. The adverse effects of brackish water are more pronounced in the desert regions. It is obvious that low salinity water due to excessive bicarbonate ion concentration have appreciable Residual Sodium Carbonate (R.S.C.) which is responsible for development of alkalinity in the soils, viz. permeability, porosity and infiltration rate etc. The areas of Barmer, Jalore, Nagaur and Pali are prone to high R.S.C. induced problems.

In the desert areas of Rajasthan prolonged irrigation with saline or sodic water has decreased the fertility status and nutrient availability in the soils and consequently affected the yield of crops to the great extent.

Management of Ground Water Resources

Since the demand for water has been continuously increasing with the ever increasing population, increased demand for food and fibre and the development of urban and industrial areas. Therefore, there is utmost need for adopting measures for management and conservation of water resources. Following are managemental strategies for protecting ground water resources :

1. Surplus water during monsoon must be conserved and artificially recharged to groundwater system.
2. Roof top rain water harvesting be made mandatory in urban areas by laws be amended.
3. Recharge storm water in urban areas municipal bodies be made responsible.
4. Revival of traditional water conservation structures like Nadi, village pond, Baori and dugwells etc., panchayat be made responsible.
5. Percolation tanks, sub-surface barriers, check dams be constructed on large scale wherever feasible. Ground Water Department of Rajasthan has constructed 52 sub-surface barriers and 2 percolation tanks under various schemes between 1986 to 2000.
6. Recycle urban waste water after secondary treatment.
7. As the desert areas receive scanty rainfall and thus it needs to be conserved in *Tankas,* so that it can be used in time of water crisis.
8. Industrial and urban effluent treatment to match quality of ground water in the areas be made mandatory.
9. Identify and delineate suitable area/aquifer near urban complexes and declare these as Ground Water Protection Zones and all industrial activities be banned in such areas.

10. Development of ground water in command area be promoted to control water quality deterioration due to water logging.
11. Water conservation—This include modern agriculture practices like dry farming, sowing of salt tolerant crops, laying efficient irrigation network by using cemented distributories, sprinkler and drip irrigation etc. conventional tanka system in use since ancient period in parts of western Rajasthan is most suitable method for collecting meagre rainfall in the area. This structure can be used to collect roof top rain water in other parts of the state to reduce pressure on water supplies of all major cities and towns. Crops which requires less water be encouraged.
    In Rajasthan, the State Government is very keen in taking initiation for water conservation devices. State Government has already started loan and subsidy facilities for purchasing water conservation devices. Recently, the government has taken a decision to include construction of underground tanka, a mandatory requirement while approving a plan for residential houses as a part of water conservation programmes.
12. Pollution of water bodies due to effluents of factories and textile industries, sewage of major cities and towns needs to be checked. Such discharges should be either properly treated and harmful constituents be removed before its release to the natural drainage courses or it should be recycled.
13. Economic water rates be charged in urban areas.
14. Regulatory Aspects : It is mandatory to regulate ground water withdrawal in over exploited critical areas, it includes—

    (i) Ban on construction of new wells,
    (ii) Register all existing structures,
    (iii) Regulation of new power connection for pumps in such areas.

Ministry of Forest and Environment, Government of India *vide* its Notification No. 38(F) dated 14 January, 1997 constituted a Central Ground Water Authority under Environment Protection Act, 1986 and assigned function to check in discriminate boring and withdrawal of ground water in the country by issuing necessary regulatory directions for preserving and protecting ground water. The Central Ground Water Authority has framed Ground Water (Development and Protection) Rules, 1998, for protecting the vital resource.

In Rajasthan, the State Government after considering "Model Bill to regulate and control Development of Ground water" circulated by Ministry of Water Resources, Government of India in June, 1996, has also drafted "Rajasthan Ground Water (Regulation) Bill, 1997" to regulate and control ground water development in the state.

15. Public Awareness : The success of the water conservation programme also depends on people's participation. Therefore, in the local languages of the areas, seminars, posters, nukkad sabha, drama be arranged using audio-visual aids for educating the people of urban and rural areas for judicious use of water, water and health, hygiene, pollution etc. NGO's are also performing excellent job of water conservation and protection, be encouraged for sustainable development.

Role of ISHEER

The registered Indian Society of Health, Environment, Education and Research (ISHEER) is also performing excellent work for creating awareness amongst the urban and rural masses regarding conservative and socio-health aspect. The society has arranged several water education camps in various schools, NGO's offices and at other places and obtained enthusiastic results.

REFERENCE

1. Ground Water Department, Rajasthan, Jodhpur various Technical Reports.

SECTION-V

BIODIVERSITY

28

Desert National Park

A Unique Biosphere Reserve for Conservation and Development of Biodiversity in India

VED PRAKASH MEENA

Severe environmental problems are waiting at the door of 21st century world humanity. In the last few decades man has caused major destruction by his malactivities. Because of this life on earth itself is endangered. Therefore, presently worldwide efforts are being made to protect environment and its factor. One of them is known as "conservation of biodiversity". In order to conserve biodiversity many plans have been executed around the world. In India too many national parks have been established under Wildlife Protection Act,1972. The Desert National Park of Rajasthan is one of them. It has also been made biosphere reserve in the year 1986.

In the present paper an attempt has been made to explore the inter-relationship between animals and plants of arid India and portray the plant communities and associations present here. In this respect the Desert national park provides a rich feast to the ecologists to unfold the intricacies of a natural ecosystem.

Park is an enclosed tract of land held by royal (government) grant or prescription for keeping beasts of the chase. Distinguished from a forest by being enclosed, and having no special laws or officers, National Park is distinct from a public park. It is an extensive area of land set apart as a national property to keep in its natural state for the public benefit.

National park are areas usually of great natural beauty and interest, now established in many countries and serve specially for conservation of native flora and fauna, as well as, for recreation of human beings—specially those interested in wild life.

In addition to those in the United States and Britain, especially famous are the Canadian National Park in the Rockies, the Kruger Park in the Eastern Transvaal, Serengeti in Tanzania, Mt. Fuiji in Japan.

In Britain, National parks were setup in England by an Act of 1949. They are areas of natural beauty, but for the most part the land is farmed or otherwise used, and the parks include numerous village and other settlements and the normal life of the countryside goes on in them. This is because the country is small and large chunk of land cannot be spared. They are, however, specially protected from building development. Lake District and North Park Moors are the two of the National Parks in Britain. The same Act setup Areas of Outstanding National Beauty (A.O.N.B.), also protected.

In United States there are mainly extensive wild areas, often largely uninhabited and serve both for recreation and for conservation of wildlife. The first to be set up was the Yellowstone Park in 1872, including Parks of the States of Wyoming, Montana and Idaho, with an area of over 3200 sq miles. It is said to have more geysers than all the rest of the world. Its creation really started the National Park Movement all over the world. The United States has 350 other National Parks some of the famous being the Crater Lake, Grand Canyon, Mamoth Caves.

In order to retain the natural environment and to conserve natural habitat, no settlement has been allowed for scores of miles around the Crater Lake in the State of Oregon, not even a hotel or lodge. The tourists have to drive long distances to arrive at the lake. They can go round it and learn about it only through a local guide. They are prohibited even to feed the brids or squirrels with a grain of food. There is not a single fish of any sort in the 2000 feet deep lake. Thus is conserved the natural habitat.

National Parks programme is today an extended activity of conservation of representative ecosystem in the form of Biophere Reserves or nature reserves. Although initiated under the auspices of the UNESCO during 1973-74 the programme did not gain momentum nearly a decade. Gradually 62 countries selected 223 areas to be brought under Biosphere. Preservation Programme but each, country interpreted the philosophy of the programme as found convenient to it. After the promulgation of Wild Life Protection Act, 1972, Government of India, Wildlife areas came to be reserved as National Parks, Sanctuaries, or Closed Areas, thus giving due importance to wildlife conservation. India so far has selected 13 sites as Potential Biosphere Reserves. The number of such areas has kept on increasing year after year looking into their importance upto 2003 when India had a total of 93 National Parks and 485 Sanctuaries making a total of 578 wildlife protected areas, some Parks and some Sanctuaries specialised for protecting various forms of wildlife as brids, tigers, lions, rhinos, elephants, deer, Great Indian Bustard and others. In the present context some National Parks are of international significance to the birds like Siberian cranes and the Great Indian Bustard. The former is called the Keoladev National Park white the latter the Desert National Park, one each in the semi-arid and arid regions of Rajasthan respectively.

Undoubtedly the incessant exponential growth of human race upon the earth, and more strictly in certain harsh areas, has brought about slow but subtle changes in the natural eco-systems, which are inimical to the humans to cause large-scale turbulence in nature and jeopardise the laws of nature. Hence it is of paramount importance to study the nature's reserves against the back drop of environmental degradation. It is this purpose that the present paper on Desert National Park is attempted.

The numerous life forms, both floral and faunal that have adopted themselves to survive and propagate under harsh habitat conditions of the hot desert, constitute an invaluable stock of rare germ plasmas, which are too valuable from the biological point of view. The desert being a fragile ecosystem, its further denudation and deterioration is likely to remove, for all times, one of the world's most engaging environments. It is imperative, therefore, that at least a part of the Great Indian Desert, inasmuch of its pristine conditions as possible, is kept further unmolested and restored to its original status to serve as a benchmark model of the desert as it once existed over a vast territory. This could be done only by establishing a Desert National Park.

The Great Indian Desert of the pre-partition period of the eastern section of the Saharan Arabian desert and was earlier known as "Tharparkar" desert (Tod-Rajasthan). The name persisted for a long time as the name of a district in Sind province of Pakistan. It was later on known as the Thar desert or the Great Indian Desert. Part of it is now in Pakistan and a larger part is in India. Though named Desert National Park it is indeed a sanctuary. The Park area is situated between 25° 47′ N and 26°46′ N latitude and between 70°45′ and to 70°45′ E longitudes and extends into 3162 sq. km, partly in Jaisalmer and partly in Barmer districts of Rajasthan. (Fig. 28.1). Its northern part is under the influence of the Indira Gandhi Canal Command and the southern part is under the influence of the Narmada Irrigation Project (under contemplation). The famous bulge of Jaisalmer is the core of the desert ecology and is under the defence establishment as well as under the command of the IGNP. The result is that there is hardly any sizeable area left for establishing a conservation unit even in the vast wasteland. The conservation unit is bounded by the Jaisalmer-Sam area in the north, the Gadra Road-Barmer rail track in the south, the Gadra Harsani-Satto defence road in the east and the International boundary in the west. This is the only available land. Its western boundary had to be shifted 20 km towards the east from the International boundary, on instructions from the Defence Department.

It is the solitary example of a desert park in the world. Its global reputation is due to its existing desert land eco-system though developed in isolation, yet manifests a large variety of consequential changes in its habitat, seasonal variations in faunal and floral composition and diverse vegetation types. It is a unique ecosystem also because it is both man-made and well maintained.

Conservation of Wildlife

Since the Desert Park is under the administrative control of wildlife wing, for planned working but no sacrosanct management plan has so far been prepared for the Desert National Park for wildlife preservation and its management.

The Present Position of Desert National Park

The total area of the Park being 3162 sq. km. It is the biggest sanctuary in the arid zone, all the rest being in the semi-arid and sub-humid. It has over 60 species of mammals and 6 species of reptiles.

The entire area was notified as a sanctuary under the Wildlife Protection Act (1972) and still has natural landscape, original vegetation, landforms, traditional land use, animal husbandry, flower, people and it is large enough to be an effective conservation unit.

Presently the land of desert national park is being used in diverse manner. Complete data of this is shown in Table 28.1.

Since a single unit of 600sq. km representing all the landform found within the desert was not available, two units of 300 sq. km each have been selected, one in Jaisalmer and the other in Barmer district for the purpose of total conservation. They serve the purpose of core area of the National Park. The two totally fenced and protected areas are not located at the centre of the National Park nor are these insulated by buffer zones, as is generally the case with National Parks and Sanctuaries. Since the areas outside the Reserves, especially towards the international boundary in the west are heavily guarded and are likely to remain so indefinitely, the protected area will always be insulated from outside influence.

The Desert National Park has no ecological boundary. Agricultural fields, Pastures, revenue land and sanctuary land all are conjoint and at some places barbed-wire fencing is the significant marking of boundaries, making pillars to show its actual status.

It is the holistic and exploratory study unveiling the cause and effect relationship which are of great importance to unfold the cause of degeneration in a particular ecosystem and to suggest measure to infuse the basic components to revive it.

In order to study the ecology of biotic variation of this park a team under the leadership of K.S. Sankhala and another team under the Bombay Natural History Society had conducted research projects. K.A. Shankaranarayan (1987) had worked on Eco-regeneration of Desert Reserve in the Thar desert.

Geology and Landforms

It is surmised that the desert area was submerged under sea during the Jurassic, Cretaceous and Eocene ages. It was uplifted probably sometimes in the Upper

Table 28.1 : Land use in Desert National Park

District	*Total number of villages*	*Total area in ha*	*Forest in ha %*	*Unirrigated area in ha and per cent cultivated land*	*Culturable waste land in ha* per cent*	*Area not available for cultivation*	*Grazingland including oran, culturable waste and barren land*	*Grazing land as per cent of total area*	*Available grazing land/ Livestock*	*Total live-stock popula-tion*
Jaisalmer	28	275272		10715 (3.89%)	250921 (91.15%)	13636 (4.95%)	264513	96.09%	1.96 ha	135241
Barmer	9	203788	17.8 (0.84%)	54769 (26.88%)	108898 (53.44%)	38413 (18.84%)	165375	81.15%	1.73 ha	95812
Total	37	479060		65484 (13.76%)	359819 (17.11%)	52049 (10.86%)	429888	89.73%	1.86 ha	231053

* Use as grazing land

Source : Project Document-7, Ministry of Environment and Forests, Govt. of India, New Delhi.

Table 28.2 : Population and occupational patterns in Desert National Park

District	*No.of villa-ges*	*Area in ha*	*Population and density/ ha*	*No. of workers and per cent of total population*	*Cultivators and (per cent)* to total workers*	*Agricultural labourers (per cent)*	*Household industry (per cent)*	*Other workers +*	*Marginal workers +*
Jaisalmer	28	275272 (0.04/ha)	11938	3785 (31.0%)	2039 (54.00%)	191 (5.00%)	68 (1.80%)	1487 (38.20%)	309 (7.76%)
Barmer	9	203788	19227 (0.09/ha)	6948 (36.14%)	6370 (91.68%)	19 (0.27%)	21 (0.30%)	538 (7.75%)	228 (3.18%)
Total	37	479060	31165 (0.06/ha)	10733 (34.43%)	84.09 (78.35%)	210 (1.95%)	89 (0.83%)	20.25 (18.87%)	537 (4.85%)

* Including graziers and pastoralists. + Mostly graziers and pastoralists

Source : District Census Handbook, Jaisalmer and Barmer.

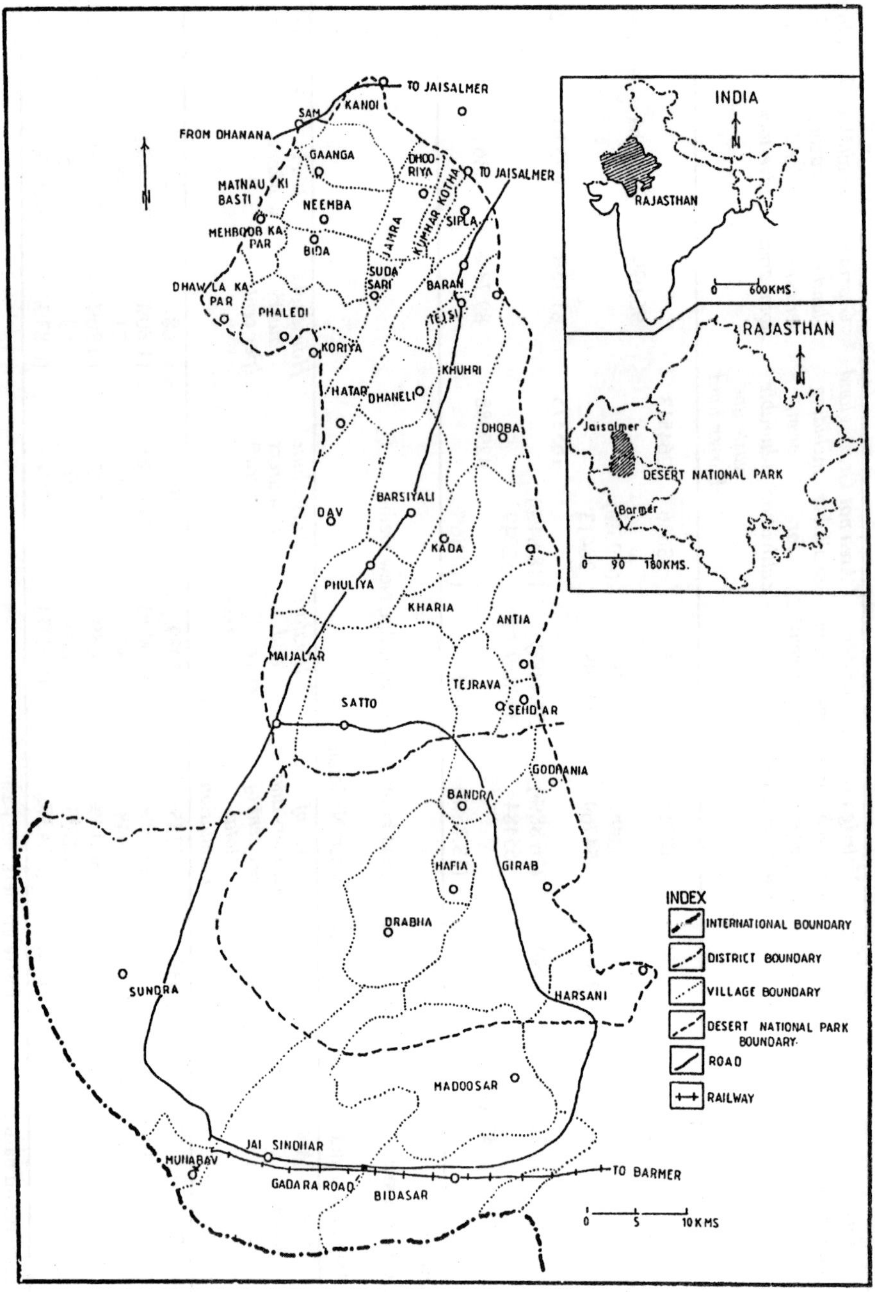

Fig. 28.1 : Key Map of Desert National Park

Tertiary. It was during the late Pleistocene after the last Phase of Glaciation that the desertic conditions were set in small outcrops of Jurassic rocks comprising limestone, sandstone and shales are found in the midst of desert sands and alluvium. Marine deposition resumed in the Eocene period before a climatic accident created the desert.

Geomorphologically, it is largely sandy with shifting sand dunes, fixed dunes and interdunal flat lands, interdunal depressions, and saline flat lands. It slopes towards the south and north-west without any perennial river. The seasonal rivulets are mostly nallahs with limited catchment areas. The little flow of water is often locked by the inhabitants as *'Khadins'* used both for their cultivation and for drinking purposes.

The soil are mostly sandy with low water holding capacity, 50 to 150 cm deep in the interdunal depressions, lacking in nitrogen, carbon and mineral salts, not suitable for cultivation. Its strong and gravelly areas are *'banjar'* lands.

Climate

The climate is typically arid with 4 to 6 rainy days. Droughts occur quite frequently. January is the coldest month affected by the western disturbances when temperatures often go 3 to 4 degrees below the freezing point. Forests damage the vegetation. May and June are the hottest months when the temperature may rise to 50° C and dunes travel long distances during the stormy summer days.

Settlement Pattern

The settlement pattern of the sedentary population in the area is of three types: (i) Entirely compact, (ii) Entirely scattered and (iii) Villages having both compact settlement and scattered homesteads to which the majority of villages belongs. The rural people mostly live in hamlets, small villages and *'dhanis'* or homesteads. The dunal topographic conditions, the formation of *'dhani'* type settlements characterised by non-accessibility, scattered location of water sources, the socio-cultural heritage of certain castes, the lately prevalent *Jagirdari system* and the customs of bringing tenants for use of land and the practice of subsistence farming. There is a tendency among people belonging to the same caste to live together in the *dhanis* or in the village core. Historically, the sequence of settlement in the villages has been first viewed families who settled there. Later people moved seasonally to their distant commuting. People living in *dhanis* take grater care of their crop land from close quarters while the people living in the core compact settlements enjoy a richer social life and make better use of the community Park. The occupation of the people residing in these are shown in Table 28.2 and Fig. 28.2.

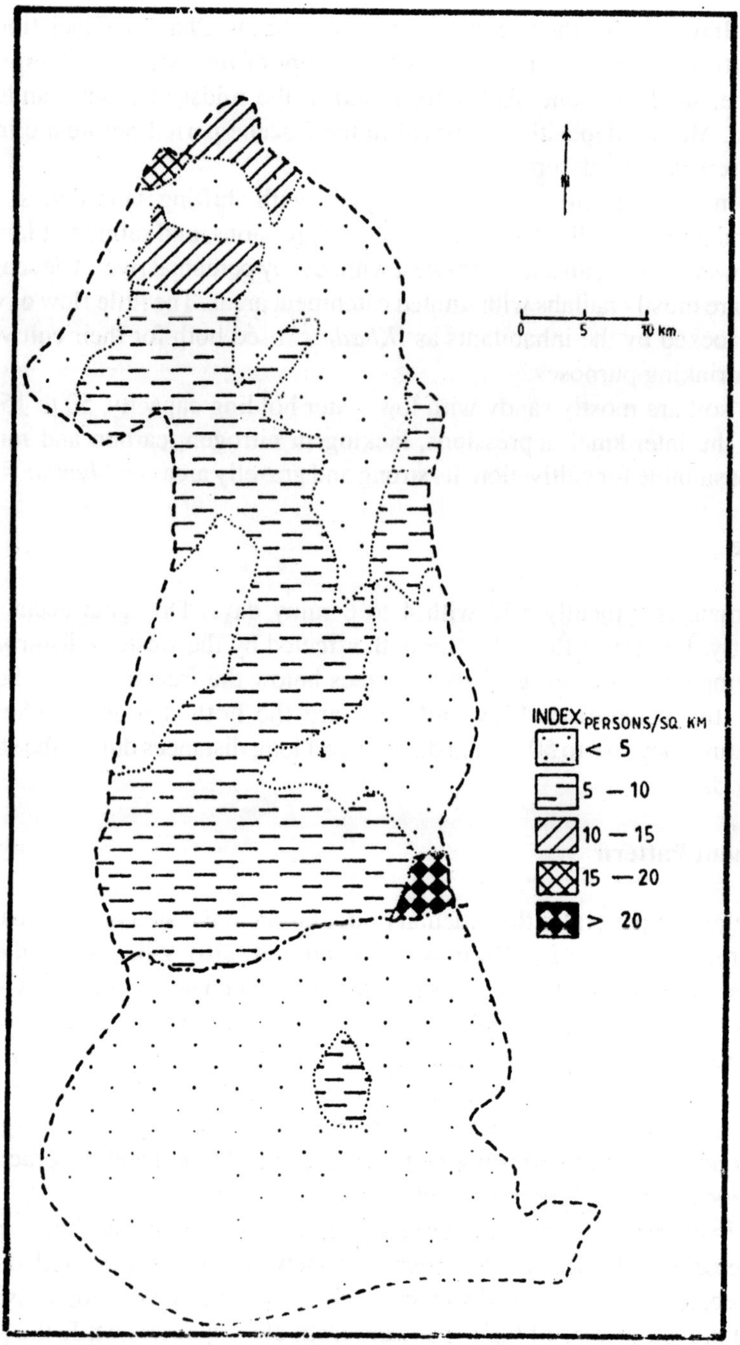

Fig. 28.2 : Desert National Park—Population Density

According to landscape modification, the development of the Park chronologically may be divided into three main periods :

1. The period of inception.
2. The period of development.
3. The period of maintenance.

The period of inception started in the 19th century and earlier when the occasional floods made this area of the desert Park natural boggy and alkaline, with layers of saltpetre overlying the soil. After the retreat of the monsoon and drying up of the marshland, it was used as pasture for cattle, abandoned only during the famines and droughts. It was also a refuge area for the wild cattle of the Jaisalmer State. The State had been robbing this area of its vegetal cover by lopping trees, foliage cover, shrubs and hay.

It is interesting to find that the construction of *khadins* which took place in gradual stages affected the development of the D.N.P. The work used to be done in years of famines when a large number of peasants and landless laboures used to be provided with relief work for their sustenance.

The management plans of the desert area were first prepared in the 1970s. Though it is named Desert National Park, but in fact, it is a Sanctuary as declared under the Wildlife (Protection) Act, 1972. Besides the main areas, other eight Closed Areas were selected for wildlife preservation and declared as Closed Areas under the Act. Today these are called satellites of this biosphere reserve under the Director of Desert Afforestation and Pasture Development, Jodhpur till 1986, after which it came under the administrative control of the Wildlife wing.

With the abolition of feudal kingship ushered an era of freedom from exploitation of vegetation and shooting animals. But during the transition, the neighbouring farmers encroached upon the peripheral lands of the Park and its allotment to refugees by local administration. But these activities were prevailed upon by K.S. Sankhla with the concurrence of the then Prime Minister Mrs. Indira Gandhi.

Broadly speaking 16 types of land use are most significant. The practice of shooting the Great Indian Bustard, the State National Bird was banned in 1972, cutting of trees for fuelwood and cattle grazing was banned in 1982 and also the temporary human settlements which existed for limestone making were also abolished. Grazing practices have been banned except in years when due to overflood in of the Chelasari *Nadi* and *khadin*, there is immense growth of grasses. Establishments of human settlements inside this park and development of tourism has been continuing and thus the human population has been increasing which is detrimental to the desert ecosystem.

The management of the dessert park maintains wireless sets, water resources through tanks and tube-wells for both the animals and staff posted at remote places and vehicles (including camels for mobility of the staff). It is good that there is no separate rest house of Desert National Park, nor their should be any if the ecosystem of the Park is to be maintained.

Fossil Wood Park at 'Akal'

About 15 km from Jaisalmer town on the NH-15 on the Jaisalmer-Barmer Road near the village Akal, is a scientific wonder protected in a small park named as 'Akal Fossil Wood Park'. There are several tree trunks lying prostrate on the ground in an area of 21 hectares. Twenty-five of the trunks are visible on the surface, the largest is 7 metre long and 1.25 metre in girth. About 180,000,000 years ago these tree trunks were real trees with leaves, roots, standing upright which grew in this part of Rajasthan and formed thick jungles. During submergence and emergence more than once, these trees were petrified. One can see large number of petrified logs and innumerable pieces of woods scattered on the surface in the area.

Water Resources

Water is the most precious natural resources of the region. The government has recently drilled tube-well Kuri, Miaji-alaar, Hasrani and grab villages. Raw water is stored in tankas (covered tanks) in almost every house and is the only sources of sweet water for drinking. In the past years, the gazzalars have been made for water supply for the staff posted at remote places. The Indira Gandhi Canal ends up at Gadra village in Barmer district. Its Sagarmal Gopa branch would enter the DNP area right from 15 km from Sam village and traverse a distance of 125 km in the notified DNP area.

Environmental Adaptation of Animals

Several animals in this part acquire necessary water from their food. The gazelle and the hare can live long without water. They generally migrate to water points in the area. The gazelle, blackbuck, nilgai and the wolves, etc. congregate around large perennial rain pools. The hare, gerbil and gazelle lick dew on grasses. Dove, common vulture, sand grouse, bulbuls, partridge, peafowl and the jungle babbler, etc. migrate in the vicinity of perennial rainpools during the drought season. Many of the mammals and birds of the desert drink considerable saline water having above 3000 ppm salinity. Some scientists mention nasal salt excreting glands is some birds whereas other animals excrete extra salts with their urine.

Socio-economic Aspect

The park is a valuable asset for harvesting fodder. In contemporary times, the human needs weighed a bit more than those of animals. So the villages small and large, surrounding the park had an access to the park to graze their domestic animals and have fuel wood. In modern times tourism industry is flourishing and generating local employment.

A few houses of the *ex-jagirdars* and some others in villages within the park are built of sandstone brought from the towns. Otherwise, mostly the villagers build mud houses with roofs of brushwood. They put 2 m high thorn fencing around an extensive compound to ward off animals, thieves and also as protection against strong winds. Inside the compound, two or three *jhupas* (huts) are built depending upon the size household. Bullocks and camels are tied in one corner of the compound while rest of the livestock are confined in separate enclosures. Fodder and grains are also stored here in specially built conical structures in *jhupas* of *khejri* wood and shrubs, locally available. The entire floor within the huts and compound is *kuchcha*, coated with cow dung, water and earth. Water is stored in earthen pitchers. Some families store rain water in tanks or pits. The members of the extended families or disjointed families build separate hutments constructed in the compound the original dwelling. Shops, community facilities and even government functionaries are housed in the *jhupas* in the village core. In general the people here are vegetarian. The desert people have developed certain peculiar food habits. Seeds and pods of several trees are harvested, dried and used as delicacies. Almost all the fruits of *Zizyphus nummularia (jharberi)* growing in accessible parts of the desert are harvested and converted into flour for human consumption specially during drought years. The grass seeds add nutrition to their foods. Cattle grazing is allowed free except inside the enclosures. Sewan grass is of high nutrition value, which grow during monsoon period.

Activity have been mainly concerned with providing absolute protection to the flora and fauna within a particular territory against degradation by people living within or outside the territory. It has been recognized that this approach antagonizes the local people. There have been conflicts, sometimes resulting in loss of life. The latest thinking is to involve the people living within the reserve in its management while ensuring sustained betterment of their lot. This philosophy of man being the hub of the biosphere will be the guiding principle in the operation of the Thar Biosphere Reserve. Rotational grazing of livestock will be the pivot around which the whole scheme will revolve. Many kinds animals are kept in Desert National Park, which are shown in Table 28.3.

Table 28. 3 : Livestock population in Desert National Park

District	*No. of villages*	*Livestock population, total and per cent*						*Total*
		Camel	*Sheep*	*Goat*	*Cattle*	*Donkey*	*Buffalos*	
Jisalamer	28	5359 (3.96%)	55631 (41.11%)	38638 (28.56%)	30423 (22.50%)	5081 (30.75%)	100 (0.07%)	135241
Barmer	9	-	30157 (31.47%)	43605 (45.51%)	17357 (18.12%)	4626 (8.82%)	67 (0.07%)	95812
Total	37	5359 (2.34%)	85788 (37.12%)	82243 (35.59%)	47789 (20.68%)	9707 (4.20%)	167 (0.7%)	231053

Source : Office of Dupty Director, Desert National Park, Jaisalmer and Barmer.

There is great scope for tree planting at suitable sites in the village and desert pots and on water points primarily to provide shade. Shady and fast growing species like Neem (*Azadirachta indica*), Siras (*Abizzia*), Rohida (*Tecomella indulata*), Khejri (*Prosopis cineria*) and Peepal (*Ficus religiosa*) may be recommended. Gas supply and energy can save trees from felling for fuel.

Vegetational History of the Park

Several studies have shown that the grasses, the shrubs and trees in the DNP area are well adapted to the environment. These are deep rooted tenacious enough to survive expanded drought and yet efficient in putting up a good biomass during favourable interlines. These are quite palatable and rich in mineral matter. What is more they occur in great variety, there being as many as 700 species, amongst which grasses alone account for 107 species. All told the arid zone if Rajasthan has a well defined vegetation which is described as Tropical Dry Thorny Forest Type. The DNP consists chiefly of scrubby and thorny forest, the main dominant species are as follows: Kikar (*Acacia decidua*), Pilu (*Salvadora Persica*), Jal (*Salvadora oleoides*), Ber (*Zizyphus mauritiana*), Khejri (*Prospopis cineraria*), Chonkra (*Acacia leucophloca*), Karil (*Cappries Decidua*). The dominant species of grasses like sewan, bhurat, phog grew in parts where land became usually dry just after the monsoon. Today, the grassy patches of the Park are the remnants of the extensive closures, which once existed in the Park stretching upto Sundasari village. The effects of climate variations, droughts and floods, etc. also brought about considerable change in the natural biomass composition of the DNP. It is interesting to note that this ecosystem was biologically less diverse prior to 1990 AD as compared to the letter restricted exploitation of biomass and that the ecosystem of area has matured and diversified over the years. This is in contrast to the general observations of other desert ecosystems which have degraded due to the onslaught of both human and animal population. About 2000 sq. km of area of the park was once densely furnished with tanks in which rain water used to collect for some months during the year and a splendid thick forest existed. This is the famous GIB shooting area of Jaisalmer. But as there was no restriction on cutting trees and grazing cattle, that serene verdant has disappeared. However, after 1980 some restrictions were imposed and development processes began to take place for the conservation of the natural environment after the recommendation of K.S. Sankhala in 1985. A detailed classification of desert species based on their adaptive characteristics is as follows:

Drought Escaping Plants

These include monsoonal plants which appear through 'seeds' just after the rain, grow profusely, flower and produce seeds within a duration of 4-6 weeks in which situation plants do not have to face harsh conditions of the desert species like *Mollugo cerviana, Gisekia pharaceoids, Indigofera cordifolia, Tributus terrestris* etc.

Draught Evading Plants

These evade drought by conserving scanty moisture, e.g. *Pulicaria crispa, Digera muricata, Celosia agrenta, Tephorosia purpurea.*

Drought Enduring Plants

They are small sized plants tolerant of drought for longer period. Examples are *Convolvulus, Microphyllus, Launea sp.* etc.

Drought Resistant Plants

They have adaptive features through which they are able to face extreme conditions of desert. These plants may be further classified as (i) succuluents (*Cassigonum Polygonoides*), *Europhorbia caducifolia, Opuntia sp., Aloe barbedense, Agave sp.,* (ii) Non-succulents are those that are true *Xerophytes Acacia senegal, Prosopis cineraris decidua, Salvadora persica, Calotropis procera.*

Grasslands *(Lasiurus scindicus)*

The grasslands which vary from dense tall sewan grass to low grasses, occupy large parts of the park area, mostly in the extreme north and extreme south, chiefly confined to the closures 1, 2, 3, 4 and 5. This *sewan grass* also grows more profusely in sandy soil areas. In the last over ten years these grasses occupied an indomitable area from Jaisalmer to Barmer. The grassland have been constantly shrinking over the years due to encroachment and agricultural expansion. Thus most of the closures have disappeared.

Ten large area of woodlands in the park also have a continuous cover of tall grasses ganging from 50 to 135 cm in height interspersed with continuous growth of trees like Khejri *(Prosopis cincraria), babul* with occasional growth of shrubs like ber and kair. The tall sewan grass and fog grow profusely during the rainy season. Owing to the diversity of vegetation and a profuse landcover of grassy patches, these areas are the most favoured sites to *Chinkara,* land birds and the mammalian fauna of the park. These tall grasslands have a predominance of shrubs like *Salvadora spp., vilayati babul, kari, ber,* amidst the array of grasses like *dhaman* and *bhural*. Micro-variation in the characteristics of vegetational diversities are also noticed on varied landforms in the park region. Flora of Desert National Park which are shown in Table 28.4.

Table 28.4 : Flora of Desert National Park

No.	*Local Name*	*Botanical Name*
A.	**Trees**	
1.	Khejri	*Propis Cineraria*
2.	Kikar	*Acacia Nilotica*
3.	Neem	*Azadirachta Indica*
4.	Hingota	*Balanites Egyptican*
5.	Ker	*Capris Decidua*
6.	Jal	*Salvadora oeloides*
7.	Jal (Khara)	*Salvadora persica*
8.	Rohira	*Tecomela undulata*
9.	Juliflora	*Prosopis juliflora*
B.	**Shurbs**	
1.	Khimp	*Leptadema prypechnica*
2.	Aak	*Kalotropis procera*
3.	Senia	*Crotoleria burilia*
4.	Thore	*Euphorbia royalena*
5.	Ber	*Ziziphus mauritiana*
6.	Murali	*Lycium auropoeum*
7.	Lana	*Heloxylon selicornicosa*
C.	**Grass**	
1.	Bhurat	*Cenchurus callaticus*
2.	Siwan	*Lasicirus sindicus*
3.	Bura	*Cymbopogom jwarincosa*
4.	Lamra	*Aristada mutica*
5.	Kucha	*Sasccharus griffinir*
6.	Murat	*Panicum turgidum*
7.	Dhaman	*Samccharus ciharisis*
8.	Bui	*Areva aomentosa*
9.	Phog	*Calligomun polygonoides*

Source : Meena, V.P. 2000, *Desert Ecology,* Printer Publisher, Jaipur.

Rajasthan 'state tree': Khejri, the Lifeline of the Desert and 'State Flower': Rohaida

The Khejri *(Prosopis Cineraria)* has been one of the evergreen and long lasting friends of the dwellers of arid zones even in the period of scarcity such as drought and famines since time immemorial. According to the Central Zone Research Institute, Jodhpur. There are only five important trees that grow in the Indian desert and Khejri is the most useful of all of them. It has the greatest viality and widest ecological amplitude in the arid and desert condition. Khejri is the state tree of Rajasthan. Rohida is the state flower of Rajasthan. These are normal and protected trees in the park. Desert National Park is an important research centre for them.

Faunal Community and Diversity

The Park lowland has remained a battleground of many animal species which on account of vagaries of climate, frequent floods, drought and natural catastrophies coupled with human invasion on grassland induced few migrations of terrestrial animals. It is, therefore, significant to consider the following faunal aspects of this Park :

(i) Food habits and breeding biology of the main wildlife species.
(ii) Case studies of rare and vanishing species like Great Indian Bustard Chinkara, desert fox and migratory Imperial sandgrouse.
(iii) Migratory phenomenon of birds and reptiles.
(iv) Behavioural study of all the wild species with special reference to reproduction.
(v) Monitoring the whole desert National Park from interrelationship of agricultural field produces and existing wildlife.
(vi) Study and monitor the reintroduction programmes.
(vii) Evaluation of population dynamics.

In the 1920s, blackbuck were abundant in the Jodhpur region. Upto the thirties of the century, the wild boar was so prolific around Jodhpur that people used to shun walking in the outskirts of the town after dark. The Asiatic wolf is also facing extinction in the desert region. The wild ass, now restricted in the Little Rann of Kuchch was well distributed in parts of Rajasthan desert. What remains now of the mammals wildlife of the desert is a dim reminder of the past glory.

Almost birds, some 300 species were reported by Huge Whistler. The Great Indian Bustard, Houbara, Imperial sandgrouse are still in fair numbers in the desert.

The developmental activities during the years 1972 to 1990 in DNP improved its ecological conditions. By all these improvements a fenced areas came up with artificial water and pasturage even in summer season for herbivores. Gradual growth in new vegetation and change in climo-vegetational factors the neighbouring population of herbivores ushered in the Park closures of the DNP.

This region is rich in fauna, including bird life. It is the habitat of common sandgrouse, Imperial sandgrouse and the Great Indian Bustard. The spiny tailed lizard (*Uromastix hardwickil*) are common. The Indian fringe-toed sand lizard (*Acanthodactylus cantories*) and Bronze grass skinks (*Maboya marcularis*) are common, movement markings of the snakes are seen on bare sand dunes. Ring doves (Streptopeloa decaocto) are very common. So are the granivorous pigeon (Columbia livia) and the sparrow (*passer domestics*), white cheeked bulbul (*pycnonotus leucogenys*), scavenger vultures (*Neophron percnopterus*), pariah kite (*Milvus migrans*), gray parrodges (*Francolinus pondicerianus*) are conspicuous. Crested larks (*Galerida cristata*), small Indian skylark (*Alauda gulgula*), the black bellied finch lark (*eremopetrix grisea*) and white throated munia (*Lonchura malabarica*) are found. Among predatory birds shikara (*Accipiter badius*), sparrow hawk (*accipiter virgatus*), tawny eagle (*Aquila raapx*), short toed eagle (*circaetus gallicus)* are seen perched on trees. The common sandgrouse (*Plerocles exustus*) are conspicuous (Fig. 28.3).

The Indian Gazelle are all over in small family groups. At times there 8-10 individuals all may group, whereas at some places or two buck hold as many as 8-10 females. Sometimes only feminine herds are observed. The desert fox (*Vulpes pusilla*) become conspicuous in the late afternoon.

This is the land of desert rats. There are two species, the hairy footed gerbil (*Ggrbillus gleadowi)* and the desert gerbil (*Meriones hurrianae)* are seen in the winters. The Imperial sandgrouse comes in thousands and their number may exceed over a million birds.

The rich insect fauna, which is a characteristic of this desert plays a remarkable role as the micro-consumer in the existing ecosystem. Although most elements of this fauna area are found to be present almost throughout the year, they exhibit a high population build up particularly during the rainy season (July to September). They cause much loss to the growing vegetation. The termites attack the roots of plants. Among the termites, the most destructive species are *Psammotermes rajasthanicus, Anacanthoyterms macrocehalus, Microterms spp.* and *Odontotermes sp.* In certain pockets heavy losses of kharif crops due to the white grub (*Holotrichia sp.*) have been observed. Locusts and grasshoppers breed in the region specially in the rainy season when sufficient moisture is available in the soft, sandy soil. These pests do much loss to the growing vegetation. Swarms of the desert locust (*Schistocerca gregaria*) visit

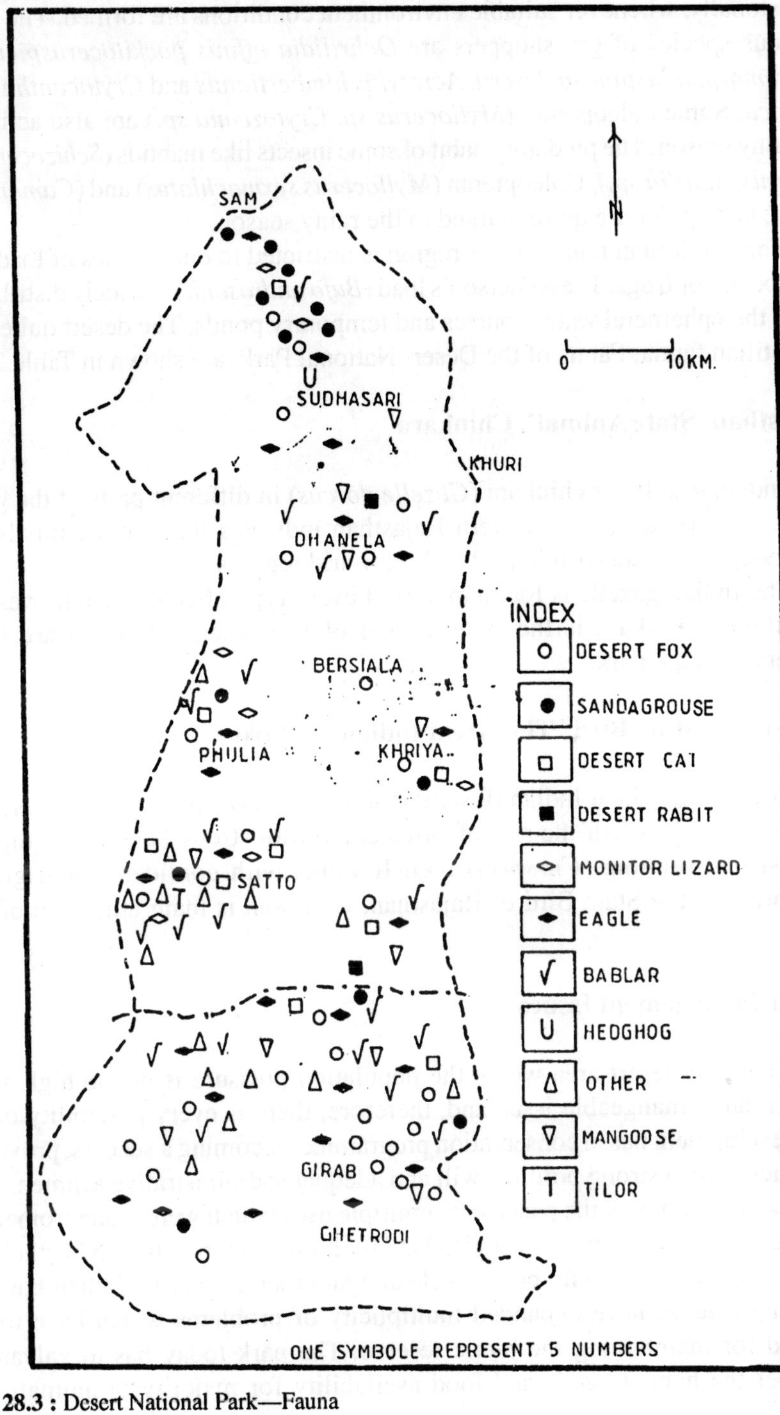

Fig. 28.3 : Desert National Park—Fauna

occasionally, whenever suitable environment conditions are formed. The most noxious species of grasshoppers are *Ochrilidia affinis pockiloceruspicticus, Pygromorpha bispinosa deserti, Acrotylus himbertianus* and *Crytacantharacis tetarica.* Some coleopteran *(Myllocerus sp. Crytozemia sp.*) are also active in the rainy season. The predatory habit of some insects like mantids (*Schizocenutus bicornis, Statilia sp.),* Coleopteran (*Myllocerus sexmachlatus*) and (*Camonutus gigas*) and spider are quite marked in the rainy season.

The amphibian fauna of this region is restricted to one species of load and five species of frogs. The Anderson's load (*Bufo androsonil*) is widely distributed along the ephemeral water courses and temporary ponds. The desert quite rich in reptilian fauna. Fauna of the Desert National Park are shown in Table 28.5.

Rajasthan 'State Animal', Chinkara

The Indian gazelle or chinkara (*Gazelle dorcus*) in different parts of the great Indian or Thar desert in western Rajasthan indicates that these animals are well-adapted for survival in arid and semi-arid regions.

The Indian gazelle is found in almost every type of ecosystem in western Rajasthan. Chinkara is the state animal of Rajasthan. Chinkara are fully conserved in the park.

Rajasthan 'State Bird,' The Great Indian Bustard

Godawan or the Great Indian Bustard (*Choriotis nigriceps)* is one of the rarest bird of India. It is half the size of *saras* crane with strong bare legs equipped with strong running (*Cursoria*) feet. It walks with confidence and grace. *Godawan* is the State Bird of Rajasthan. Godawan is main attraction of the Park.

Major Management Issues

The Thar is a desert area where the population pressure is not so high as to make it an unmangeable issue and, therefore, there is every possibility of an eco-development based conservation programme becoming a success, provided it is backed by a strong political will and adequate administrative acumen. The park, with a legacy of the past for its multiple uses which were made from time to time, is at the moment a forbidden land for human intervention. Nonetheless, its former usage as a wild cattle enclosure and then as a great Indian bustard shooting reserve have expanded multiplicity of problems which have to be tackled for maintaining the biotic reserve. The park today has to galvanize together the habitat needs and food availability for majority of animals the

Table 28.5 : Fauna of the Desert National Park

No.	*Name*	*Zoological Name*
A.	**Animals**	
1.	Chinkara	*Gazella gazella*
2.	Jackal	*Canis aurens*
3.	Hare	*Laus nigricollis*
4.	Hedgehog	*Parachinus micropus*
5.	Jungle cat	*Felis chaus*
6.	Fox	*Vulpus bengalensis*
7.	Mongoose	*Herestes edwards*
8.	Wolf	*Canis lupus*
B.	**Reptiles**	
1.	Common lizard	*Calotus variciclor*
2..	Common lizard monitor	*Varanus bengalensis*
3.	Dhaman	*Ptyas mucosus*
4.	Krait	*Bungarus caeruiens*
5.	Russel's viper	*Vipera russeli*
6.	Spiny-tailed lizard	*Uromaxtyx hardwicki*
C.	**Birds**	
1.	Babbler, common	*Aggya caudata*
2.	Babbler, jungle	*Tudoeds somervillai*
3.	Bee eater, blue-checked	*Merops superciliosus*
4.	Bee eater, small-green	*Merops orientalis*
5.	Babbler, red–vented	*Molpastes cafer*
6.	Babbler, red–whiskered	*Otocompsa jocasa*
7.	Crow, house	*Corvus splendense*
8.	Crow, jungle	*C. macrophychos*
9.	Cuckoo	*Euchyanamis scolopaceus*
10.	Dove, little brown	*Streptolia bengalensis*
11.	Dove, red turtle	*S. tranquibarica*
12.	Dove, ring Indian	*S. decaotao*
13.	Drongo	*Diruus macrocerous*
14.	Eagle, short-toed	*Circatus gallicus*
15.	Eagle, tawny	*Aquila rapax*
16.	Great Indian Bustard	*Choriotis nriceps*
17.	Falcon, larger	*Falco jugger*
18.	Flycatcher	*Rhepidura aureola*
19.	Kite, common	*Malvus mighans*
20.	Lapwing, red wattled	*Lobivanellus inducus*
21.	Lark finch	*Eremopterix grisea*
22.	Myna, common	*Acridotheres tristes*
23.	Owl, great horned Indian	*Caprumulgus asiaticus*
24.	Owlet, spotted	*Athene brama*
25.	Partridge, gray	*Francolnus pondicerainus*
26.	Robin, Indian	*Saxecohdes fulicate*
27.	Sandgrouse, common	*Pteroceles Exustus*
28.	Sandgrouse, common	*Tringa glareola*
29.	Shrike	*Asturbadius sp.*
30.	Shrike, grey common	*Tephordornis Pondiceranus*
31.	Stilt, black-winged	*Himantopus himantopus*
32.	Tailor bird	*Orthotimis sutorius*
33.	Vulture, white backed	*Pseudogyps bengalensis*
34.	Vulture scavenger	*Neophron peonoterus*
35.	Woodpecker	*Brabnchyternus bengalensis*

Source : Meena, V.P. 2000. *Desert Ecology*, Pointer Publishers, Jaipur.

large mammalian fauna, reptiles, birds and other animals, as well as the human population in village within the park area.

The isolated location of the park is both a boon and a disadvantage. Boon because there are not many human habitations in the immediate proximity of the park to distrub its natural eco-system. Disadvantage because only a few manager and experts are available to take care of this national reserve far away from the urban metropolises. Every issue of this reserve or every component of this fragile ecosystem like wind, water, soil texture and biotic life on such land has to be studied thoroughly in scientific pattern, like the ten year project at the Keoladeo National Park by the Bombay Natural History Society. Neglect of a single component may give irreparable loss to the nature of the park. The management objective of the park should aim at :

(i) Preservation of a fragile and unique ecosystem as a natural heritage, and study the ecological behaviour of the desert life.
(ii) Provision of healthy outdoor recreation of the people of urban areas.
(iii) Involving the people living inside the park in the maintenance of the ecosystem and the safety and preservation of biotic life of the DNP.
(iv) Developing domestic and international tourism in desert ecosystems, restricted to the carrying capacity of the park.
(v) Developing interesting and unique destinations for educational tourism.
(vi) Programmes of research projects for study of the Indian Desert ecosystem for their conservation.

The desert park management has to maintain a control on pollution from mining activities, grasses and weeds, management of wastelands, woodlands, grasslands, water resource points and by ban on poaching solution.

The mistique of desert national park is its bio-diversity within a big area, exhibited in a myriad forma of animals, birds, insects, aquatic and tree vegetation which has developed by relentless efforts, of man over the year. In their continued efforts the people living within the ecosystem should be educated and made aware of the great advantages of maintaining the quality of the desert ecosystems by dedication to the natural system would also improve the quality of their life within the desert. They may be provided gas for their kitchens from the oil fields which have been explored in the vicinity in recent years. This would save the trees from cutting for fuel wood. Efforts may be made in scientific harvesting of the sewan grass of the upkeep and earning revenue to the park. The indigenous system of water harvesting and utilization system of water harvesting and utilization of the water of the intending I.G. Canal is a prime requisite as the biotic life may vanish with water famine.

The glory of this natural desert ecosystem can be enhanced if the aforesaid points are taken into consideration.

There is a need to save this great project of Indian planners. A final notification regarding declaring it a national park should be released without delay. A National Desert Museum should also be developed here. This project should be kept in natural ecosystem which is helpful in the prevention of international smuggling and infiltration experiences of this project tell us that in this region permanent greenery should be developed in place of cultivating heavy grains. This desert national park needs to be saved from the dangers of Indira Gandhi Canal. The area of this desert national park should not be reduced. A research institute for desertic wildlife creatures should be opened here. On account of these grounds desert national park will prove out to be unique project and it will also become a multipurpose project.

REFERENCES

1. Anonymous. 1977. *Desertification and its Control.* ICAR, New Delhi, pp. 1-358.
2. Anonymous. 1983. *Progress in arid zone research :1952-1982.* CAZRI, Jodhpur, pp. 1-65.
3. Anonymous. 1984. Action plan for Biosphere Reserves, In: *Natural and Resources,* Vol. XX (4) , October-December 1984, pp. 11-22, UNESCO, Paris.
4. Bhandari, MM.1978 *Flora of the Indian Desert.* Scientific Publishers, Jodhpur, pp. 1-471.
5. Department of Science and Technology, Government of India, 1981. *Report of the working group on Arid Zone Research.* Mimeo. pp. 1-231.
6. Gupta, R.K. and Prakash, I.1975 *An Environmental Analysis of the Thar Desert.* English Book Depot, Dehradun, pp. 1-484.
7. Maan, H.S. (ed.) 1980. *Arid Zone Research and Development.* Scientific Publishers, Jodhpur, pp.1-531.
8. Meena, V.P. 2000. *Desert Ecology.* Pointer Publishers, Jaipur. pp. 1-192.
9. Paroda, R.S., Mann, H.S. and Verma, C.M. 1980. *CAZRI,* Jodhpur, pp. 1-381.
10. Project document, 7. 1988. Ministry of Environment and Forests. Govt. of India, pp.1-100.
11. Rajasthan District Gazetteers—Jaisalmer. 1973. Directorate, District Gazetteers, Government of Rajasthan, Jaipur. pp. 1-381.
12. Rajasthan District Gazetteers—Jodhpur. 1979. Directorate, District Gazetteers, Government of Rajasthan, Jaipur. pp. 1-562.
13. Rajasthan District Gazetteers—Barmer. 1962. Directorate, District Gazetteers, Government of Rajasthan, Jaipur. pp. 1-227.
14. Sen, A.K. 1972. *Agricultural Atlas of Rajasthan.* ICAR, New Delhi, pp. 1-52 (Plates 34).
15. Shankarnarayan, K.A. 1987. Eco-regeneration and sympathetic resource utilization in Desert Biosphere Reserve in the Thar desert In: *Proc. First National Symp.* pp.195-214.
16. Tod, James. 1914. *Animals and Antiquities of Rajasthan.* Vol. I and II. Routledge and Kegan paul Ltd. London.
17. Whistler, H. 1938. The Ornithological Survey of Jodhpur State. *J. Bombay Natural History Society,* 40 : 213-235.

29

Pesticides are Killing our Birds

PRAKASH P. BAKRE

Where have all the crows gone? Cried a headline in one of the news paper in Jaipur recently. During the last 'Sharadha Paksha', the fortnight when pious citizens feed the crows so that their forefathers live in peace in the Moksha, people suddenly realized that the omnipresent crow that frequented our cities and town has suddenly disappeared and only few are left and were in great demand during the said fortnight. It is not only the crow but even vultures, sparrows, saras and many more birds are gradually declining from our cities and towns. And even if we ignore the religious aspect of it, the decline of bird species can have disastrous effects on the ecosystem. Birds have played a significant role not only as our allies in our war against insect pests of agriculture, but also as pollinators of our crops and controlling the insects that are vectors of many diseases. They have been variously described in our literature and have their presence in art, sculptures, carvings, paintings, and poems and prose as messengers. Life without birds is just not imaginable as we have even worshipped them as Jatayu, Hansa, Mayura or Garuda and their colourful beauty has brightened our lives. And if they are declining, as has been observed in case of vultures, can one think of ecosystem without birds!

The problem is there are many reasons for gradual decline of avifauna and most are of anthropogenic origin. When the vulture decline was reported some five years back initially it was taken as a problem related to a single species because of their peculiar habits. But the study sent scientist scurrying to look for similar declines in other avian species and it was soon discovered that the phenomenon is not restricted only to vultures but *sarus* cranes and sparrows as well. This necessitated the need for in-depth study of the possible changes in the environment which these feathered fauna share with us. When the three realms of the environment, the hydrosphere, the lithosphere and the atmosphere are all contaminated due to human activities other denizens of biological world are bound to have adverse effects.

Traditionally Indian farmer depended totally on the feathered fauna for the control of insect pests that attacked his crops. Birds have been frequenting agricultural areas and nesting in hedgerows since optimum supplies of soft-bodied forms of pest insects were at hand in the nearby agricultural fields to feed the young ones. No doubt, when insect pests were not there, they did consumed grain and seeds. Many of these seeds were in fact of weeds in the agricultural areas. But the total loss to grain to birds was much low in true economic terms compared to total number of insects consumed by a single bird and the grains saved in the process. Consequently, the loss to grain was taken as a good price to pay for the service rendered by birds towards insect control.

However, with the dawn of "Green Revolution" this scenario changed and the traditional work of birds of controlling insect pests, as natural predators, was then assigned to insecticides. The results of this modern change were quickly evident in the near-total elimination of insect pests from agriculture fields. In fact, accompanied occasional "Kills" of large number of birds was taken a bonus to the farmer adding further economic gains because of prevention of losses due to birds. Average farmer who would normally shoo away the birds when the crop was ripe was now content that insecticides not only controlled insects but birds too.

Yet another aspect of this revolution in the Indian agriculture was that newer and newer hybrid varieties of crops were introduced. These heavily depended on extensive use of synthetic manures and modern farming techniques and also extensive use of water. A heavy use of insecticides was also a must for protection against insect pests, particularly in crops like paddy and cotton. The brighter side of the story is that through the tireless efforts of our agricultural scientists we were able to produce bumper crops to feed our ever-increasing population of millions. However, the consequences of heavy dependence on insecticides and other varieties of pesticides were to follow thereafter. The darker side started making itself apparent with occasional mass killing of birds in fields. However, these were too infrequent and insignificant and restricted to certain geographic areas only.

Although the first indication of possible adverse effects on birds were evident within two years of use of DDT in United States, the science of wildlife toxicology that deals with the adverse effects of environmental contamination on wildlife has developed recently. But, before it could develop standard methods to assess the impact of agricultural uses of pesticides much damage had already taken place. And today we find ourselves in a no win situation. Since even though the indications are clear and possible future hazards are known we cannot halt the use of these perilous pesticides in agriculture. Since it might spell a doom and our import bills may swell many folds if enough production is not maintained to feed our ever-increasing population. Currently we are using around 131 pesticides in various formulations at an annual rate of more than hundred thousand metric tonnes. Most of these are the insecticides and the fungicides primarily used in agricultural fields to control a particular insect pest a target organism. But there are many "non-target" organisms inhabiting the area. Since basically the pesticides are poisons developed to control a form of life having deleterious effects on other life forms. If the non-target is physiologically similar to that of the target organism the possible ill effects can be predicted. But, the situation is totally different and unexpected when the non-target happens to be a physiologically totally different organism, say like birds.

It is to be expected that when pesticide is used in agriculture it will be aimed at the target organism. But how much of total pesticide used actually reaches the target organisms is a moot question. As per the estimates of Pimental and Levitan (1988), only less than one per cent of the pesticide reaches target organisms, normally an insect. Caterpillars targeted were present on different parts of crop plants and not necessarily at the top to receive the following spray of pesticides. Thus, almost more than 99 per cent of pesticides used goes into general environment and may either enter biological realm or may fall on inanimate objects and get absorbed on them. However, if the pesticide happens to be the persistent one, it may remain in the environment over a very long period and may disperse over varying distances and has chances of being incorporated into biomass and further biomagnify in the food chains in the habitat. Earlier Joyce (1982) had reported that only 0.000,0001 per cent DDT applied for control of Heliothis reached the pest species. Unfortunately, the application of these toxic substances for insect control is not uniform on different crops. The situation is worst in Indian context. Average land holding of our farmers is very small and everyone is free to grow a crop of his choice and use any pesticide thought/advised proper. As a result a bird species inhabiting agricultural area and with a wide home range, is likely to be exposed to a variety of pesticides in these small fields, in one single day. Application of various classes of pesticides may result in birds consuming seeds coated with pesticides, insects and other smaller invertebrates impaired or killed as a result of pesticide exposure. Substantial dependence of some bird species on insect larvae is well known. Some species of birds seem to follow the rains, and have more than one wintering quarter that assures them of optimal food supply. With the onset of rains, organic production increases enormously and the vegetation developed and consequently abundance of invertebrates feeding on green matter.

Opportunistic feeders exploit such food sources of seasonal and local abundance and are very important agents of insect pest control. Unfortunately, with the increasing uses of pesticides these seasonal sources of food are lost and this has ultimately resulted in reduced reproductive success of birds (Heath *et al.* 1969; Jefferies, 1971; Bitman *et al.* 1970). Unfortunately, practice of coating of seeds with fungicides may result in many seed eating birds that normally follow the plough or a tractor, consuming these dressed seeds. Organophosphates and carbonates used for seed dressing may particularly prove to be highly hazardous. What the average farmer is not realizing is that these birds inhabiting agricultural area are his allies in the war against insect pests.

Massive uses of insecticides and herbicides in rice fields may result in decline of insects and other invertebrates on which birds like herons depend. Further, all fishes, amphibians, crabs killed may attract storks, kites, eagles etc. which could give rise a case of secondary poisoning (Roux, 1980; Thiollay, 1980; Curry Lindahl, (1981). The final impact at population level like local extermination or mere sub-lethal effects like impaired cholinesterase inhibition or reduced reproduction may be evident, depending on species involved. Young (1968) has suggested that direct mortality generally affect long-lived slow breeding bird species. However, reduced fertility has more profound effect on rapidly breeding short-lived species.

Fry (1975) had opinion that the reproductive success in birds of tropical area is low, but adult survival is excellent, implying that the destruction of the adult population will take many years to recover. Large scale killing of birds are not that common although there has been a recent spurt increase in killing of peacocks in and around Jaipur and many other areas of Rajasthan. Direct kill of a bird is just one of the 'effect' possible but in most cases birds carry heavy loads of pesticides in their body systems and may suffer from a variety of physiological and behavioural effects which are rarely detected. One of the most well studied subtle effects has been thinning of eggshells, which may prove disastrous to a species. Particularly vulnerable are the species occupying the apex of the food chains. Thin eggshells were first reported in double crested cormorants and white pelicans in Wisconsin (Anderson et al.,1969) and DDE was identified as the culprit responsible for the condition. Thereafter large number of cases of thinning of eggshell were reported. Normally a shell thinner to 80 per cent of the normal thickness was shown to result in "dead in shell" syndrome. Thinning of the shell was produced by a variety of organo-chlorines but was species specific and chemical dependent.

Other effects reported included reduced clutch size, reduced hatchability, reduced fertility, lesser fledging success. In some cases there was longer period of ovulation, aspermia, longer periods of incubation and rearing, stunted young ones, less heat being applied to egg during incubation, abnormal behaviour of chicks, abnormal building of nest etc. Apart from these visible effects pathological damage has been reported in liver, kidney, gastrointestinal tract, spleen, pancreas, testis, ovary, adrenal, apart from a variety of hematological parameters studied as a result of exposure to pesticides. Most prominent physiological effects were inhibition of the enzyme acetyl-cholinesterase, which manifested in a variety of behavioural abnormalities including delayed response to a stimulus. As a result of these adverse effects the birds are continually under pesticidal stress and this has resulted in decline in the feathered fauna frequenting our agricultural fields and even our own backyards.

With present trend in pesticidal usage are we heading towards total extermination of some of these birds, which are our allies in the war against insect and rodent pests. This needs to be critically analysed since one of our common bird in the villages and cities alike, the vulture, is facing the problem of population decline while the number of *Sarus* crane too is heading for a quick depletion. In both cases pesticides are suspected to be the factors responsible and there may be many more species of birds that are not as prominent but heading for population decline. Unfortunately, in our own situation the pesticide usage is indiscriminate and most of our farmers who are using them are illiterate to follow the instructions given on the label of the container. Often they end up using totally irrelevant pesticide for a given pest since he is wrongly advised/guided by unscrupulous seller who takes undue advantage of farmer's ignorance. Maybe we are heading for a situation similar to that of Thailand where there is relative lack of birds both in number and variety that were once encountered in their marshes, swamps and paddy fields. Contamination of habitat has been reported to be cause for this decline (IUCN, 1979) and what that country is left with insect pests with immunity to a variety of pesticides and no natural predators, the birds, to control them.

In our own situation there have been reports of birds carrying pesticide loads as early as 1981 when Kaphalia *et al.* (1981) reported residues of BHC, lindane and DDT in various organs of few species of birds like pigeon, kite, crow, vultures and cattle egrets. Pesticide residues were also reported by Misra and Bakre (1994) in a variety of birds inhabiting an aquatic habitat near Jaipur. Murlidharan (1993) had reported presence of dieldrin in *Sarus* crane in and around Keoladeo National Park, Bharatpur. Thus as the situation stands today the draconian hands of pesticidal pollution is gradually engulfing the avian fauna of our agricultural fields where they are primarily used and our aquatic habitats where the pesticides end up ultimately. Unfortunately the science of wildlife toxicology is still in infancy in our country and mere disappearance of vultures and sarus has not set the alarm bells sounding loud enough for our planner to take note of the trend of vanishing of our feathered fauna.

REFERENCES

Anderson, D.W., Hickey, J.J., Risebrough, R.W., Huges, D.F. and Christensen, R.E. 1969: Significance of chlorinated hydrocarbon residues to breeding Pelicans and cormorants. *Can.Field-Nat.*, 83, 91-112.

Bitman, J., Cecil, H.C. and Fries, G.F. 1970: DDT induced inhibition of avian shell gland carbonic anhydrase: a mechanism for thin eggshells. *Science* N.Y. 168, 594-6

Curry-Lindahl, K 1981: *Bird migration in Africa. Movements between six continents.* Vols. 1 and 2. Academic Press, pp. 444 and 695

Fry, C.H. 1975: *Bird conservation and prospects in Nigeria.* Ardeola, 21(esp) pp.993-1015.

Heath, R.G., Spann, J.W. and Kreitzer, J.F. 1969: Marked DDE impairment of Mallard reproduction in controlled studies. *Nature, Lond.* 224,47-8.

IUCN 1979: *Conservation for Thailand.* IUCN Working Document. IUCN/UNEP.

Jefferies, D.J. 1971: Some sublethal effects of pp DDT and its metabolite ppDDE on breeding passerine birds. *Medel. Fakult.Gent.* 36, 34-42.

Joyce, R.J.V. 1982: A critical review of the role of chemical pesticides in Heliothis Management. pp.173-188 in *International Workshop on Heliothis Management, Patancheru,* Andhra Pradesh, India.

Khaphalia, B.S., Husain, M.M., Seth, T.D., Kumar, A. C.R. Krishna Murthy 1981 : Organochlorine pesticide residues in some Indian wild birds. *Pesti. Monit.J.,* 15, 9-13.

Misra,Vipul and Bakre, Prakash P. 1994: Organochlorine contaminants and avifauna of Mahla water reservoir, Jaipur, India. *The Science of Total Environment* 144 pp. 145-151.

Murlidharan, S. 1993: Alderin poisoning of Sarus crane (*Grus antigone*) and a few garnivorous birds in Keoladeo National Park, Bharatpur, India. *Ecotoxicology.* 2; 196-202.

Pimental, D. and Levitan, L. 1986: Pesticides—Amounts applied and amounts reaching pests. *Bio-Sciences,* 36(2), 86-91.

Roux, F. 1980 : On the wintering habitats of Palearctic migratory birds. Interview by C. de Jong-Boon.

Thiolly, J.M. 1980: On the wintering habitats of Palearctic migratory birds. Interview by C. de Jong-Boon (1980).

Young, H. 1968: A consideration of insecticide effects on hypothetical avian populations. *Ecology,* 49(5), 991-994.

Our Contributors

Professor Savindra Singh
Department of Geography
Allahabad University,
Allahabad (U.P.)

Dr. Amal Kar
Senior Scientist
Central Arid Zone
Research Institute, Jodhpur

Dr. Ajay, Group Director
Forestry, Landuse Planning and Photogrammetry Group (FLPG)
Space Application Centre,
Ahmedabad

Dr. P.S. Dhinwa
Senior Scientist Landuse Planning and Photogrammetry Division,
Space Application Centre,
Ahmedabad

Mayank Bhavsar
Remote Sensing Signal and Image Processing Area, Space Application Centre, Ahmedabad

Professor S. Padmaja
Dept. of Geography,
Osmania University,
Hydrabad (A.P.)

Dr. Kausalya Ramachandran
Senior Scientist CRIDA (ICAR)
Santosh Nagar, Hydrabad

G. Somiah
Project Fallow, SAP
Department of Geography
Osmania University,
Hydrabad (A.P.)

Dr. Surendra Singh
Former Head Resource Survey & Monitoring Division
CAZRI, Jodhpur

Prof. R.K. Rai
Former Head, Department of Geography
North Eastern Hill University,
Shillong, Meghalaya

Dr. K.M. Haroni
Department of Geography
Govt. College, Kota

Dr. Meenakshi Vijai
Environmental Chemistry Laboratory
Department of Chemistry
Govt. College, Kota

Dr. Ashu Rani
Environmental Chemistry Laboratory
Department of Chemistry
Govt. College, Kota

Rajesh Bhakar
Department of Geography
Govt. Dungar College,
Bikaner (Rajasthan)

Prof. S.R. Basu
Department of Geography
35, Ballygunge Circular Road
Calcatta University, Kolkata-19

Dr. Anusuya Ghosh
Department of Geography
Calcatta University, Calakuta

Dr. Sunil Kumar De
Department of Geography
UCT College Berhampore,
Murshidabad

Prof. S.C. Mukhopadhyay
Former Head, Department of Geography
35, Ballyagange, Circular Road
Calcatta University, Kolkata-19

Dr. Jayanti Das
Senior Lecturer in Geography
Shri Shikshayaton College, Kolkata

Dr. Lakshmi Shivaramakrishnan
Senior Lecturer
Gokhale Memorial Girls Collage, Kolkata

Dr. L.C. Agrawal
Department of Geography
Govt. College, Kota

Dr. D.N. Sathi
Former Head, Department of Geography
Govt. College, Ajmer

Dr. Ved Prakash Meena
Department Of Geography
Government M.S.J College, Bharatpur

Prof. Prakash P. Bakre
Director, Indira Gandhi Centre for Human Ecology Environment and Population Studies University Rajasthan, Jaipur

Dr. T.I. Khan
Indira Gandhi Centre for HEEPS
University Rajasthan, Jaipur

Dr. S.K. Wadhawan
Geological Survey of India
Western Region, Jhalana Area, Jaipur

Prof. Ashok Kumar Singhvi
Senior Professor
Physical Research Laboratory, Ahmedabad

Prof. Santosh Shukla
Department of Geography
Dr. H.S. Gour University, Sagar (MP)

Dr. B.L. Sharma
Head, Department of Geography
Govt. College, Kota

Prof. S.K. Sharma, Director PSC
Department of Geography
Dr. H. S. Gour University,
Sagar (MP)

Dr. Azra Akhtar
Department of Botany
Govt. College, Kota

Dr. R.B. Singh
Department of Geography
Delhi School of Economics
Delhi University, Delhi

Dr. S. Sen Roy
Department of Geography
Delhi School of Economics
Delhi University, Delhi

Pooja Puar
Research Scholar
Department of Geography
Govt. P.G. College, Kota.

Dr. M.L. Sharma
Principal, Bhawani Niketan
College, Jaipur

Prof. Baleswar Thakur
Department of Geography
Delhi School of Economics
University of Delhi
Delhi

Dr. Punyatoya Patra
Department of Geography
Aditi Mahavidhyalya, Bawana
Delhi University
Delhi

Prof. S.C. Kalwar
Department of Geography
University of Rajasthan
Jaipur

Dr. H.V. Sharma
Department of Geography
University of Rajasthan
Jaipur

Avinash
Research Scholar
Department of Geography
University of Rajasthan
Jaipur

Dr. J.K. Sharma
Department of Geography
Bhawani Niketan College
Jaipur

Dr. R.D. Doi
Associate Professor
Department of Geography
University of Rajasthan
Jaipur

Dr. D.D. Ozha
Groundwater Department, Jodhpur

F.M. Golani
Ground Water Department, Jodhpur

Dr. Ripunjaya Singh
Lecturer in Geography
Government Collage, Ajmer

Dr. Suresh C. Sharma
Department of Botany
Government College, Kota